KNOWLEDGE ENCYCLOPEDIA
HUMAN BODY

© Wonder House Books 2024

All rights reserved. No part of this book may be reproduced or transmitted in any form by any means, electronic or mechanical, including photocopying and recording, or by any information storage and retrieval system except as may be expressly permitted in writing by the publisher.

(An imprint of Prakash Books)

contact@wonderhousebooks.com

Disclaimer: The information contained in this encyclopedia has been collated with inputs from subject experts. All information contained herein is true to the best of the Publisher's knowledge.

ISBN : 9789354404115

Table of Contents

Brain & Nervous System

Making Sense of Who We Are	7
Command and Control: The Brain, Spine and Nerves	8
Action and Reaction: Sense and Motor Organs	9
Fitting Together: Moving Pictures Sent on Radio Waves	10–11
Putting Matters in Grey and White	12–13
Voluntary and Involuntary Movements	14
Moves Like Lightning: The Autonomic Nervous System	15
Breaking News to the Brain: The Senses I	16–17
Sight and Sound: The Senses II	18–19
Nerves and Neurones: How They Work	20–21
Your Body's Other Brain	22–23
Switching Your Nerves Up and Down	24–25
From Thought to Action	26
Making Memories	27
Higher Thought: From Animal to Human	28
The Pituitary and Pineal Glands	29
Higher Thought II	30–31
An Attack of the Nerves: Epilepsy & Multiple Sclerosis	32
Sleep Healthy and Wake Up Healthy	33
A Healthy Mind in a Healthy Body	34
Improve Your Mental Health	35

Heart & Circulatory System

The Amazing Human Body	37
At the Heart of Everything	38–39
Inside Your Heart	40–41
Arteries & Veins	42–43
Feeding the Organs: Capillaries	44
Keeping the Beat	45
Heart Troubles	46–47
Transplants: Gifting A New Life	48
Know Your Blood	49

Blood Cells: Little Transporters & Defenders	50–51
Platelets & Plasma	52
Donate Blood, Save a Life	53
Know Your Blood Type	54–55
Know Your Blood Pressure	56
Lymph: The Second Circulatory System	57
Lymphatic System: Your Body's Janitor	58–59
Blood Disorders	60–61
Life Cycle of Blood Cells	62
Watch Out	63
A Healthy Heart Makes a Healthy Body	64-65

Immune System & Common Diseases

Our Body's Army	67
Our Skin: The Great Wall	68–69
The Organs that Protect Us	70–71
Immune Cells: Tireless Soldiers	72–73
Phagocytosis: Eating Our Body's Enemies	74
Shoot at Sight	75
You are Under Attack	76–77
Clotting and Wound Healing	78
Platelets: Our Body's Repairmen	79
Keeping Our Gut and Lungs Safe	80
Our Microscopic Allies	81
How Our Immune System Fights	82–83
Our Immune System's Chemical Toolkit	84–85
Our Soldiers Never Forget	86
Why We Get Allergies	87
Fighting Cancer	88
When Our Soldiers Turn Against Us	89
Fighting Common Diseases	90–91
Vaccines: A Jab of Safety	92–93
Immunity in Plants & Animals	94
Helping Our Immune System	95

Lungs & Respiratory System

The Breath of Life	97
Nose: Your Breath's Gatekeeper	98–99
Taking Air to the Lungs	100–101
The Last Mile: Bronchi and Bronchioles	102
Destination: Lungs	103
Inside Your Lungs	104–105
The Respiratory Cycle	106
Taking Oxygen to the Tissues	107
Oxygen and Carbon Dioxide Cycles	108–109
Cellular Respiration	110
Breathing Underwater	111
Breathing at High Altitudes	112
Asthma	113
Poison in the Air	114–115
Under the Weather	116–117
Invaders of Your Lungs	118–119
Exercises to Breathe Better	120
Your Body's Sound Box	121
Making Words, Making Sense	122–123
Fluid in Your Lungs	124
Bringing Someone Back to Life	125

Skeletal & Muscular System

The Bones that Make Us	127
Protecting the Brain: The Skull	128–129
Stiffening the Back: The Vertebral Column	130–131
At the Heart of It: Shoulders and Ribs	132–133
Inside Your Bones	134–135
The Bones that Keep Us Going	136–137
In Your Hands and Feet	138–139
Some Very Hip Bones	140
Helping You Move: Skeletal Joints	141
How Joints Work	142–143

Breaking Point: Bone Fractures	144
Moving Our Bodies: Muscles and Tendons	145
The Muscles We Can Control	146
The Muscles We Cannot Control	147
How Muscles Work	148–149
Keeping the Heart Going	150
Exercise in Moderation	151
Getting Older	152–153
Osteoporosis	154
Muscular Diseases	155

Stomach & Digestive System

We Are What We Eat	157
The First Bite	158–159
Teeth & Gums	160–161
Hunger & Thirst	162–163
The Stomach	164–165
The Things that Make Us	166–167
The Intestine	168–169
The Liver	170
The Pancreas	171
Enzymes: How We Digest Food	172–173
Large Intestine & Appendix	174
Excretion: Eliminating Waste	175
Taking Food to Those Who Need It	176
Metabolism: Make It or Break It	177
The Urinary System	178–179
Tummy Troubles	180–181
A Healthy Mind & Healthy Body	182–183
Gall & Kidney Stones	184
When Organs Give Up	185

Word Check — 186–190
Credits — 191–192

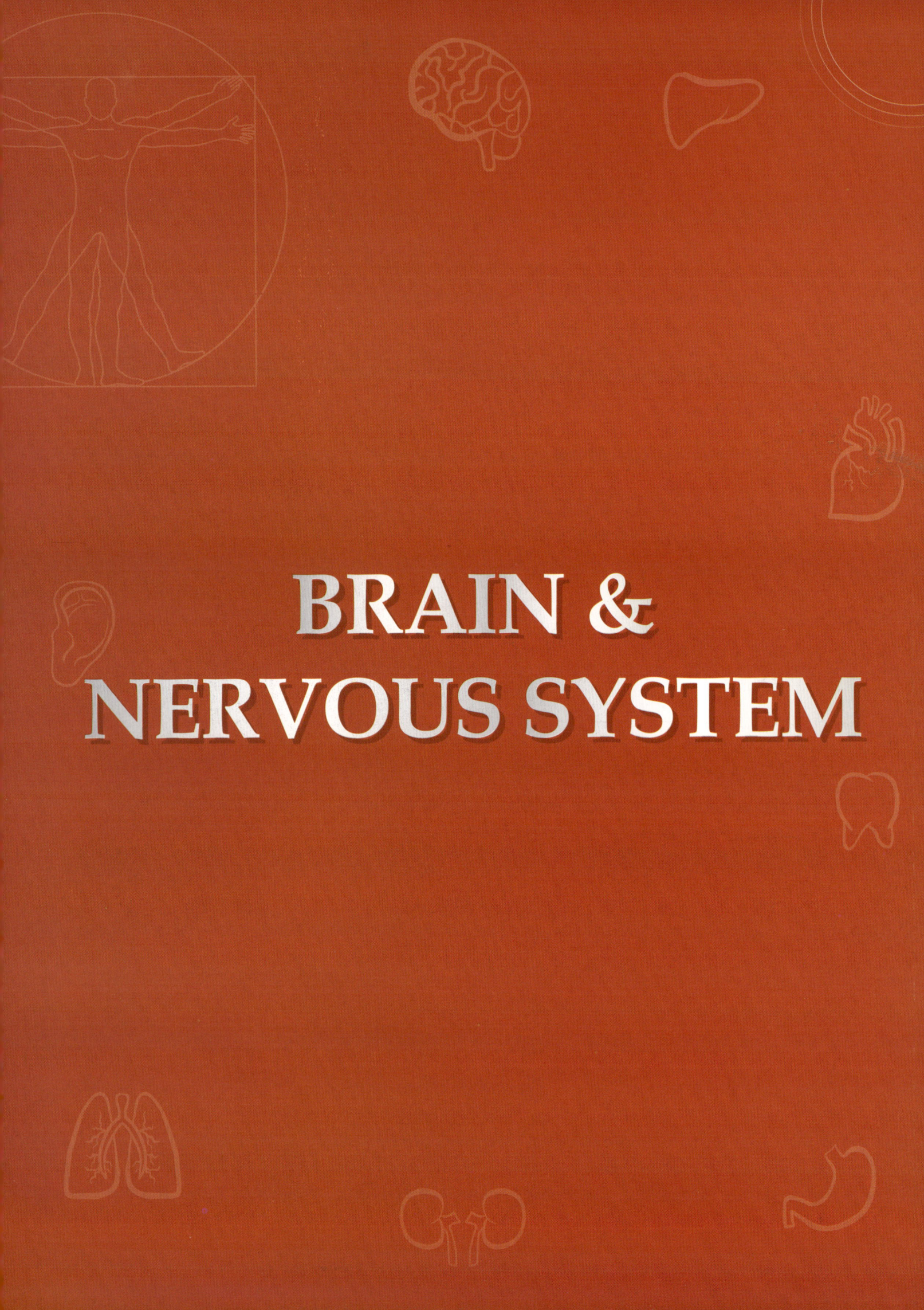

MAKING SENSE OF WHO WE ARE

Our brain is a complex organ. Scientists are still understanding how it works. We know that the brain takes information from the eyes, ears, tongue, nose, and skin to the spinal cord. It sends this information back to the brain through the sensory nerves. Through many steps, the brain puts together the information from the senses and makes up the world that we live in—our thoughts, memories, likes, and dislikes.

Based on what it sees, hears, tastes, smells, touches, and feels, the brain passes information through the motor nerves to the muscles and organs. Which is why we feel hungry, frightened, angry, thirsty, etc. For instance, when we are faced with a difficult situation, our brain tells us whether we should run away or stay and fight.

The nervous system also controls what we do without being aware of it, like breathing, circulating blood, digesting food, and removing waste from the kidney. Doctors call this part of the nervous system the autonomic nervous system.

▶ Our brain remakes the universe inside our heads

Command and Control
The Brain, Spine, and Nerves

The nervous system is to our body what the government is to a country. The nervous system takes the information coming from various sensory organs, processes it and makes decisions, which are then carried out by other organs. The nerves, which are a part of the nervous system, are made up of special cells called **neurones** that take tiny electrical signals from the sense organs to the brain and from the brain to the motor organs. Then the motor organs, mainly the muscles, act together to do what the brain wants them to. The neurones communicate through tiny switches called **synapses**, and chemicals that doctors call **neurotransmitters**.

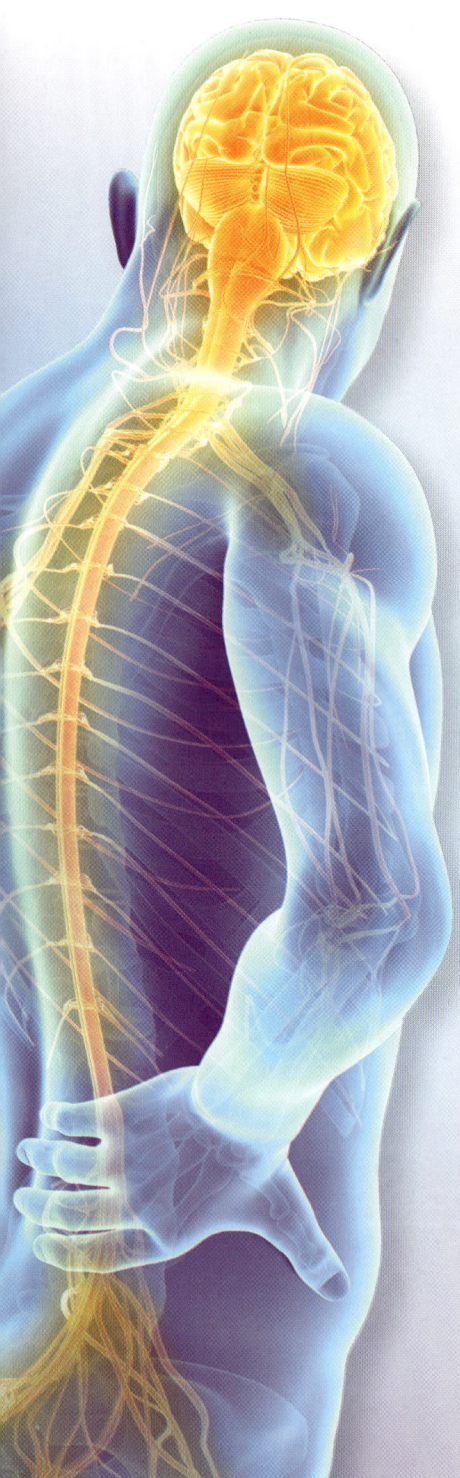

The Brain

The human brain is made of **white matter** and **grey matter**. As human beings evolved, they grew bigger brains with more neurones. To fit them all in, the grey matter had to fold itself into wrinkles. The neurones have connections with each other that run all over the brain. This makes up white matter. Inside the spinal cord is the central canal which contains the cerebrospinal fluid. This supplies the brain and spinal cord with nourishment.

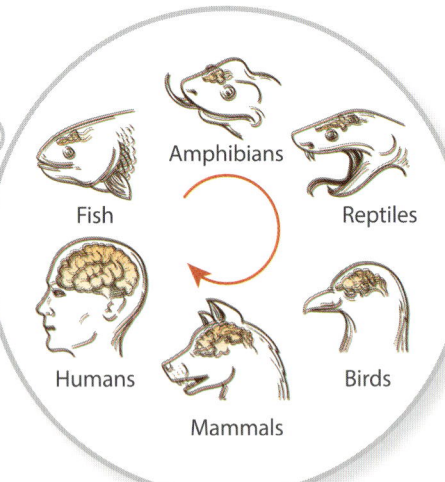

▲ *The grey matter does the processing, whereas the white matter provides communication*

The Spinal Cord

The spinal cord starts at the back of the neck and extends to below the last rib. It takes information to the brain, and brings back orders from the brain to the body. It also takes care of the body's reflexes. The spinal cord is protected by the backbone which starts at the base of the skull and ends above the hip area.

The nerves work like telephone cables that carry messages from the organs to the brain and back. The organs in the head are directly connected to the brain, while those in the body connect to the spinal cord.

◀ *The brain and spinal cord govern the rest of the body. Nerves connect them all together*

In Real Life

Look closely at the pictures of a walnut and brain. They look a little alike because like the brain, the walnuts has folds and is divided into two halves. Additionally, eating walnuts is good for the brain. Walnuts are rich in minerals and lipids that help your nerves and brain function well.

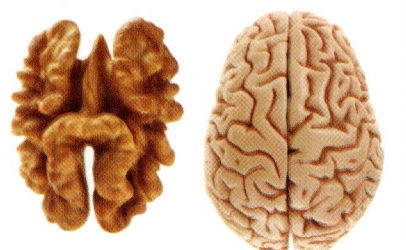

▶ *Some scientists claim that walnuts are the healthiest nuts of them all*

Action and Reaction
Sense and Motor Organs

Imagine a pineapple cake in front of you. Your eyes give your brain a picture—it is big, round, and white. Your nose tells your brain that it smells of pineapples and cream. These are called stimuli. Your brain puts all the stimuli together and makes your hands reach out to grab a piece of the cake! As your fingers touch it, your brain knows that it is creamy on the outside and crumbly on the inside. You put it into your mouth, and your tongue tells you that it is sweet to taste. How did it all happen?

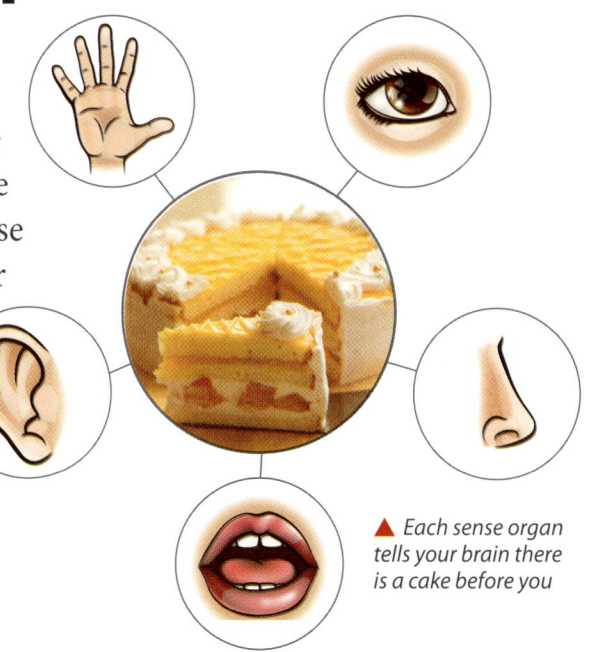
▲ Each sense organ tells your brain there is a cake before you

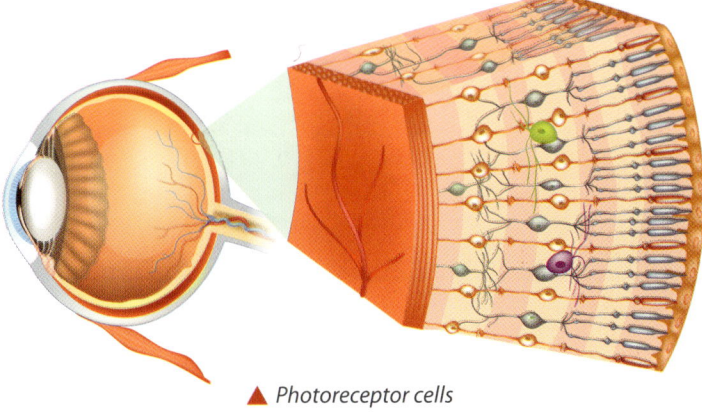
▲ Photoreceptor cells

Sense Organs

Each of our senses works in their own way. Chemoreceptors in the tongue and nose can sniff out the smallest amounts of chemicals in our food and air. Photoreceptors in the eyes pick up colours and light. Auditory receptors in the ears pick up the tiniest vibrations in the air. All of these turn into tiny electric signals that go through the nerves to the brain, which turns them into tastes and smells, sights and sounds.

Isn't It Amazing!

We cannot sense electricity by looking at a wire. But bees can detect the tiniest of electric currents from the flowers they visit. A bee can even sense whether another bee has already been to a flower that it is visiting. What's more, it comes to know whether the first bee took nectar from the flower. This saves the worker bee's time!

▲ Bumblebees can sense electricity

Motor Organs

When your sense organs tell your brain that there is a cake in front of you, your motor organs take action to grab a slice of the cake. Your brain sends signals through your nerves to your hand muscles, so they reach out for the cake. It also sends signals to your mouth and your voice box, so you can ask the grown-up in the room for permission to grab a slice!

▲ Your motor organs act on the information they receive from the brain and the sense organs

Fitting Together
Moving Pictures Sent on Radio Waves

The brain is divided into many parts. These include brain stem, little brain, midbrain, and higher brain. They work like a jigsaw puzzle where all the pieces make sense only if they fit together.

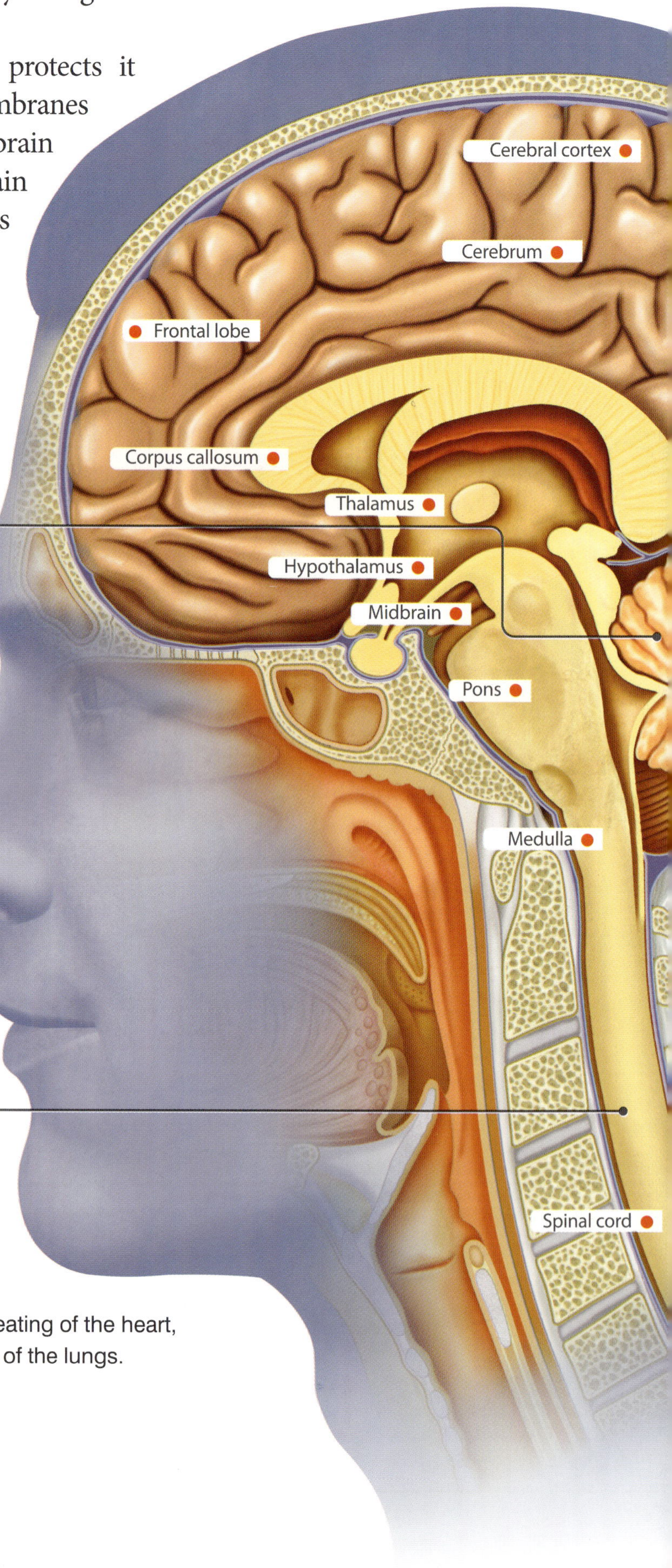

▼ The diagram shows the parts of the inner brain

The brain is enclosed within the skull, which protects it from injury. It is also covered by three membranes called **meninges**. The meninges help keep the brain cushioned, so even if you hurt your head, the brain does not smash against the inside of your skull. Veins and arteries criss-cross the meninges, bringing nutrition to the brain, and taking away any waste.

👤 The Little Brain

This part of the brain is just above the back of your neck and makes up about one-tenth of the brain. Doctors call it the cerebellum. It acts like a checking centre for the brain, helping muscles correct themselves. For example, if something seems heavier than it looks, your cerebellum will get your muscles to put in more force. The little brain also helps you learn to avoid things that cause pain.

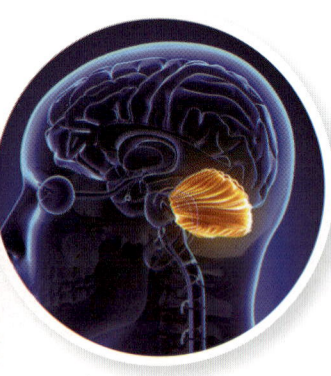

▲ The cerebellum contains significantly more nerve cells than the cerebrum

👤 The Brain Stem

The brain stem is the first part of the brain. It is made of two parts—the **medulla oblongata** connects the spinal cord and the brain, while the pons connect the cerebellum and the cerebrum to the medulla. Between them, they control our daily activities like the beating of the heart, the circulation of blood, and the working of the lungs.

◀ Structure of the human brain stem

HUMAN BODY | BRAIN & NERVOUS SYSTEM

The Higher Brain

The rest of the brain makes up the cerebrum. Most of what you see is the outer **cortex**, divided into two brain hemispheres (right and left) by the longitudinal fissure. The cortex is made up of four lobes, divided by inward folds called **sulci**. The outward folds of tissue that make up each lobe are called **gyri**. Each lobe has a function and also helps the other lobes.

The temporal lobe is responsible for hearing, memory, and learning. The occipital lobe is responsible for the function of sight. The parietal lobe is responsible for touch and movement. The frontal lobe is for thinking, acting, language, and personality.

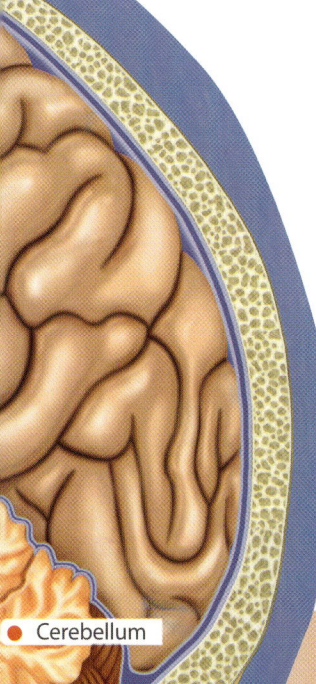
● Cerebellum

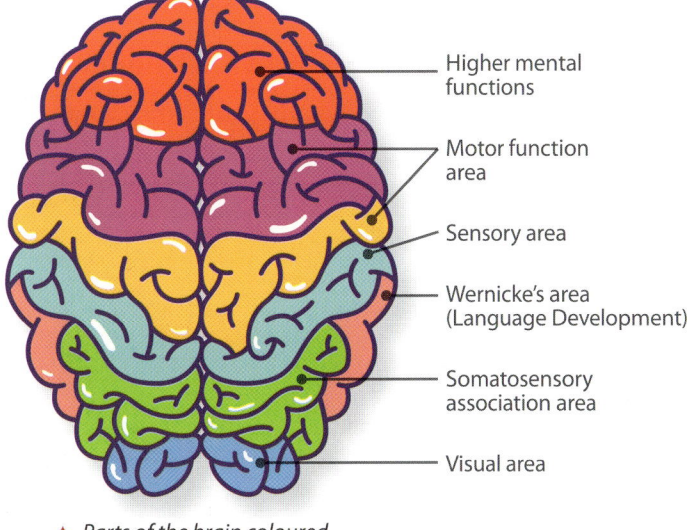
- Higher mental functions
- Motor function area
- Sensory area
- Wernicke's area (Language Development)
- Somatosensory association area
- Visual area

▲ Parts of the brain coloured according to their functions

In Real Life

A few decades ago, doctors would prescribe cutting off bits of the brain to cure moral deviancy and brain diseases like epilepsy. They called this lobotomy. More often than not, it left the patients worse than before. Often, after a lobotomy, a patient was unable to show emotion. Modern doctors think it was a cruel and useless practice.

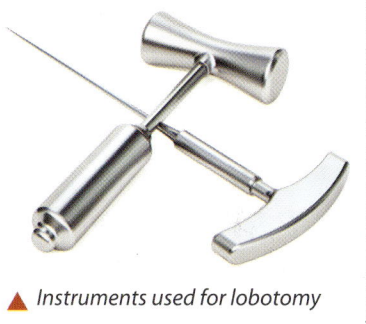

▲ Instruments used for lobotomy

The Inner Brain

The inner brain is made up of the **thalamus**, the **hypothalamus**, and the midbrain. The thalamus is the part that decides which stimuli to give attention to. This is the one that says, 'If you see a tree, do nothing. But if you see a lion, you better run!' The hypothalamus takes care of things like feeling hungry or thirsty, tired or fresh, sleepy or awake. The midbrain connects the senses, if you hear mum's voice, you immediately turn to see where she is.

Incredible Individuals

The procedure of lobotomy was developed by Antonio Egas Moniz but it was American neurologist Walter J. Freeman II, who popularised its use despite widespread criticism. He approached the media to promote the procedure, even giving it the name 'prefrontal lobotomy'. During his time, he performed around 3,500 lobotomies. Thankfully, he was banned from doing it after the last lobotomy, which happened in 1967

▲ Antonio Egas Moniz

▲ Walter J. Freeman II

Putting Matters in Grey and White

Ever heard of the phrase, 'use your grey matter'? That is the part of the brain responsible for thinking, remembering, and imagining. The brain is made of billions of tiny cells called neurones, that act like electric cables. Each neurone has a cell body, and long, wire-like dendrites and axons. The cell bodies of neurones clump together into tissues called grey matter.

The axons of neurones, tied into long bundles called tracts, make up white matter. They connect the different parts of the brain so that the grey matter can do all of its work.

Grey Matter

In the cortex, grey matter is present as gyri. In the inner brain, it is made of clumps known as **basal ganglia** or **nuclei**, surrounded by white matter. Each of these has a name and a specific function to perform. These nuclei let your brain decide how to act once you have a stimulus, like which way to turn when you want to catch a ball. They are also important for learning and memory and to control emotions and behaviour.

The most important is perhaps the claustrum. Famous scientist Francis Crick compared it to the conductor of an orchestra as it connects with all parts of the brain.

◀ Grey matter is made of cell bodies of neurones while the thread-like white matter is made of their axons

Nucleus	What It Controls
Caudate nucleus	Planning movements, memory storage, learning, and decision-making
Putamen	Voluntary and **involuntary movements**
Globus pallidus	Voluntary movements
Subthalamic nucleus	Voluntary and involuntary movements, learning, emotional reactions
Substantia nigra	Eye movements, mood, learning, likes, and dislikes
Claustrum	Consciousness

White Matter

The biggest tract of white matter joins the right and left halves of the cortex. Doctors call it the corpus callosum. This lets the nerves in each half talk to each other. Other tracts connect the basal ganglia with the cortex, the midbrain with the basal ganglia, and the hindbrain with the midbrain. The pons is also made of white matter. One important tract connects the eyes with the occipital cortex, crossing over in the middle, so that the left eye connects to the right side and the right eye to the left side!

In Real Life

Atrophy means the shrinking of a tissue. People who are addicted to alcoholic drinks get affected by atrophy of the white matter in their brains. This makes them forget things and they are unable to move their arms and legs properly.

▶ Alcoholism doubles the risk of brain shrinkage in the 30s to 50s age groups

Brain Ventricles

Like the heart, the brain also has four '**ventricles**' filled with cerebrospinal fluid that keep it nourished. There are two lateral ventricles on the sides, and the third ventricle and fourth ventricles connect them to the central canal of the spinal cord. A thick network of blood capillaries in the choroid plexus supplies the cerebrospinal fluid (CSF) with the minerals and nutrients that it needs.

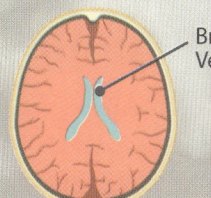

▶ The ventricles are hollow spaces that contains the CSF

Incredible Individuals

Many mental and nervous diseases like Tourette (pronounced as too-ret) syndrome happen because of damage to the basal ganglia. A patient of this disease appears quite normal, but may repeat themselves while speaking or make sudden sounds while talking, including grunts, whistles, or swears, or make sudden movements called tics. A lot of what we know today about this syndrome comes from Oliver Sacks (1933–2015), a British neurologist who researched on unusual and rare neurological disorders. He received the Guggenheim fellowship for his research on Tourette syndrome.

▶ Oliver Sacks' research helped many patients with this syndrome. David Beckham and Wolfgang Mozart have suffered from Tourette syndrome

Voluntary and Involuntary Movements

Did you know that our body is in motion all the time? But we are aware of only some movements, like walking and eating. We can control these movements when we like. When you are lagging in a race, you can make your leg muscles work harder so you run faster.

▲ Voluntary movements done without thinking are called reflexes

 ## Understanding Body Movements

Some movements in the body happen on their own and we do not have control over them. For example, our heart beats and our stomach churns. These are called involuntary movements. Your heart beats faster when you are excited or frightened, but you cannot make it beat slower even if you want to. Conversely, some voluntary movements can become involuntary when you have to react very fast. For example, if you burst a balloon behind someone, they cover their ears. Some involuntary movements can become voluntary. Breathing, for example, is an involuntary action. Your brain regulates your breathing; specifically, the part of your brain called medulla oblongata. However, if you run too quickly, you might need to take in greater gulps of air to feel better. At this point, your breathing becomes voluntary. This is controlled by another part of the brain called the precentral gyrus.

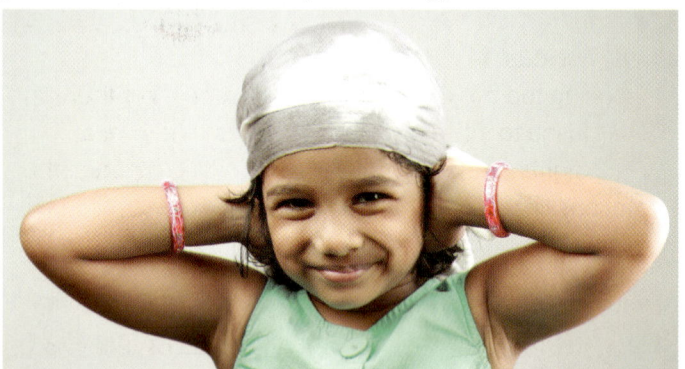

▲ When you hear loud or sudden noises, your hands might reflexively cover your ears

Isn't It Amazing!

Gurning is the British slang word for making faces. If you can make a lot of faces, you should head to the Gurning World Championships in Egremont, UK. The top prize goes to men, women, boys, and girls who can make the most faces.

▲ Voluntary movements are controlled by the higher brain

 ## Voluntary Movements

These are movements that are controlled by the higher brain, especially the sensorimotor cortex in the parietal lobe. These include control of the muscles of the legs and arms, fingers and toes, face and jaws, and the abdomen. That is why you can smile at friends or draw your stomach in when you have to stand in attention.

 ## Involuntary Movements

These are the movements controlled by the cerebellum and lower nuclei of the brain and also by the autonomic nervous system. These include the contraction of the stomach and intestines to move food forward, the beating of the heart, the blinking of the eyes, and the contraction and expansion of the lungs.

▶ While closing her eyes and sniffing, this girl has made her involuntary muscles become voluntary

Moves Like Lightning
The Autonomic Nervous System

Sometimes you need to react so fast that there is not enough time for your brain to think. That is when the autonomic nervous system takes over. For example, when you accidentally touch a hot pan, your hand instantly withdraws away from it; or when ambient light shifts, the dilation of the pupils will change accordingly.

It also takes care of the things which would otherwise take too much of your time, such as digesting food, making blood flow through your body, helping your tired muscles become relaxed, and keeping your urinary bladder in check when you are sleeping. When you wake up, it lets the brain take over. The automatic nervous system is divided into two parts: the sympathetic and parasympathetic nervous systems.

In Real Life

Sympathy has nothing to do with the sympathetic nervous system! Instead, it is controlled by a part of the brain's temporal lobe called the **amygdala** and special kinds of neurones called **mirror neurones**.

▲ Did you know that children can feel sympathy from the age of two?

Fight or Flight

The spinal cord and a set of nerves, which constitute the sympathetic nervous system take care of most situations when you need to act without thinking too much. It prepares your body to either run away from a dangerous situation, or to turn around and fight. So, if a mosquito sits on your hand, you might swat at it without thinking. In more serious situations such as people getting into fights or discovering a fire, they might punch the opponent or douse the fire. When this system is active, you feel emotions like excitement, anger, fear, and disgust.

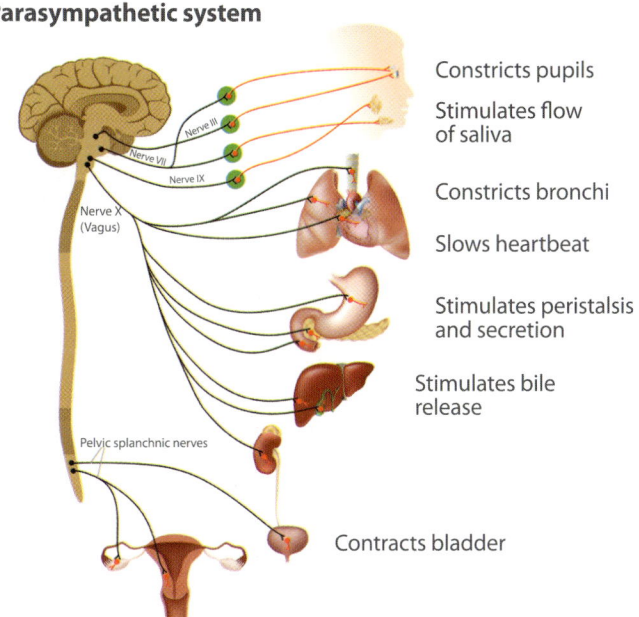

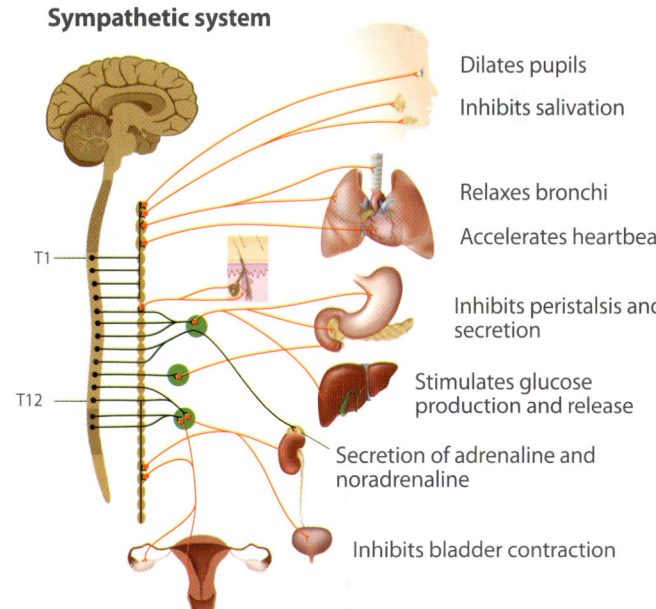

▲ The parasympathetic nervous system prepares your body for rest and digestion

▲ The sympathetic nervous system prepares your body to either fight something or run away from it

Rest and Digest

The parasympathetic nervous system makes you feel tempted to eat when you see the food you like. It makes you calmer and happier, while your digestive system gets ready to digest. You start drooling, your stomach and small intestine start releasing digestive enzymes, and your large intestine prepares to get rid of your last meal so that there is space for the new meal. When this system is active, you feel satisfied and sleepy.

Breaking News to the Brain
The Senses I

All our senses can be divided into two. There are senses that need to be felt, such as touch, taste, and smell. Then, there are senses like sight and sound that are not felt. You can also divide them into those that help you ascertain the direction of the sensory stimulus (as shown on this page) and those that do not (see *pp 14–15*). The sense organs in the head communicate directly to the brain, while those in the rest of the body (mostly in your skin) communicate with the spinal cord. The sense of balance is unique as it does not depend on external stimuli.

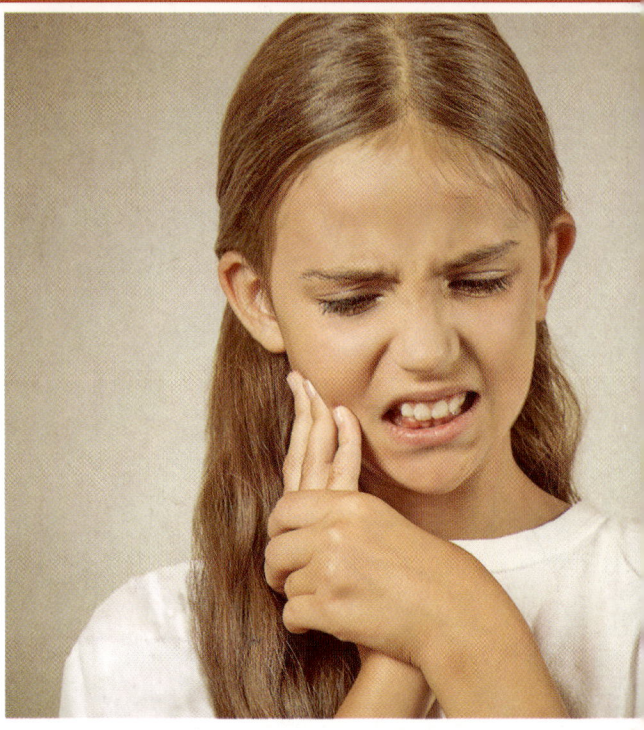

Taste

If you look at your tongue in the mirror, you will see that it is made up of hundreds of grainy bumps that doctors call **papillae**. Each papilla has a number of special nerves called **gustatory receptor cells** (GRCs). Different papillae have different kinds of GRCs for things that taste sweet, sour, bitter, or salty.

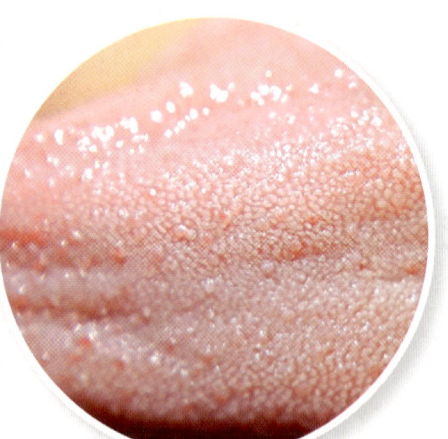

◀ *The tongue has around 3,000 to 10,000 taste buds*

Smell

Deep inside the nose is an organ called the **olfactory bulb**, which gives us the sense of smell. It is made of thousands of **olfactory receptor cells** (ORCs), which catch chemicals in the air we breathe and tell us whether they are nice, like oranges, or pungent, like garlic. The nose also has pain receptors, which help pick up very strong smells that suggest that you may be exposed to a dangerous chemical like ammonia.

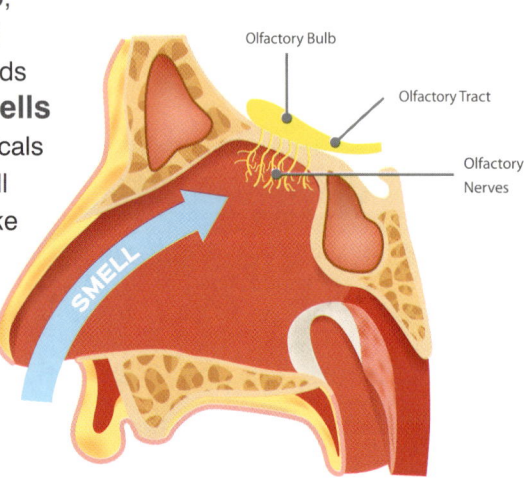

HUMAN BODY | BRAIN & NERVOUS SYSTEM

Isn't It Amazing!

Butterflies have their taste buds on their feet!

▲ *Butterflies put their feet on the flower to quickly decide if it has nectar*

In Real Life

A lot of what we think of as 'flavour', is actually both smell and taste together. The brain treats them together. That is why, when you have a cold and your ORCs are blocked due to a runny nose, food tastes odd even though your GRCs are not blocked.

▶ *Having a cold prevents you from tasting or smelling things properly*

Touch, Heat, and Pain

Our skin is a giant sensory organ that can feel many things. Different kinds of nerve endings in the skin tell the brain and spinal cord different things. Each part of the skin is represented in the brain in a map called the homunculus.

▶ *Skin is the largest organ in the body*

Cutaneous receptors

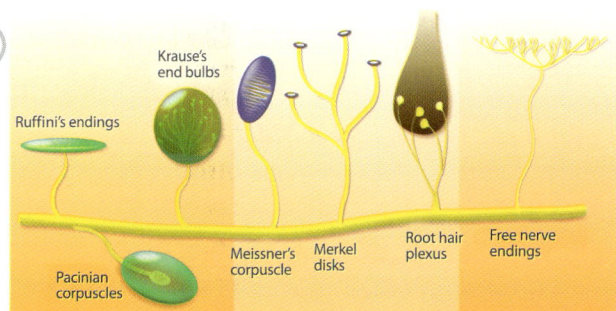

Sensory Ending	Place in the Body	What it Senses
Free nerve endings	All over the body	Pain, heat/cold, pressure, or twisting
Merkel's discs	All over the body	Very light touch (like a mosquito landing)
Ruffini's corpuscle	Skin and joints	Stretching the body
Meissner's corpuscle	Fingertips and lips	Light touch
Pacinian corpuscle	Inner skin	Strong pressure (like a tight hug)
Root hair follicle plexus	Hair follicles in the skin	Movement of hair (like an insect brushing past it)
Krause's end bulbs	Eyes, lips, and tongue	Cold

Incredible Individuals

Anaesthesia is a medical treatment given to a patient before surgery by doctors, so that one does not feel any pain at all. It was invented in England in the 19th century, but many doctors did not use it, making their patients go through surgery in great pain. Queen Victoria (1819–1901) decided to take anaesthesia when she gave birth to her eighth child, Prince Leopold (1853–1884). This made the public demand it and doctors began to use it more.

▶ *Queen Victoria was the first member of the Royal family to live at Buckingham Palace*

Sight and Sound
The Senses II

Our eyes and ears do not just see and hear. Since we have two of each, the brain can also ascertain where the stimuli are coming from, using the eyes and ears. So, if you hear something falling nearby, you know that you have to run in the opposite direction to save yourself. When you are playing cricket, your eyes tell you how far away the ball is from where you are standing, so you know when to hit it. Your sense of balance allows you to walk smoothly and correct yourself if you slip.

Hearing

The ear 'hears' just like a telephone does. The outer ear acts like a microphone, collecting sound from around you. These sounds go through your ear canal to the ear drum. This is a tiny stretch of skin that vibrates, just like a telephone diaphragm. The tiny ear bones pick these vibrations and take them to the cochlea. The cochlea is coiled like a snail's shell and breaks what you are hearing into separate sounds, so you can hear even if separate people are calling for you at the same time.

The brain can hear the same sound through both ears, so it can figure out whether the sound is coming from the left, the right, the front, or behind you.

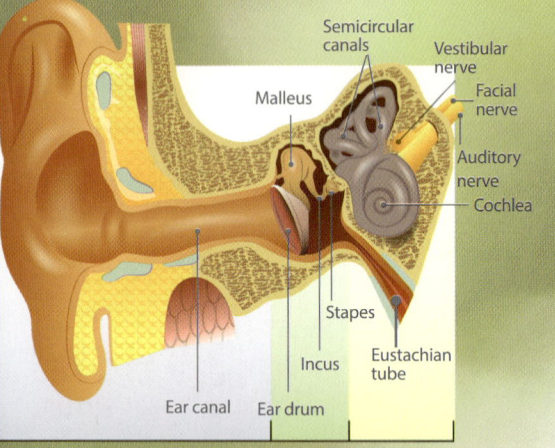

▲ Your outer ear never stops growing throughout your lifetime

Balance

Do you see the semi-circular canals in the figure above? These have nothing to do with hearing, but tell your brain about your body's balance. They are filled with a jelly-like substance and have tiny hairs inside them. If you are wobbling without balance, the jelly moves and the hairs pick up on this movement and communicate to the brain. The brain then sends a message to the relevant muscles to steady yourself, even without you realising it.

HUMAN BODY — BRAIN & NERVOUS SYSTEM

⭐ Incredible Individuals

The famous music composer Wolfgang Mozart had a rare ability called an 'absolute pitch', which helped him identify correctly a musical note and even compose the same himself.

▲ Mozart composed his first piece of music at the age of five!

Sight

The human eye works just like a camera. The pupil and iris in the front of your eye act like shutters that close and open to let in light. The lens concentrates light just like the lens of a camera. Light finally lands on the retina, which has special **photoreceptor cells** (PRCs). They are of two types—**rod cells**, used for night vision; and **cone cells** that work better in daylight. The cone cells sense only three colours of light: red, green, and blue.

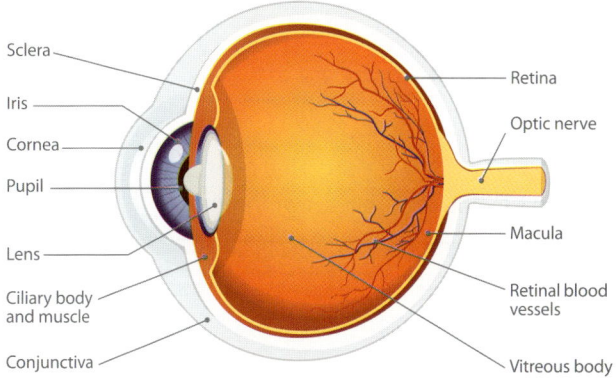

Binocular Vision

Both your eyes tell the brain slightly different things. What is right in front of you is the same, but the left eye sees things to your left, and the right eye sees things to your right. Using these differences, the brain can make a 3D map of the world in front of you, so not only do you see things, but you also know how far they are.

▶ Humans have a maximum horizontal field of 200° with both eyes

In Real Life

Can you spot the numbers in the picture? Colour blind people cannot as they have defects in their cone cells that stop them from sensing the difference in colours.

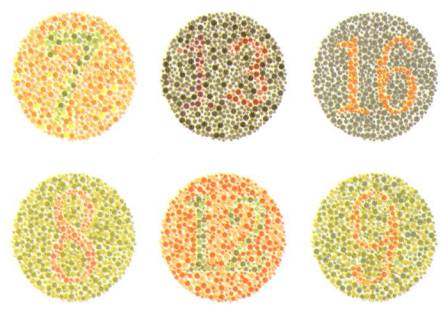

▶ The numbers are written in different colours than the circles

Nerves and Neurones
How They Work

The neurone is a nerve cell that forms the basic unit of the nervous system. Unlike other cells, neurones can stretch for several inches along the body. They clump together in the brain to make nuclei. In the rest of the body, neurones clump together to make **ganglia**. Bundles of neurones that interact with the same organ are called nerves.

▼ The diagram provides a closer look at the nerves, neurones and axons of the body

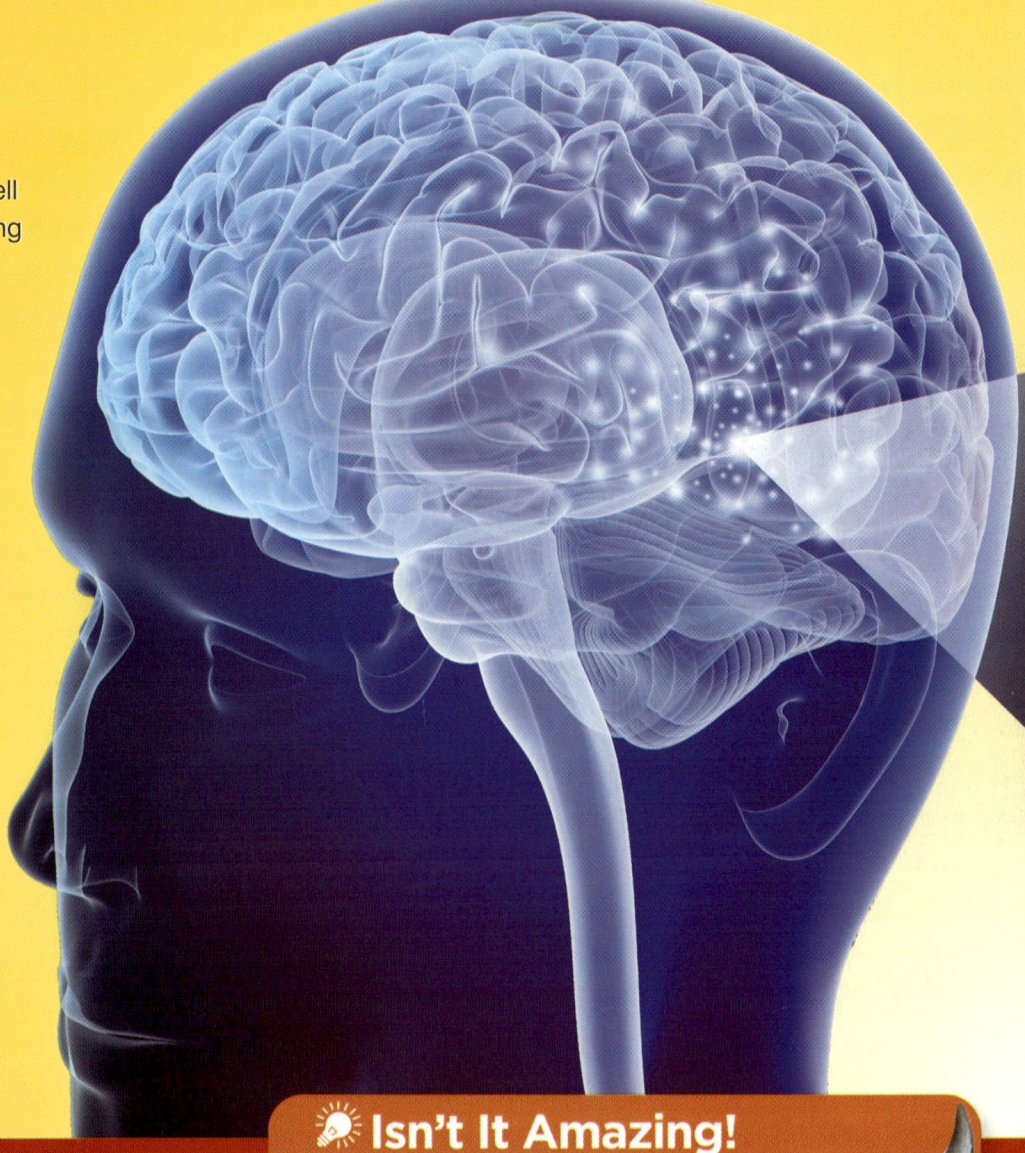

Cell Body

Most of the neurone's mass sits in the cell body. This has lots of mitochondria, giving the neurone all the energy it requires.

Dendrites

Dendrites look like plant roots and are parts of the neurone that touch other body cells or neurones. They gather information from the sensory cells and pass them onto the axon at the other end.

Axon

This is the longest and most important part of a neurone. It carries news from the dendrites to the next neurone over long distances. This information is carried in the form of tiny amounts of electricity called an action potential. **Axons** in the brain make up its white matter and are not covered by myelin sheaths.

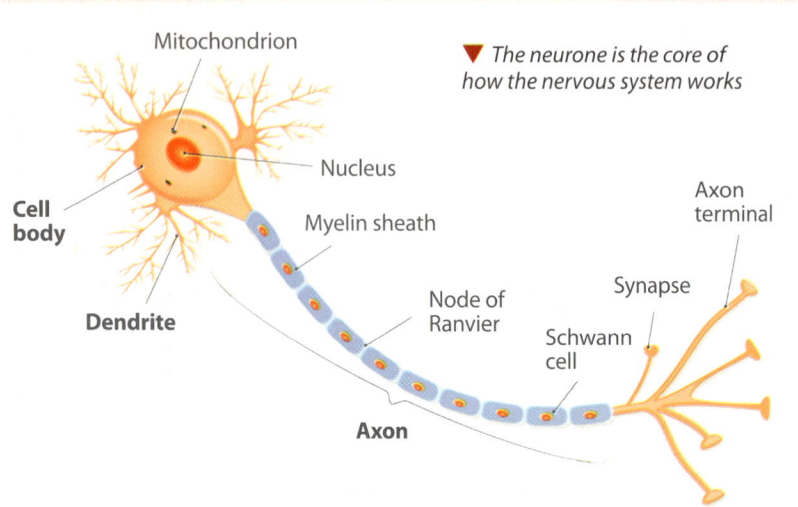

▼ The neurone is the core of how the nervous system works

Isn't It Amazing!

A blue whale can grow up to 34 metres long. Some of the neurones in its spinal cord run all the way to the tip of its tail. At nearly 30 metres, this makes the blue whale's spinal neurones the longest single cells in the world!

▲ Blue whales have the longest neurones in the world

Myelin Sheath

Like the plastic insulation that protects electric cables, the **myelin sheath** protects the axon, so that the information does not leak. It is made of a material called myelin, which is made by the **Schwann cells**. The glial cells wrap themselves around the axon, making an extra cover. We call the gaps between Schwann cells the **Nodes of Ranvier**. These allow the neurone to recharge itself.

Synapses

A synapse is where one neurone meets another to pass on the information. A neurone will have synapses with dozens of others. They also act like switches, slowing down or speeding up news travelling through the nerves. The brain makes new synapses between different neurones all the time, and that is how we make new memories and learn new things.

Action Potential

This is a tiny electric current that passes along the wall of the axon. When there is no message to be passed along the nerve, there is sodium (Na$^+$) outside the neurone and potassium (K$^+$) and chloride (Cl$^-$) inside it. This is the resting potential. When there is a message to be passed, Na$^+$ rushes in and K$^+$ rushes out, creating the action potential. After the message has passed, the two slowly switch places again, bringing the neurone back to normal.

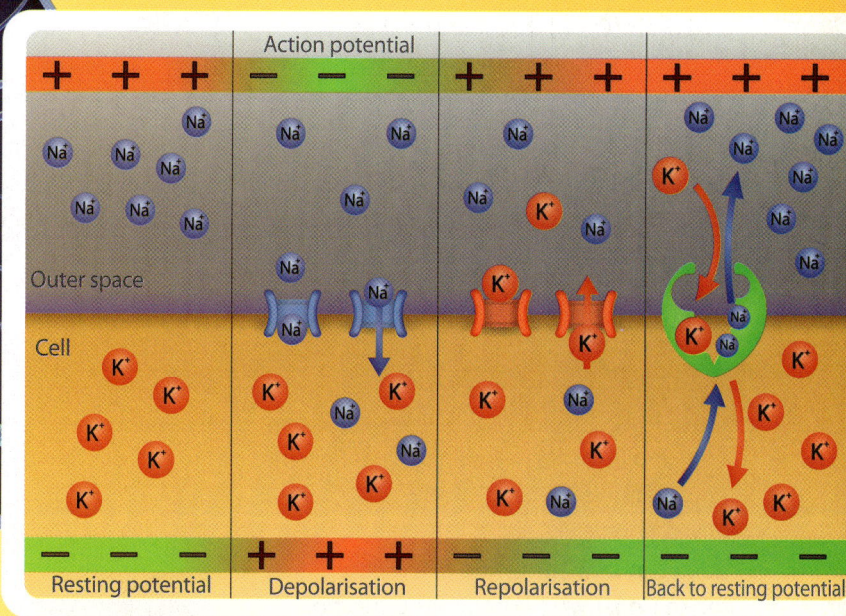

▲ Neurones take messages to and from the brain through action potentials

Mirror Neurones

These are neurones found in many parts of the brain. They help to bridge some of the senses with action. Scientists conducted experiments in which they found that these neurones become active when we see or hear something that is familiar to us. For example, these become active in ballet dancers' brains when they see other ballet dancers. These neurones help us learn things by seeing or hearing them, and also in expressing feelings like sympathy.

Incredible Individuals

Santiago Ramón y Cajal (1852–1934) was a Spanish neuroscientist who discovered that the neurone was the basis of the nerve cell. He developed techniques to 'stain' individual neurones and follow their path across the nerves. Many of the drawings he made are still used by medical teachers.

▲ Santiago Ramón y Cajal was the first Spaniard to win a scientific Nobel Prize

Your Body's Other Brain

The spinal cord takes messages from the organs of the body to the brain and back from it to the organs. It also takes messages to the muscles from the higher brain, but it does a lot more than acting like a courier. It also governs the sympathetic and parasympathetic nervous systems, which do not depend on the brain. It further handles automatic reactions of your body that the brain does not get told about, like your reflexes. That is why it acts as your body's second brain. Any damage to the spinal cord makes you unable to move some parts of the body, which doctors call paralysis.

Spinal Cord

The spinal cord runs entirely inside your backbone. Unlike the brain, the white matter is outside, and the grey matter is inside. The white matter is made of the axons and dendrites of sensory neurones coming in from the various organs and the skin, motor neurones going out to the organs and the skin, and neurones travelling through the length of the spinal cord, connecting to the brain stem. The grey matter is made of the cell bodies of these neurones, and small connecting neurones called **interneurones**.

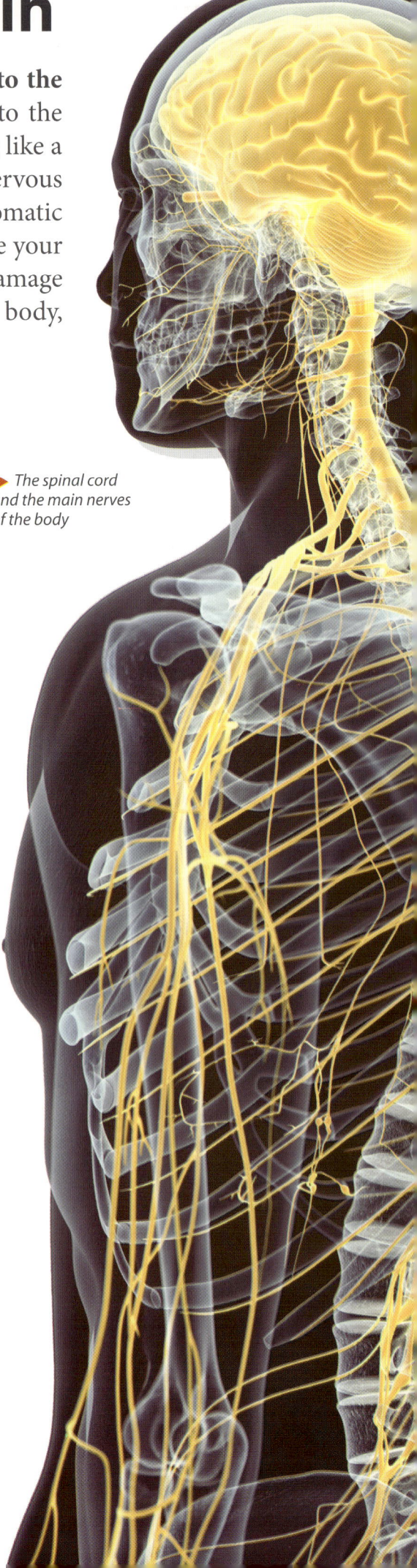

▶ *The spinal cord and the main nerves of the body*

In Real Life

Does the octopus have a backbone? No! That is precisely why it is called an invertebrate marine animal. This feature, along with them being extremely intelligent and curious creatures, makes octopi incredible escape artists. For instance, Sid, an octopus who was in captivity in New Zealand, escaped his tank multiple times before finally being released by the aquarium officials as they were fed up.

▲ *When bored, octopi are known to amuse themselves—from juggling hermit crabs to short circuiting aquarium lights*

HUMAN BODY — BRAIN & NERVOUS SYSTEM

Reflexes

Reflexes are reactions of the body that do not require you to think. These are actions you take out of fear, hunger, pain, or anything else that requires you to act quickly. During a reflex, the sensory neurone connects with an interneurone through a synapse, which connects to a matching motor neurone. For example, some neurones from your fingers that can sense very hot things are connected to neurones in your forearm that pull the hand away. So, when you touch a pie right out of the oven, you feel the heat and pull your hand away, dropping the pie. Your eye sees the mess and your hands get to work cleaning it up. That is a second reflex! But it is the one that goes through the brain.

There are two kinds of reflexes. The somatic reflexes connect the sensory organs with skeletal (voluntary) muscles. These often act as part of the sympathetic nervous system, which helps you to deal with sudden situations. You can train yourself to stop these reflexes.

Visceral reflexes connect the sensory organs to internal (involuntary) muscles and you cannot control them. These are often part of the parasympathetic nervous system, such as feeling hungry after seeing a cake.

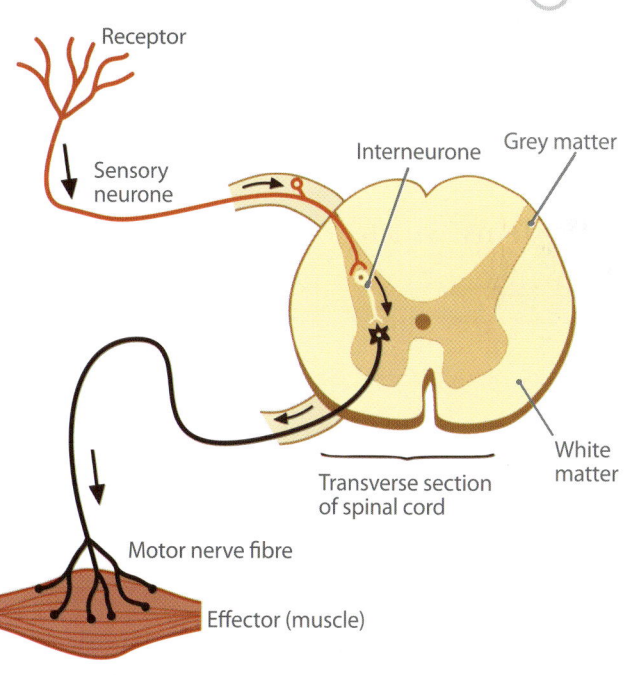

▲ Spinal reflexes allow you to act without waiting for the brain to decide

Keeping Your Two Brains Safe

The brain and spinal cord are not supplied with blood directly. Instead, the cerebrospinal fluid keeps them nourished. Blood capillaries in the brain are surrounded by tiny cells called glia. The glia makes up the blood-brain barrier and makes sure that nothing other than water and minerals can leave the capillaries and enter the cerebrospinal fluid. Proteins, large fats, and germs cannot cross this barrier. Glucose is moved into the brain by active transport.

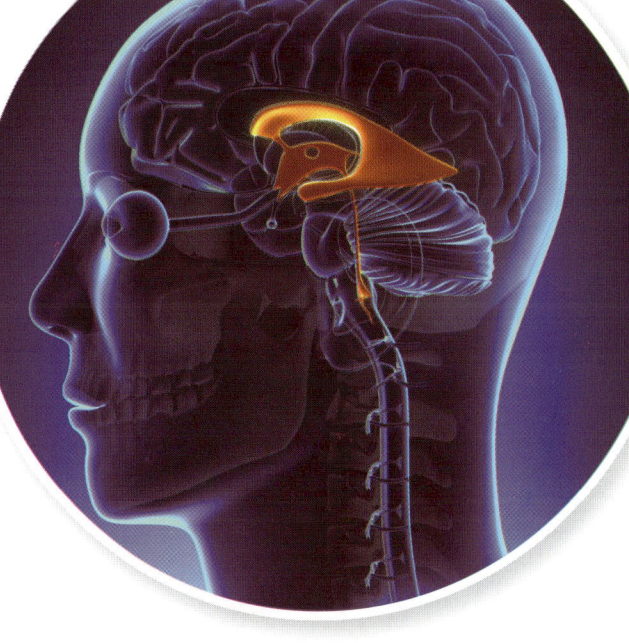

▲ The ventricles, full of cerebrospinal fluid, nourish the brain

◀ Chameleons have some of the fastest reflexes in the world

Isn't It Amazing!

Chameleon tongues have some of the fastest reflexes in the world. Once it has seen its prey, a chameleon can shoot its tongue out at 2.6 kmps to catch it.

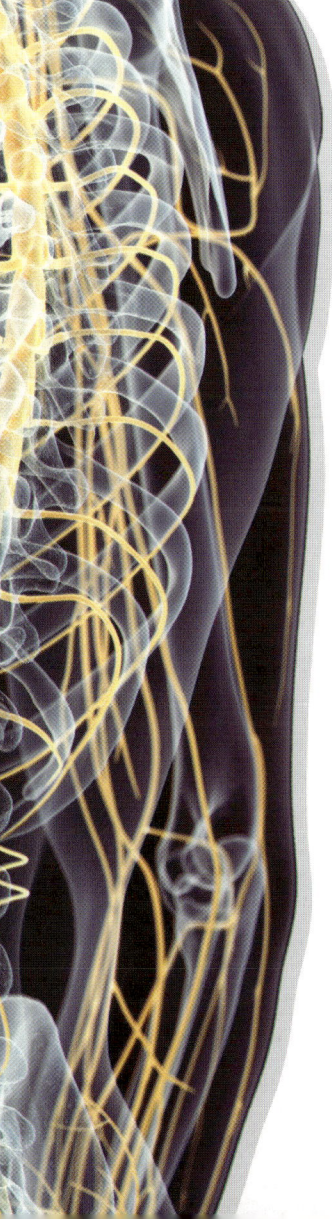

Switching Your Nerves Up and Down

If you think of the nervous system as a cable network, it has got to have switches that turn things on and off. This is done by joints between neurones called synapses and chemicals called neurotransmitters. The brain and spinal cord are full of synapses. In the spinal cord, these are important for reflexes to happen smoothly. In the brain, they help it to make up new ideas, thoughts, and memories, but we do not know how that happens. However, it does seem like the more synapses that animals have, the smarter they are.

In Real Life

The brain stem, midbrain, and frontal cortex all work together to help you decide whether you like some things or not, through the reward pathway. The synapses of the neurones in this pathway use **dopamine**. If you really like something, like chocolate, a lot of dopamine is released. Sometimes this goes out of control and becomes an addiction. Drugs that block dopamine from tagging its receptors are used to treat it.

▲ Dopamine helps control what you like and what you do not like

Synapses

Synapses link the axons of one nerve with the dendrites of the one after it. When an action potential reaches the end of its axon, it triggers tiny sacs at the tip, called synaptic vesicles, that are filled with chemicals known as neurotransmitters. These go to the end of the axon and burst, spraying the chemical into the space of the synapse. The chemicals land on the surface of the next neurone, where there are proteins called receptors lined up on the membrane. When the neurotransmitter touches the receptor, an action potential starts in the second neurone. Our brain may have nearly three hundred trillion synapses, 300,000,000,000,000 of them!

Synapses form when two nerves connect with each other. If they keep 'communicating' to each other, the synapse becomes a strong synapse, else it remains a weak synapse. If the nerves have not communicated in a long time, the synapse may be broken up. Some scientists believe that the formation of strong synapses helps in keeping memories and that the breakage of weak ones makes us forget things.

Neurotransmitters

These are the chemicals that help neurones 'talk' to each other. There are three key types of neurotransmitters based on their functions. Excitatory neurotransmitters turn up the action potential in the next neurone; inhibitory neurotransmitters turn down the action potential in the next neurone; and modulatory neurotransmitters send messages to many neurones at the same time and can also communicate with other neurotransmitters. Many neurotransmitters also act like hormones.

★ Incredible Individuals

Camillo Golgi (1843–1926) was an Italian neuro-scientist and teacher of Ramón y Cajal. He discovered many features of the nervous system, including the 'Golgi Vesicle'. This is a tiny bunch of membranes within each cell, where things that need to be sent out of the cell (like neurotransmitters) are packaged into little bags called 'vesicles'.

▲ *For their work, Golgi and Cajal were jointly awarded the Nobel Prize in Physiology or Medicine in 1906*

Neurotransmitter	Location	Nature	Response
Acetylcholine	Heart	Inhibitory	Makes the heart slow down
	Skeletal muscle	Excitatory	Makes the muscle contract
Glycine	Spinal cord and brain stem	Inhibitory	Quietens the next neurone
Glutamic acid	Brain and spinal cord	Excitatory	Turns up the next neurone
Dopamine	Some nuclei of the brain	Inhibitory	Quietens the next neurone
	Some nuclei of the brain	Excitatory	Turns up the next neurone
Noradrenaline	Skeletal muscle	Excitatory	Makes the muscle contract
	Heart	Excitatory	Increases heartbeat
Serotonin	Grey matter of brain	Inhibitory	Quietens the next neurone

Electroencephalography (EEG)

You know that little electric currents are how our nerves carry messages from the sense organs to the brain and back to the motor organs. These can be measured by neuroscientists by using a special instrument called an electroencephalogram or EEG machine. Diseases like epilepsy and motor neurone disease can be diagnosed using this machine. By attaching other machines, a process called Functional Magnetic Resonance Imaging (FMRI) is used to find out which part of the brain becomes 'electrically active' when you are thinking about something, solving a math problem, or reading.

From Thought to Action

Sometimes we do things after carefully thinking about them, like writing the answer to math problems. Often, we do not think at all, like when we need to catch a ball in a hurry. All these actions are governed by the motor nerves. They carry the brain's decision to the muscles, which do as they are told. The polio-causing virus, poliomyelitis virus (PMV), attacks the motor neurones and causes paralysis in its patients. Another disease called Amyotrophic lateral sclerosis (ALS) also affects the motor neurones.

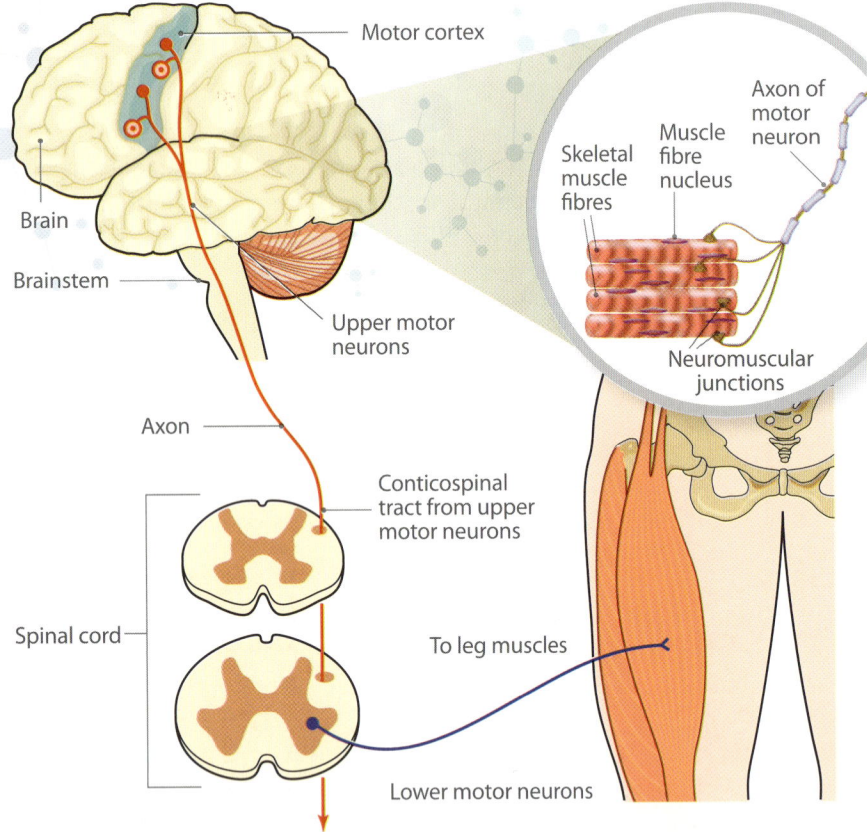

◀ The motor neurones finally tell the muscles what to do

Motor Neurones

The part of the brain that deals with actions is called the motor cortex. From there, upper motor neurones take messages to the spinal cord through the midbrain. **Lower motor neurones** start from the spinal cord and finally connect with muscles. Conscious actions need both, while reflex actions work only through the latter.

Muscles

Our muscles are made of lots of muscular fibre. Motor neurones make lots of synapses with them, that doctors call **neuromuscular junctions**. When a current comes through the neurone to the junction, it makes the muscle contract. In turn, the bone or organ that it is attached to is made to move.

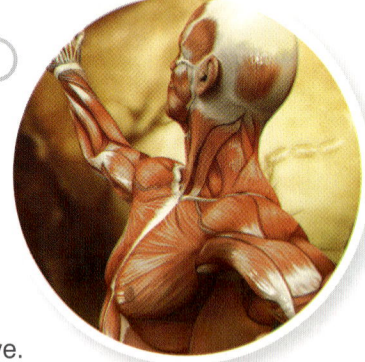

▲ Human muscles are made of lots of muscular fibres

In Real Life

An hour of play outside home helps train your nervous system to control your movements and speed up your reflexes. It also speeds up the ability to think on your feet. So, go ahead, drop that gaming console and pick up a football.

▲ Play is good for your motor nerves

Stress

Stress is how our nervous and muscular systems respond to changes in the world around us. Acute stress happens when there is immediate danger. The sympathetic nervous system wakes up the muscles, and prepares you to run away. Chronic stress happens when you cannot easily run away from a problem, like exams.

Making Memories

We still do not quite know exactly how our brain remembers things. Some scientists believe it has to do with the number of synapses in the brain. But we do know that memories come in two forms. The first are permanent ones, called engrams. You remember these for a long time, like your birthday; or a really terrible experience you might have had, like being pranked with a cake full of chilli. Others are called short-term memories, which you soon forget, like what you had for breakfast. You might forget it faster if you did not like what you ate.

Hippocampus

Did you know that **hippocampus** means seahorse in Greek? Well, whether it looks like a seahorse to you or not, it is very important in how you remember things. You have two of them, in each of the temporal lobes. They help you remember where things are—as in your spatial memory, mix many inputs into a single memory—so when you remember an event, you remember the sights, sounds, smells, and more; and also decide to remember some things for a long time such as a trip to Disneyland and forget other things, for instance, your dog having eaten your homework.

Incredible Individuals

London's taxi drivers have to learn the routes to nearly 25,000 streets to get a licence. This means the hippocampi in their brain grows bigger to help make room for all that memory.

▶ London taxi drivers have bigger hippocampi than regular people

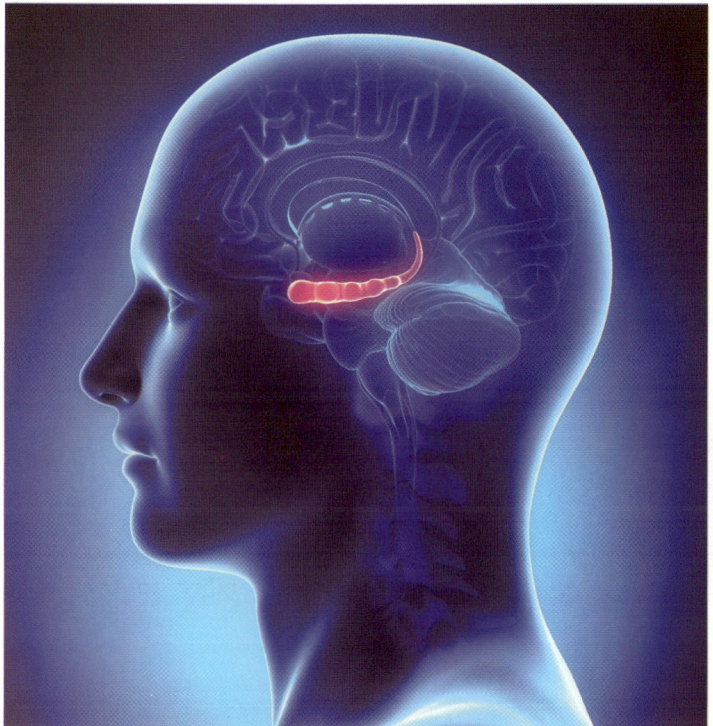

▲ Location of hippocampus

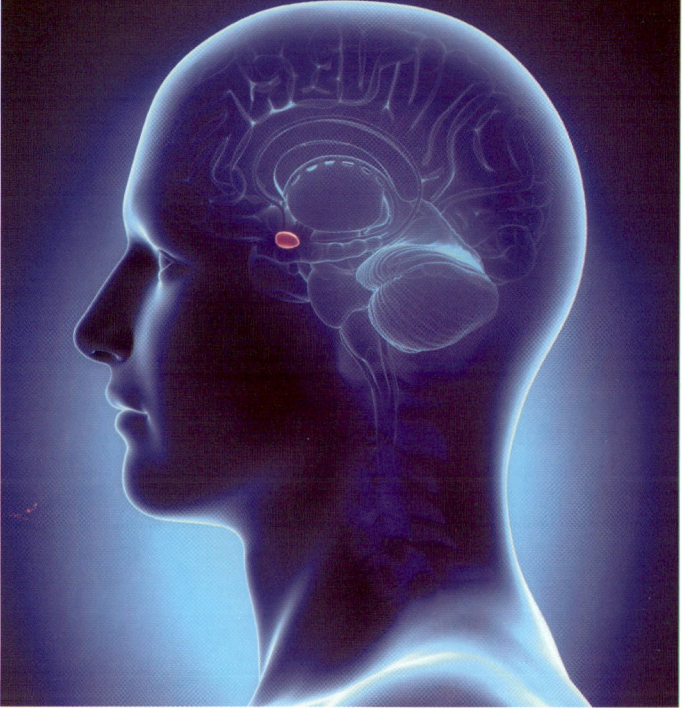

▲ Location of amygdala

Amygdala

This is a pair of tiny, almond-shaped parts of the temporal lobe, one on each side. The amygdala controls learning and memory along with the hippocampus. It also controls the reward pathway. This is the way in which the brain makes you feel good after eating something nice like, a pastry or doing something good such as helping someone in need. It is also involved in addictions to drugs and alcohol. It also makes you feel bad for things you didn't like to do, like eating broccoli!

Higher Thought
From Animal to Human

What happens when you see a delicious plate of food? How does the brain put all the senses together? How do we remember what the food tastes like, and how do we learn to like it? The answer to these questions is in the cortex. The cortex appeared only in animals such as birds and mammals. As human beings evolved, it became more complex.

Cortex

Humans have the largest cortex (*see diagram of human brain pp 6*) compared to body size. A human cortex has 16.3 billion neurones, compared to an elephant's cortex, which has only 5.6 billion neurones, even though it is twice as large.

Combining the Senses

Different parts of the cortex take care of different senses and motor actions. They connect to sense or motor organs via the inner brain and midbrain. The associated areas bring different senses together and that is how something becomes a multisensory experience. Broca's area and Wernicke's area are better developed in human beings than in any other living organisms. They help with language processing. Finally, the prefrontal lobe takes up more complex things like telling right from wrong, social behaviour, doing math, personality, and character.

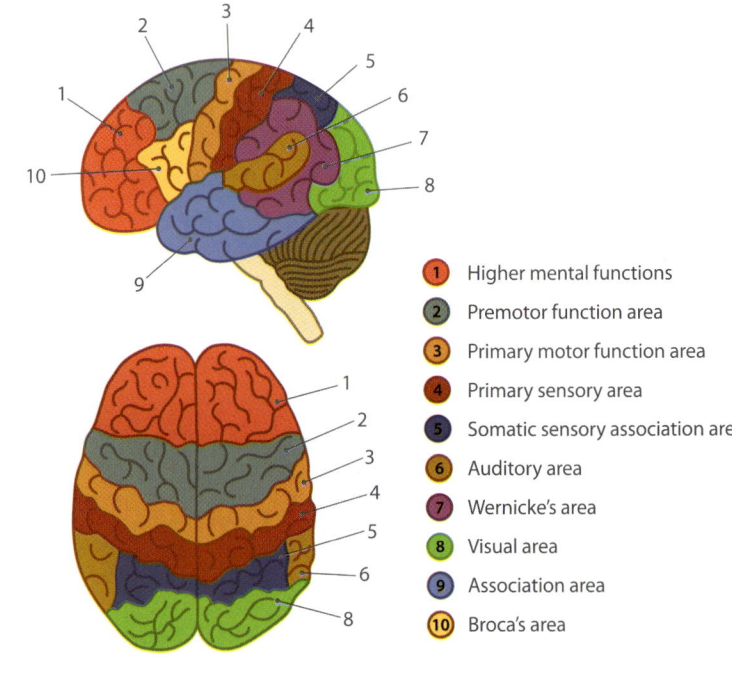

▲ *Different functional areas of the cortex based on its lobes*

1. Higher mental functions
2. Premotor function area
3. Primary motor function area
4. Primary sensory area
5. Somatic sensory association area
6. Auditory area
7. Wernicke's area
8. Visual area
9. Association area
10. Broca's area

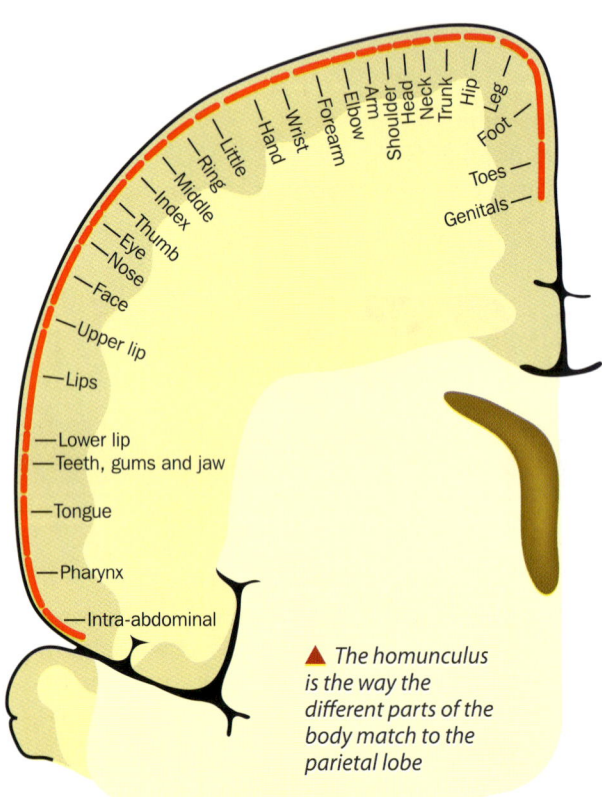

Homunculus

Homunculus is Latin for little fellow. It is a drawing made by neuroscientists that shows how much space the parietal lobe, which deals with touch gives to each part of the body. Some parts of the body, such as the lips and fingers are very sensitive to touch. Therefore, they have more nerve-endings and have a lot of matching space in the brain.

▲ *The homunculus is the way the different parts of the body match to the parietal lobe*

In Real Life

A stereotype is how we think of a group of persons or animals. For example, most people believe foxes are sly or cunning. Some stereotypes are useful, for example, if you see a lion, or a tiger, you better stay away. But many of them are harmful, and lead to negative thoughts like fear or bias against people of different backgrounds.

The Pituitary and Pineal Glands

When we are tired, we go to sleep. When we go out in the Sun, our body makes more **melanin**, giving us a tan. When we have been sufficiently rested, we wake up. How does our body know all this? This is where the endocrine system comes in. The organs of this system are called glands and make different **hormones**, which are chemicals that travel through blood and tell the other organs of your body what to do. The glands of the body, such as the liver, pancreas, thyroid, and adrenal gland are, in turn, governed by the pituitary gland. You can call it our body's chemical brain.

The Pituitary Gland

The **pituitary gland** makes hormones that control how other glands work. In turn, the pituitary is controlled by the hypothalamus, which contains centres for controlling hunger, thirst, body temperature, tiredness, etc. It makes the:

- ✔ Growth hormone, which makes your body grow.
- ✔ Prolactin, which helps a woman's breasts to make milk for her baby.
- ✔ Thyroid-stimulating hormones, which make the thyroid gland release the thyroid hormone.
- ✔ Adrenocorticotropic hormones, which make the adrenal gland release adrenaline.
- ✔ Antidiuretic hormones, which make the kidneys take up water filtered from the blood.
- ✔ Melanocyte-stimulating hormones, which make the melanocytes in the skin make melanin, which protects you from the UV rays of the sun.

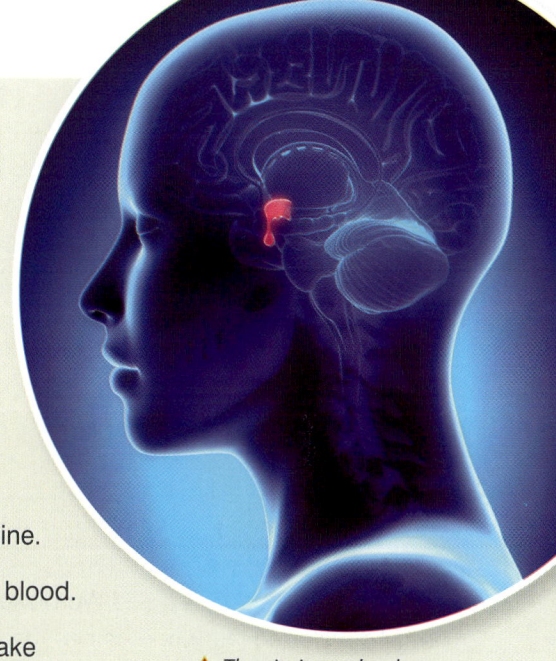

▲ The pituitary gland is called the master gland of the endocrine system

Pineal Gland

This is the brain's other gland, behind the hypothalamus. It makes a hormone called melatonin, which acts on the hypothalamus and tells it to go to sleep. You can also get melatonin from fish, eggs, nuts, and mushrooms. The pineal gland makes melatonin all through the day and releases it into the blood as daylight fades. When light falls on you, the brain and intestines make serotonin, which tell the brain to wake up. Over time, the cycle of sleep and waking becomes regular, and less dependent on light. This is called the circadian rhythm.

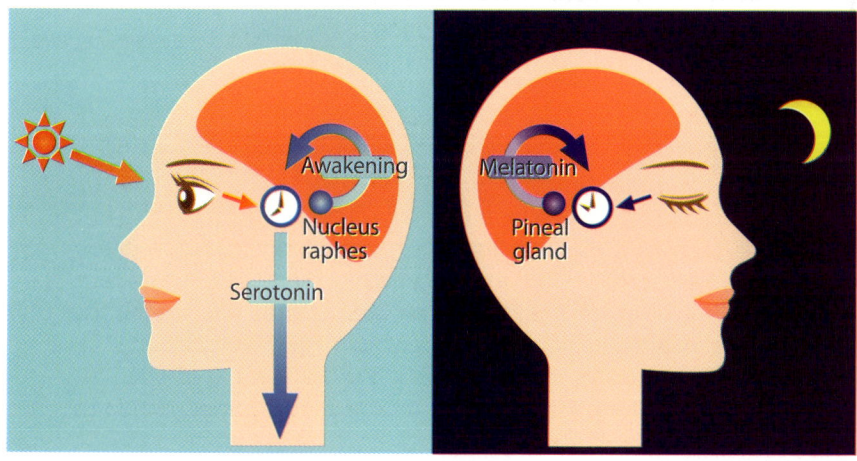

▲ Melatonin makes you sleep; serotonin wakes you up

💡 Isn't It Amazing!

In many birds, reptiles, and mammals, the pineal gland is very important. The pineal gland helps these animals to know the right season to mate and reproduce. Removing this gland makes them unable to breed in the right season. In some species, the pineal gland can also detect light and make melatonin accordingly. So it is sometimes called the 'third eye'.

Higher Thought II

The brain is an organ just like any other. It too can fall sick. But unlike other organs, illnesses of the brain manifest in three ways. The symptoms could be physical, such as uncontrollable movements or paralysis; mental, like loss of memory or the inability to pay attention; or emotional, such as excessive stress or sadness that does not go away.

The Brain and the Mind

The brain is the organ that governs your nervous system. The mind, on the other hand, is a complex set of faculties involved in perceiving, remembering, assessing, and deciding. While the brain exists as an organ with a concrete structure, the mind is invisible and thus, does not have a form like the brain. The mind is what the brain does. The study of the brain and the nervous system is called neuroscience. **Neurology** is the study of how the nerves work and what can go wrong with them. Psychology is the study of how your brain deals with the world, including how you think, your likes and dislikes, good and bad behaviour, and how you make decisions. Psychiatry is a field of medicine that studies what can go wrong with the brain, and how it can be set right. People who work in these fields are called mental health professionals.

▲ Mental health problems can be treated by counselling, forms of therapy, and medication

Depression

A depressed person tends to be sad for a long time. They think very poorly of themselves and feel hopeless, unhappy, and tired all the time. They find it hard to eat or sleep. It happens because the amygdala is not healthy and is not able to control serotonin and noradrenaline. It is treated with medicines called antidepressants.

Incredible Individuals

What is common between Winston Churchill, Beyoncé, Jon Bon Jovi, Johnny Cash, and Robin Williams? They have all suffered from depression in their lives. Remember, it is a mental health disorder and not a state of mind.

Bipolar Disorder

Someone with this disorder of the brain will feel depressed for a few days, and suddenly become very happy and excited on other days. This repeats again and again. It happens because of an underlying trouble in the release of neurotransmitters in the brain.

▲ Did you know that the music star Lady Gaga has bipolar disorder?

Anxiety Disorder

If some people worry about things too much, without any reason, they probably have anxiety disorder. Patients with this condition experience fear, sleeplessness, and mood swings. It happens because the thalamus and other parts of your brain that deal with stress, are not able to work properly. It is treated with counselling and drugs called anxiolytics.

ADHD

Know someone in school who has a hard time paying attention, can barely sit quietly, and gets bored easily? They might just have attention deficit hyperactivity disorder (ADHD). This happens because parts of the basal ganglia are unable to make enough dopamine and **noradrenaline**, which help pay attention and also act as 'brakes' on activity. ADHD cannot be cured completely, but counselling and some drugs help patients become calmer and more attentive.

In Real Life

It is important to seek clinical diagnosis of a mental health condition. However, an inability of the society to be understanding towards a diagnosed mental health patient, and discriminating or making fun of them, or associating shame with their condition causes social stigma. It is likely to make patients want to hide a mental health problem. Be nice to them and try to understand their problems.

▲ ADHD can go undiagnosed in children for years if their inability to focus on a task for long is misunderstood as a choice

An Attack of the Nerves
Epilepsy & Multiple Sclerosis

Some diseases affect the structure or working of the neurones, or both. These may not affect one's mental health, but they show up as physical symptoms like trembling and shivering or being unable to move. Some decades ago, such diseases were met with fear, scorn, or both. It was believed that such conditions were brought about by devils or the wrath of the heavens. Patients were subjected to electric shocks, lobotomies (*see pp 7*), and even tied in chains. However, modern-day medical science has evolved to help treat such conditions with medicine and surgery, if needed.

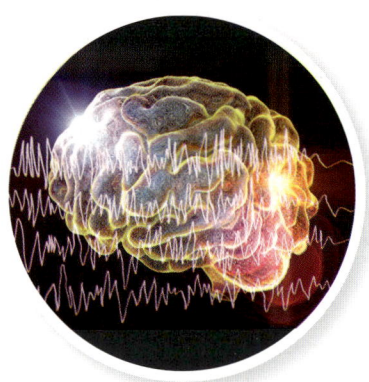

▲ Around 1 in 5 of the world's children and adolescents have a mental disorder

 ## Epilepsy

This illness affects the neurones of the brain. It may be in a small part of the brain or the whole brain. Both happen because the neurones suddenly lose control over their action potential, leading to currents running all over the white matter. During this time, the patient experiences muscle twitching, an inability to breathe, dizziness, and confusion, and sometimes fainting. Doctors call this a seizure. The patient may recover in a few minutes and may sometimes not remember what happened. Seizures may happen because of stress, sleeping poorly, or excessive consumption of alcohol, but unless they happen many times over, it is not considered epilepsy.

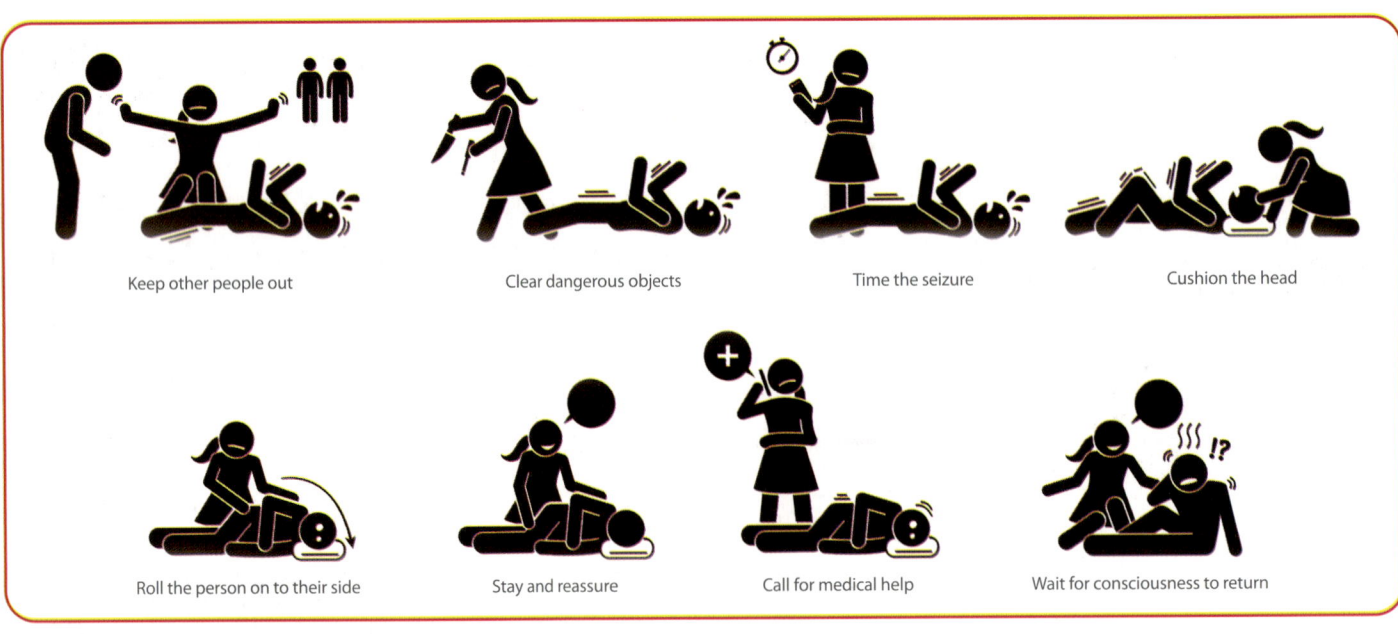

▲ Here's what you should do when someone has a seizure

 ## Multiple Sclerosis

This is a disease in which the myelin sheath of your neurones is lost over time. The neurones are not able to carry currents correctly. It causes you to feel dizzy and things to appear blurry. People can find it difficult to control their movements and can experience difficulty in remembering things. They slowly lose their ability to walk or use their hands.

In Real Life

It is believed that one should put something inside a patient's mouth when they are having a seizure, so that they do not swallow their tongue and choke themselves. However, doctors advise against this practice, so you should not put anything in their mouths.

HUMAN BODY | BRAIN & NERVOUS SYSTEM

Sleep Healthy and Wake Up Healthy

Your brain does not entirely sleep when you do. But it needs you to sleep, so that you can turn the day's experiences into memories, give your muscles and nerves some rest, and let your body grow. This rest is also important for your body to fight disease. Melatonin, made by the pineal gland, tells the brain when it is time to sleep at the end of a day. Over time, it becomes a habit, known as your circadian rhythm. Breaking the habit causes sleep deprivation. Stress or illness may make it difficult for your body to follow the rhythm.

Good Sleeping Habits

It is a good idea to start by maintaining a consistent sleep and wake-up time. Another point to bear in mind is to dedicate an hour before sleep to calming activities such as listening to soothing music or reading a book.

▶ Good sleep habits keep your body and brain healthy

▼ This girl has fallen into the habit of sleepwalking. Her motor neurones are active, but her sensory neurones are not!

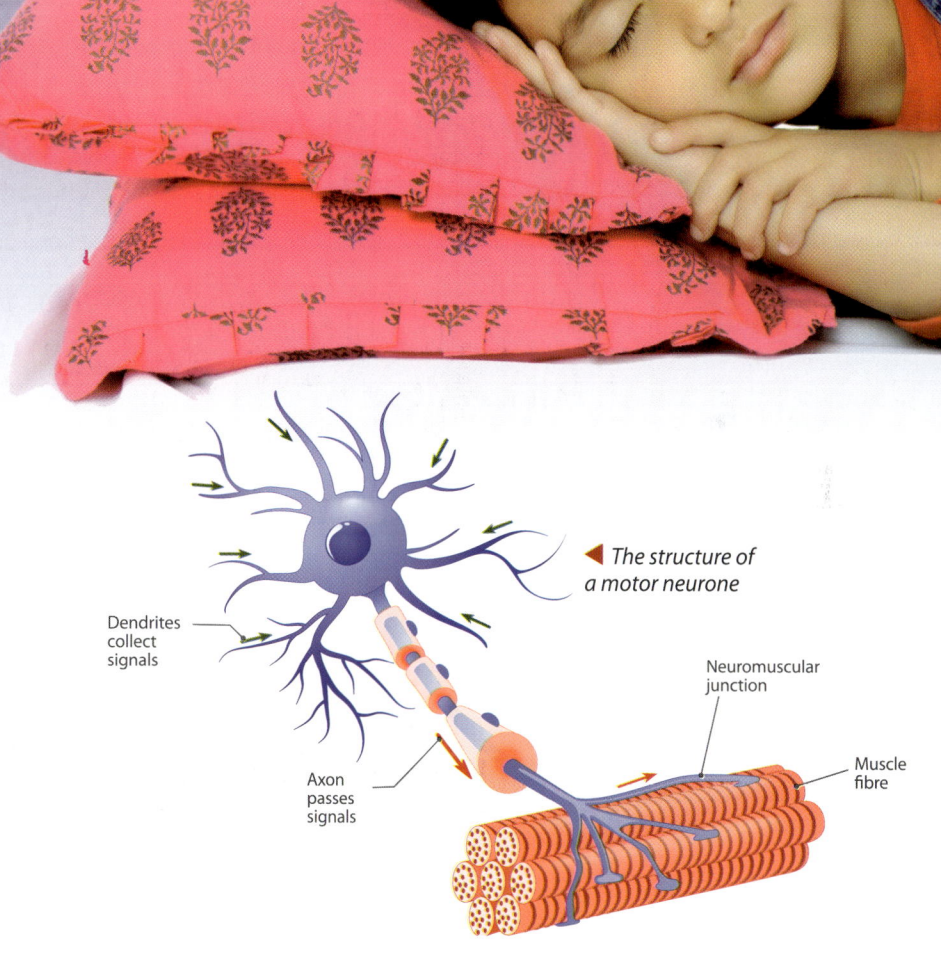

◀ The structure of a motor neurone

Dendrites collect signals

Axon passes signals

Neuromuscular junction

Muscle fibre

Dreaming and Sleepwalking

Did you know that when you are dreaming, your eyes dart about very quickly? This is called **rapid eye movement** or REM sleep. Though your senses are dormant, your higher brain centres are awake and experiencing things that are not happening in reality!

At other times, that is, in a **non-REM sleep**, your higher brain senses are also asleep. In some people though, especially children, the parts of the brain that control movement, that is, the motor cortex are awake. These children may get up from bed, start walking about, and even eat from the fridge! This is called sleepwalking, or somnambulism. It often goes away as you grow up.

A Healthy Mind in a Healthy Body

Though the brain occupies only 2 per cent of your body's weight, it burns upto 20 per cent of the calories you eat and drink. This is because your brain is always at work, even if you are sleeping, which is why you get dreams. But the brain and the rest of the nervous system can only use glucose as a source of energy. The hypothalamus makes sure that there is always enough glucose in the blood to keep the brain fed, even if other tissues have to do with less. If there is not enough glucose, the liver and muscles will turn the glycogen stored in them to glucose to keep the brain nourished.

The Nerves and Fatty Acids

Omega-3 and omega-6 fatty acids are very important to keep your neurones healthy, as they make up the myelin sheath. They are also needed in the **retina**. You get lots of them from seafood, nuts, soybean, and canola oil.

So, tell your parents to get you a big banana split with lots of nuts in it. The nuts give you vitamins, minerals, and fatty acids, the milk in the ice cream gives you calcium, and the bananas give you glucose and potassium. But remember to play and exercise a lot afterwards, so your motor nerves are in good health too!

◀ Foods rich in omega-3 and omega-6 fatty acids

Vitamins and Minerals

Your body needs a lot of vitamins and minerals to keep the nervous system working well. Iron and many vitamins are needed to make neurotransmitters. The minerals potassium, sodium, and chloride, which you get from salt, aid the normal working of action potentials. Calcium is needed for the synapses to release neurotransmitters properly.

▼ Foods rich in potassium are good for your nervous system

In Real Life

Ever called someone who acted crazy 'nuts' or 'gone bananas'? Something that is odd is called 'fishy'. But all these things are actually good for you, for they keep your nerves healthy and make you smarter.

▲ Bananas also support heart health

Improve Your Mental Health

Though we still do not have answers to all the illnesses of the nervous system, indulging in activities that help you unwind can be very fruitful for maintaining a good mental health. This could include anything: from the use of colours and paints to bring your ideas to life, to playing with your pup. Science also backs the use of varied forms of therapy to diagnose, as well as treat mental health conditions in children.

Play Therapy

Play forms a crucial part of how you make sense of your world, as a child. It is also a means for children to express their feelings. Psychologists often use play therapy to aid children in processing complicated emotions such as grief resulting from loss of a parent or caregiver. Play therapy is also great to help you learn desirable behaviours and unlearn the faulty ones.

▲ Psychologists watch children play and figure out signs related to their mental health

Art Therapy

It is not always easy to express your complicated emotions. In order to do so, you can draw art. Psychologists sometimes suggest drawing, painting, sculpting, clay modelling, etc. for children with depression, anxiety disorders, and so on. It is also used with children diagnosed with ADHD as well as low IQ.

Animal-assisted Therapy

Animal-assisted therapists have specially trained animals, who are friendly and warm, so you enjoy playing with them, and feel less worried or frightened about things. It facilitates a child in communicating their complicated emotions better with the therapist. So, when you are feeling bad, give your dog or cat a big fluffy hug!

▲ Art therapy helps you explore your inner 'experience'

Isn't It Amazing!

Catching a flight can be stressful. Before the flight, you are probably hoping you make it in time past security check or long queues. Airports around the world now have new kind of employees—dogs and cats who help passengers calm down and enjoy their wait, known as Emotional Support Animals (ESAs).

◀ Cats, dogs, horses, bunnies, and even goats help with animal-assisted therapy

HEART & CIRCULATORY SYSTEM

THE AMAZING HUMAN BODY

You might not realise it, but our body is constantly working. There are so many functions and actions that the human body needs to perform. The heart, for example, needs to keep pumping blood constantly. If it stops beating, we stop breathing. So, our body is the perfect example of a complex machine. There are many different cells and tissues that form the organs in our body. These organs work together to perform different tasks.

When different organs work together, we call this an organ system. These organ systems work together for our body to function well as a whole. Each organ system has a specific role. Do you know how many organ systems there are in the human body? Eleven! One of these is the circulatory system. The heart, blood, and blood vessels are part of this organ system. Read on to find out how our circulatory system works.

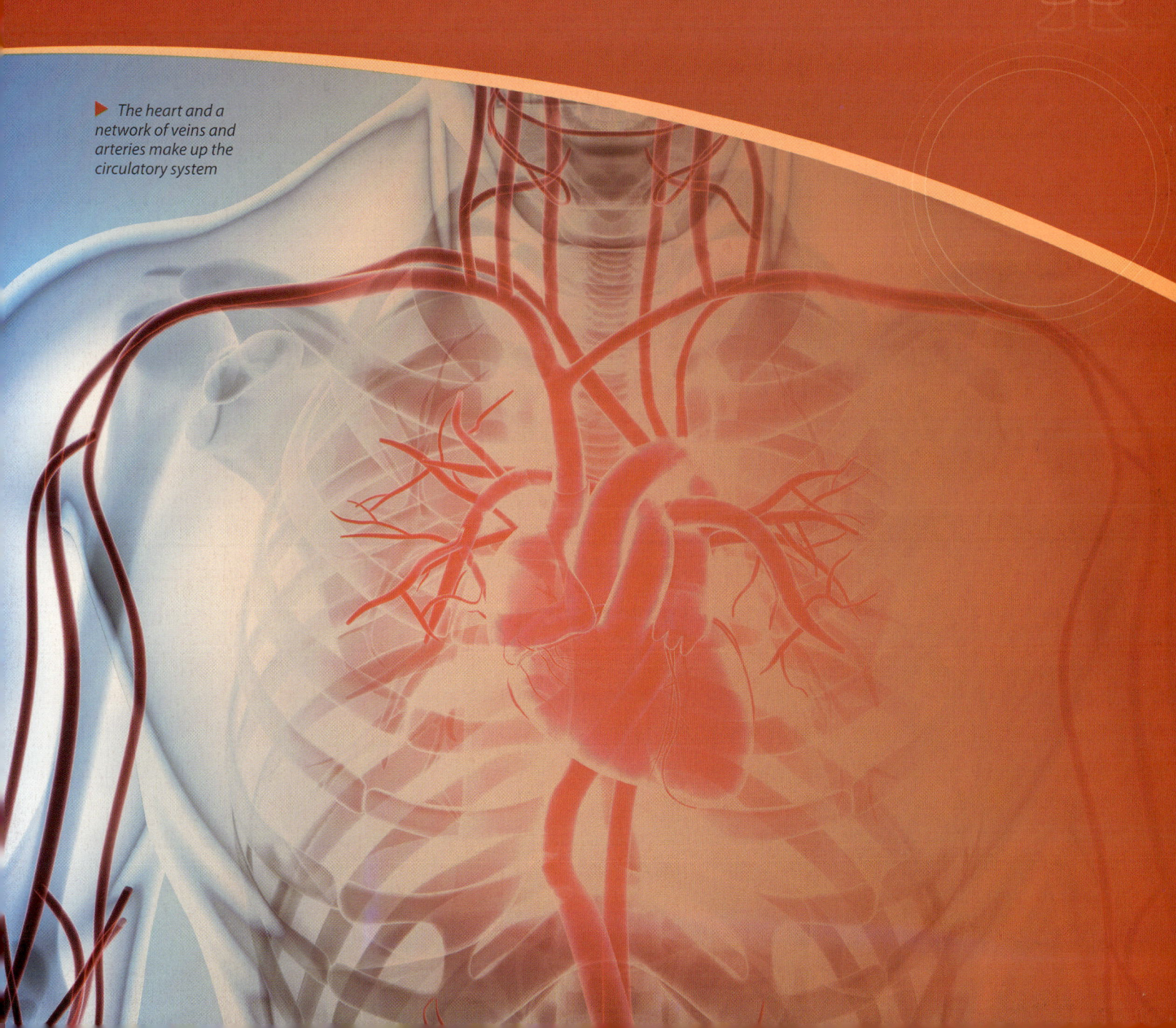

▶ The heart and a network of veins and arteries make up the circulatory system

At the Heart of Everything

You might have heard doctors say that the heart is a pump. That is true. This little organ is the size of your fist. It is the second-most important organ in your body after the brain. It pumps blood all the time, to all the organs and tissues of your body, whether you are awake or asleep, running, or sitting. This is how your organs get oxygen and get rid of unwanted carbon dioxide.

 Red Blood Cells in a blood vessel

Arteries

Your arteries carry **oxygenated blood** from the heart to all the organs. Once the organs have taken up the oxygen they need, the veins bring the **deoxygenated blood** back to the heart. Only the lungs are an exception, as they get the deoxygenated blood, and give back oxygenated blood to the body.

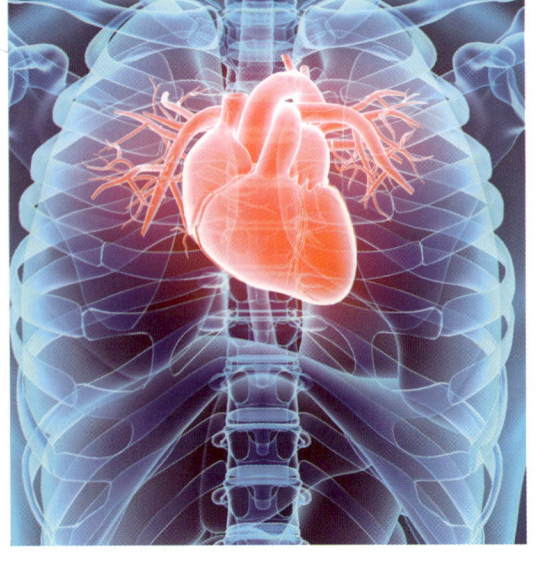

Your Body's Highways

Your heart, **arteries**, and **veins** make up the body's **circulatory system**. Think of it like the system of interstates (USA) or motorways (UK) that connect the whole country. The biggest artery is the aorta, which branches into different arteries for each organ. These, in turn, branch into tiny **capillaries** which go into each tissue of the organ.

Oxygen goes out of the capillaries and into the cells of the organ, while carbon dioxide comes into them. The capillaries join up to form veins, which come out of the organs and join the main vein, the vena cava, which takes blood back to the heart. The whole thing goes on and on, and so your blood 'circulates'.

 Put your hand over the left side of your chest. Can you feel your heart beating?

HUMAN BODY — HEART & CIRCULATORY SYSTEM

Heart Muscles

Your heart is mostly made up of muscles. As the vena cava (see pp.6) pours blood into your heart, the muscles squeeze at one go, pushing the blood out into the aorta. Your muscles require the strength to push blood with so much force that it reaches all the organs, no matter how far away they are from the heart.

The heart keeps pumping all day without rest; if your organs do not get oxygen even for a few minutes, they will collapse and you will die.

In Real Life

Did you know that your heart beats 80 times a minute? It beats faster when you are excited or exercising.

▶ Wearable pulse meters have become a go-to for everyone nowadays

Keeping the Heart Safe

As our heart is a vital organ, the body protects it in many ways. In front of the heart is a tough bone called the sternum or breastbone. The breastbone connects to your spine by a number of bones called ribs, which make up the ribcage. The ribcage keeps your heart safe, and protects your lungs.

How Your Heart is Made

Your heart is made up of 'cardiac tissue', which is laid out in three layers. The outermost layer is known as the **pericardium**, which is filled with a liquid that stops the outer wall of the heart from getting dry, so that it does not rub against the lungs. Pericardium also keeps the heart muscles well lubricated. It stops germs from attacking the heart and helps it heal quickly if they do attack it. It makes sure that the heart does not expand too much when being filled with blood, so that your **blood pressure** can be maintained.

▲ Your heart pumps about 9,092 litres of blood each day

◀ The pericardium anchors your heart to your chest

Labels: Atriums, Parietal pericardium, Ventricles, Visceral pericardium, Pericardial fluid, Endocardium, Epicardium (visceral layer of serous pericardium), Pericardial cavity, Myocardium, Parietal layer of serous pericardium, Fibrous pericardium

▲ Layers of the heart

Inside it is the myocardium, which is made of special muscles called cardiac muscles that never get tired. When it contracts, blood is pushed out, and when it expands, new blood fills in. It is the thickest layer. The innermost layer is the endocardium. It is a very thin layer that shields the heart's valves.

Inside Your Heart

Let us try and understand why the heart is so important. When a heart surgeon opens the heart, they see four chambers, separated by the inner heart wall called septum and valves. There are two on the upper, broad part called the base. These chambers are the **atria** (singular: atrium) or auricles, one to the left and one to the right. The other two chambers are the left and right **ventricles**, which are in the apex, the narrower, lower part of the heart.

There are valves between the chambers, just like there are doors between the rooms of your house. They work like gates, stopping blood from leaking out of a heart chamber once it has been filled. This makes the heart's work smooth, and also maintains your blood pressure.

Isn't It Amazing!

The left ventricle is the strongest and largest chamber of the heart, sending blood to all parts of the body. It pumps five times as much blood as the right ventricle.

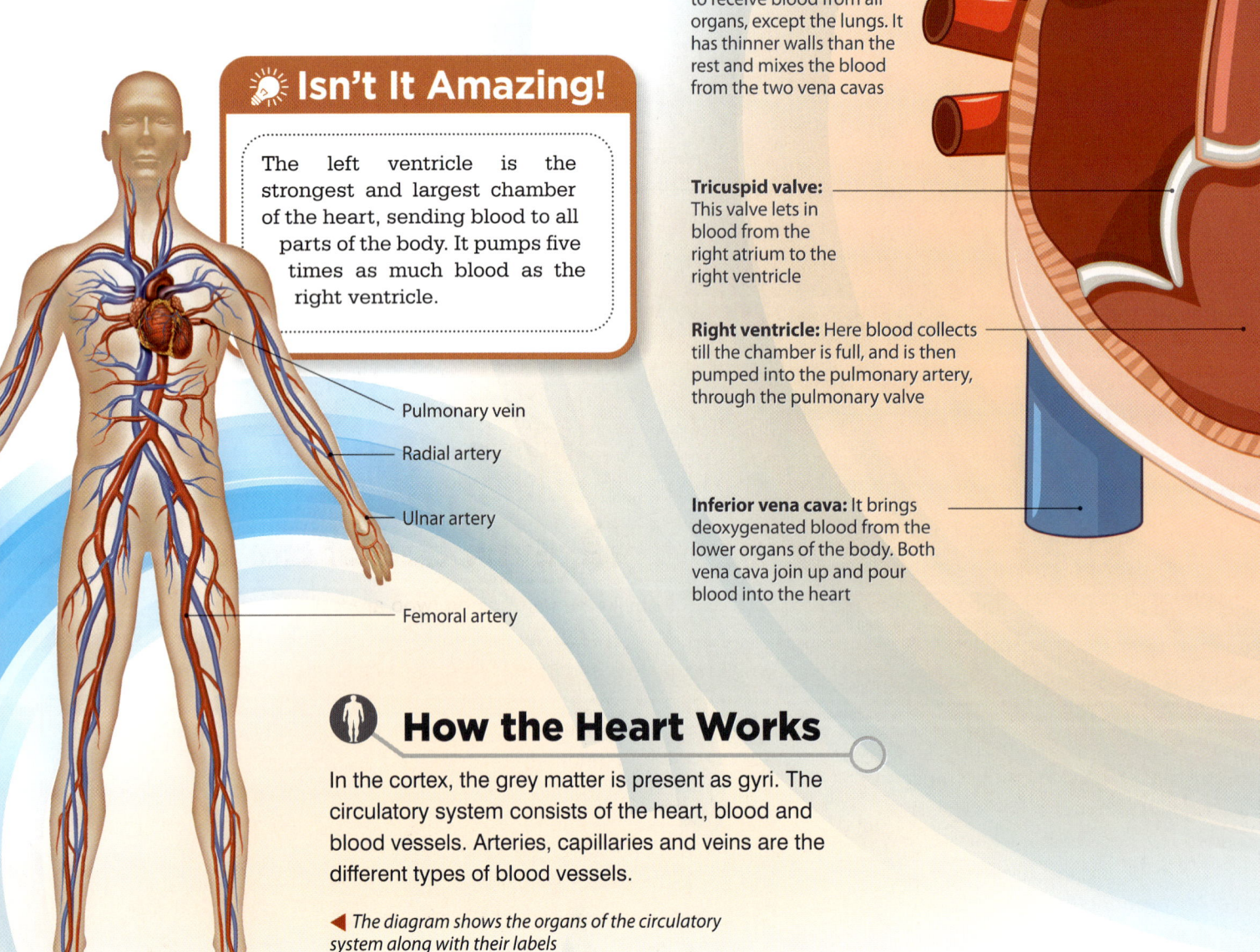

Superior vena cava: It brings deoxygenated blood from the head and upper organs

Right atrium: This is the first chamber of the heart to receive blood from all organs, except the lungs. It has thinner walls than the rest and mixes the blood from the two vena cavas

Tricuspid valve: This valve lets in blood from the right atrium to the right ventricle

Right ventricle: Here blood collects till the chamber is full, and is then pumped into the pulmonary artery, through the pulmonary valve

Inferior vena cava: It brings deoxygenated blood from the lower organs of the body. Both vena cava join up and pour blood into the heart

- Pulmonary vein
- Radial artery
- Ulnar artery
- Femoral artery

How the Heart Works

In the cortex, the grey matter is present as gyri. The circulatory system consists of the heart, blood and blood vessels. Arteries, capillaries and veins are the different types of blood vessels.

◄ *The diagram shows the organs of the circulatory system along with their labels*

Circulatory Cycle

Every drop of blood finishes one cycle when it goes out of the heart carrying oxygen, through the organs, back to the heart, into the lungs and back from the lungs to the heart, where it receives fresh oxygen. This cycle is kept going by the cardiac cycle.

Aorta: This is the body's biggest artery and takes blood to all the organs. It loops around the heart and breaks into two—one going to the head and one to the lower body

Pulmonary artery: This comes out of the heart and branches into two. It carries deoxygenated blood to the left and right lungs, where the carbon dioxide is released into air and blood becomes oxygenated

Pulmonary veins: These bring back blood, freshly oxygenated from the lungs to the left atrium

Left atrium: This has a thicker wall than the right atrium. It collects blood coming from the lungs

Mitral valve: This valve opens when the left atrium is full and lets blood into the left ventricle

Left ventricle: This pumps out the oxygenated blood to the rest of the body through the aorta. It is plugged by the aortic valve

▲ The chambers of the heart are joined by valves

 ## Pumping of the Heart (Cardiac Cycle)

The heart never rests. But it has a quiet diastolic phase when the ventricles are being filled with blood. Once they are full, the heart enters the active systolic phase. The heart muscles contract together, and blood is pumped out with force into the aorta and pulmonary artery. Together these make up one heartbeat. The force with which your heart pumps decides the pressure with which blood flows.

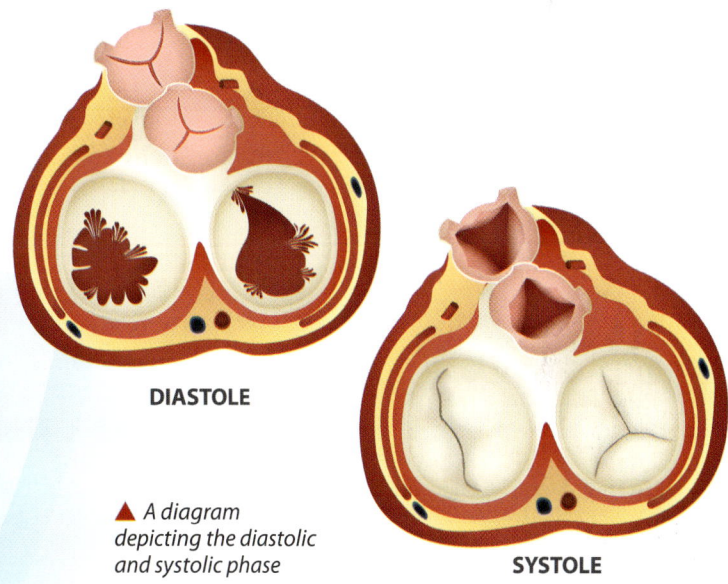

▲ A diagram depicting the diastolic and systolic phase

 ## Evolution of the Heart

Unlike human beings (who are **vertebrates**), invertebrates have no blood to transport oxygen or nutrients. Air reaches tissues directly through spiracles, while nutrients are absorbed through a fluid called haemolymph. The heart, as an organ, originated in fish—with two chambers: an atrium to receive blood from the gills, and a ventricle to distribute the blood to tissues through arteries. After lungs evolved in amphibians and reptiles, the atrium split into two—one for deoxygenated blood from the tissues, and one for oxygenated blood from the lungs. But the two get mixed in a single ventricle. In crocodiles, birds, and mammals the heart finally becomes four-chambered, and deoxygenated blood is completely separated from oxygenated blood.

Arteries & Veins

Did you know that within a minute, blood from your heart will have reached every cell of your body? That is because our body has an efficient transport system made of arteries and veins, together called blood vessels. Let us take a look at their similarities and differences. The arteries and veins are tube-like, so they ensure that blood flows in a direction. Valves in the arteries and veins make sure that blood does not flow backwards. Blood flows in the arteries away from the heart towards the organs. The oxygenated blood in them makes them look red. In the veins, blood flows towards the heart from the organs. The deoxygenated blood in them makes them look blue.

What Do Arteries Do?

Oxygen is very important for your body's cells as they need it for making energy from the food you eat. As they keep using it, they need more. It is the arteries that bring them the oxygen from the heart, which it gets from the lungs. Other than oxygen, your arteries bring hormones from various glands in the body. Hormones help communicate to the cells what to do and when. Arteries also bring nutrients like glucose and amino acids from the intestines, which digest the food you eat. The pulmonary artery is the only one that's different, as it carries deoxygenated blood from the heart to the lungs for oxygenation.

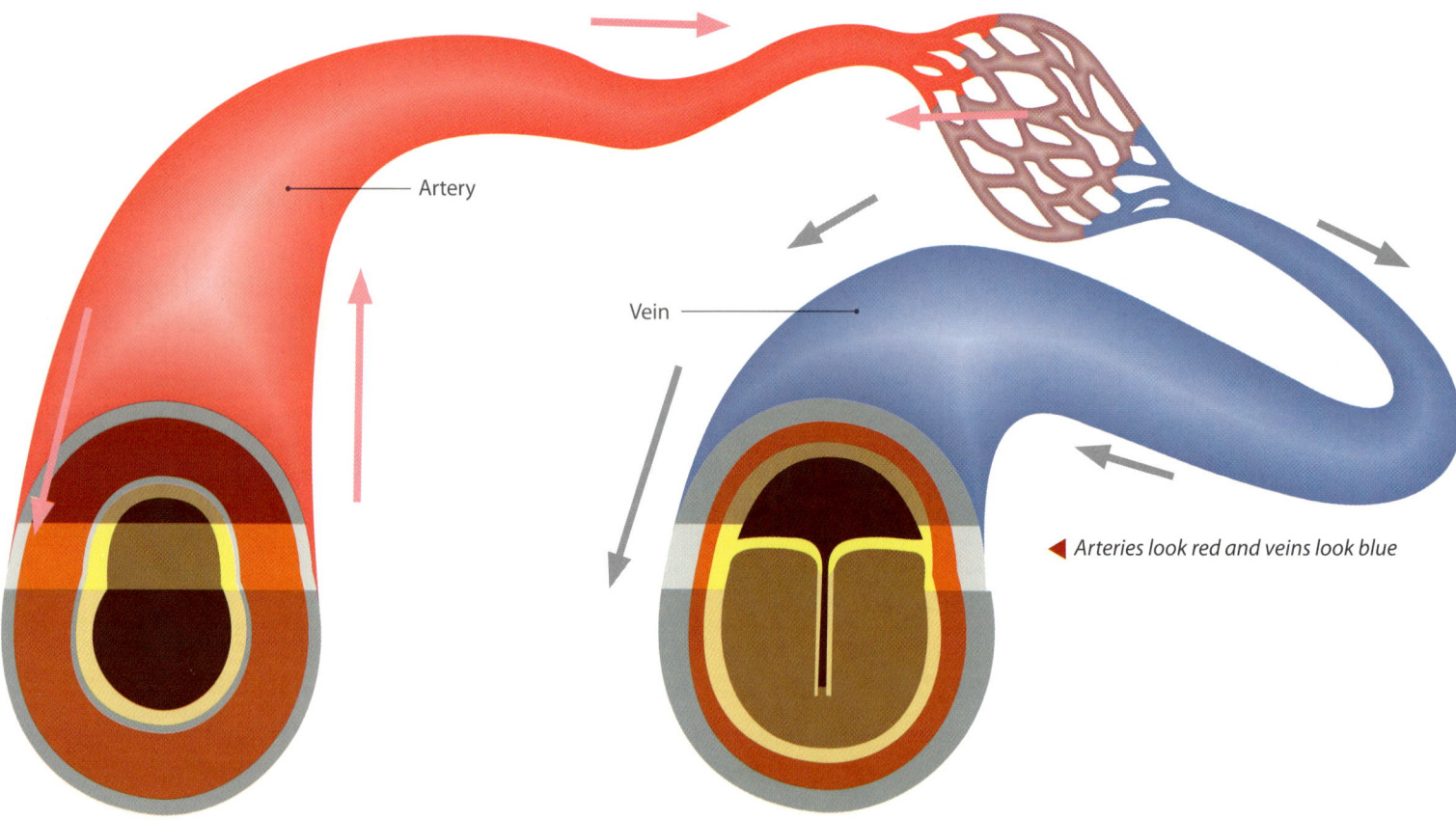

◄ Arteries look red and veins look blue

What Do Veins Do?

Carbon dioxide is what is left once food is turned into energy (or used for making proteins needed by your tissues). It must be removed from the body so that fresh oxygen can be brought in. This job is done by the veins. They also carry away other wastes from the cells, like urea. Two veins are different from the others. The pulmonary veins bring blood rich in oxygen from the lungs to the heart. The hepatic portal vein takes blood with glucose from the intestine to the liver, where the extra glucose is stored.

HUMAN BODY | HEART & CIRCULATORY SYSTEM

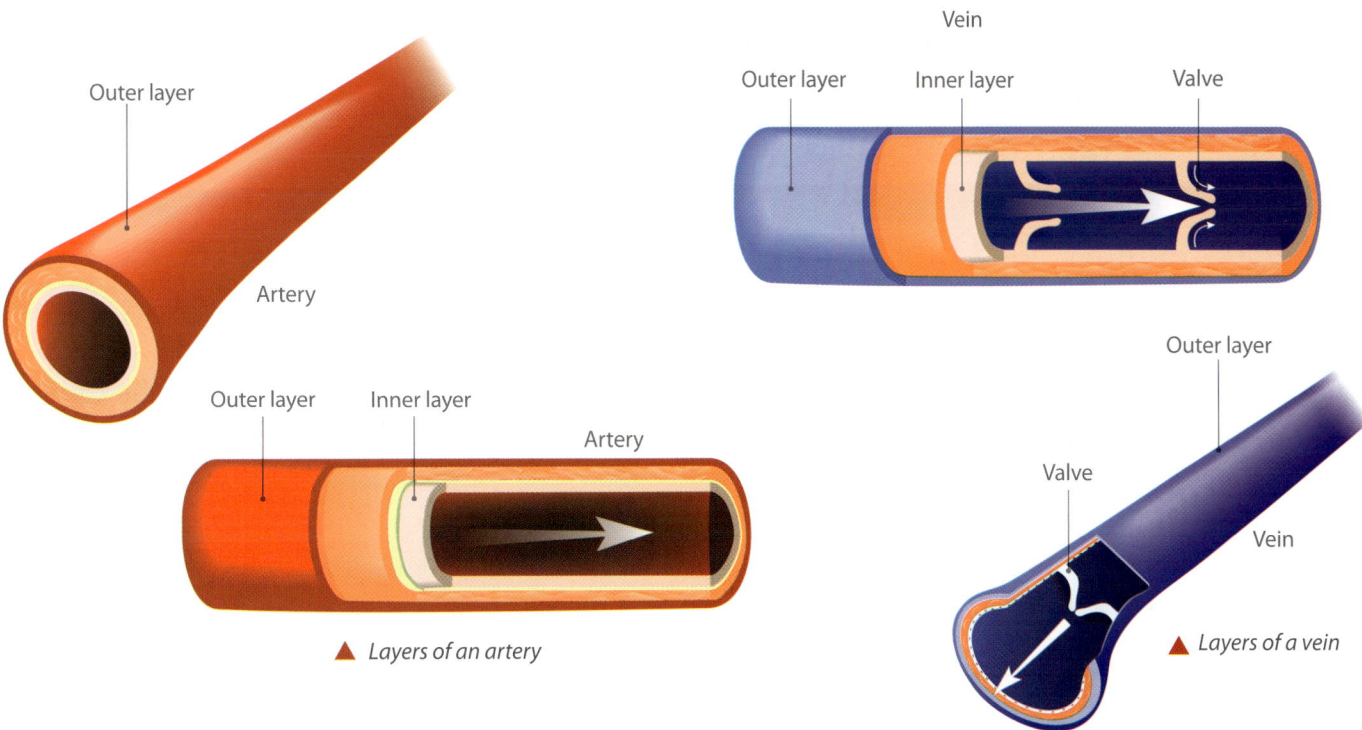

▲ Layers of an artery

▲ Layers of a vein

How are Arteries and Veins Made?

Like your heart, your arteries and veins also have three layers wrapped around each other. The outermost layer is the 'tunica adventitia' which is made of loosely bound cells called 'connective tissue'. Next to it is the 'tunica media', which has tiny muscles that help push the blood along. It is thicker in arteries and thinner in the veins. Innermost is the 'tunica intima', made of a wall of cells that are tightly bound to ensure no leaks occur.

Incredible Individuals

William Harvey (1578–1657) was the personal doctor to King James I of England. He demonstrated that the arterial and venal blood formed a single system and that the heart was the pump that kept blood flowing. Though he discovered circulation in 1618, he waited till 1649 to publish his results. Doctors did not believe him, because capillaries, which connect arteries and veins, could not be seen and the microscope was not invented at the time.

▶ It took 20 years for William Harvey's theory on blood circulation to be accepted generally

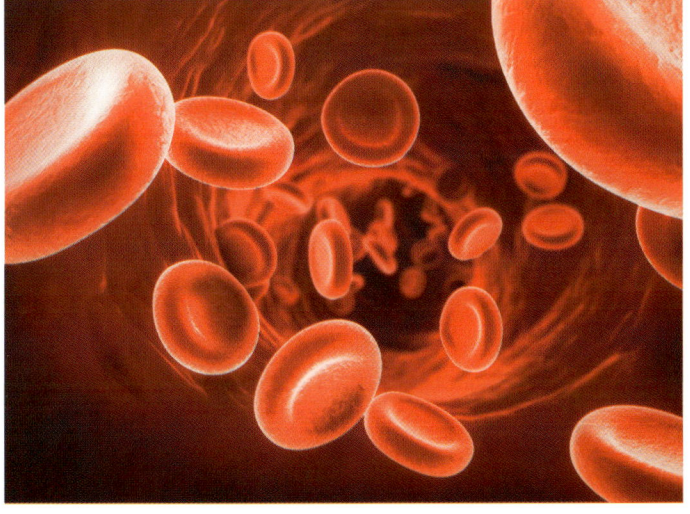

Deep & Superficial Veins

The inferior vena cava is so called that because it lies posterior to the heart. It is the largest vein in the body, collecting deoxygenated blood from all the organs. The veins that bring blood to it are called deep veins, because they are deep inside your body. The veins on your skin are called superficial veins. You can see one stick out of the insides of your elbows. These veins connect to the deep veins.

◀ Red Blood cells are responsible for the transportation of oxygenated and deoxygenated blood

Feeding the Organs

Your aorta starts from the left ventricle of the heart, and gives out branches or arteries that go to each organ. Inside the organ, they branch into smaller arterioles. As they go deeper into the organ, they branch out into really thin blood vessels called capillaries. Did you know that the heart has its own artery that supplies blood to each of the heart's muscles?

Isn't It Amazing!

Look at one of your hair strands. Some blood vessels are extremely small; that the smallest blood vessels measure 5 mm, that is less than one-third of a hair strand!

▲ Human hair under a microscope

Capillaries

Capillaries act like both arteries and veins. They do not just supply oxygen and nutrients, but also pick up waste and carbon dioxide from the organs including the heart. They join to form venules, which come out of the organ to form the main vein, which in turn joins the vena cava. You can see a network of capillaries if you look at your eye in a mirror by gently pulling down an eyelid.

Precapillaries are located between the capillaries and the smallest arterioles or arteries. These are considered to be intermediate vessels. Precapillaries control how the capillaries are emptied and filled. They have muscle fibres, unlike the capillaries.

Types of Capillary

Your body has capillaries of three types. They are the continuous capillary, the fenestrated capillary, and the sinusoid or discontinuous capillary.

Continuous Capillary

These capillaries have thin gaps between the cells that make up its walls. This lets the fluid part of blood go out into the tissue, and come back in. This kind is found in the lungs, muscles, and nervous system.

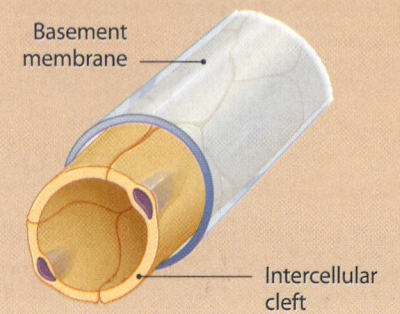

Fenestrated Capillary

The fenestrated capillaries have little holes (like a shower head) that let blood go out and come in. This kind runs in the kidneys, intestines, and glands.

▼ Capillaries are usually 10 μm in size

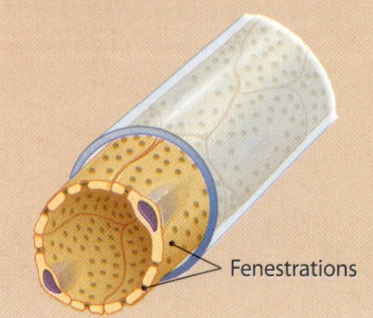

Sinusoid Capillary

Lastly, the sinusoid or discontinuous capillaries have really big gaps that allow blood cells to get into the tissue. You see them in the liver; spleen, where damaged blood cells are destroyed; and bone marrow, where new blood cells are born.

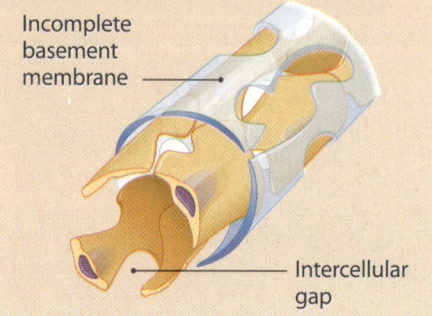

Keeping the Beat

As the heart beats, it pushes blood into the arteries. The blood pushed in with each beat is called a pulse. The number of pulses every minute is therefore called your 'pulse rate'. Doctors measure your pulse rate when you go to them to find out what is making you ill and whether the cause of your illness is affecting your heart. Patients with arrhythmia have an irregular pulse and may need a pacemaker.

Know Your Pulse Rate

Take the index and middle fingers of one hand and press them softly on the thumb-side of your wrist on the other hand. You can feel the pulsations. Look at a clock and count how many beats you feel every ten seconds. Multiply by six—that is your pulse rate.

▲ *Your pulse rate can be quite different from someone else's*

##

Did you know that Wilson Greatbatch (1919–2011), who invented the pacemaker, was not a doctor, but an electrical engineer? His device has saved millions of lives.

▲ *Greatbatch had over 325 patents to his name!*

Keeping the Rhythm

The beating of the heart is controlled by two nerves. The accelerans nerve speeds up the heartbeat and the vagus nerve slows it down. Arrhythmias occur when one or both of these nerves do not work properly, or when the muscles of the heart do not correctly respond to these nerves.

A normal pulse rate is an indicator of good health, and hence it needs to be maintained properly. When this rhythm is adversely affected, doctors try to fix it using a **pacemaker**. Artificial pacemakers help correct signals to the heart muscles by giving them tiny electric pulses.

Patients who wear a pacemaker need to be careful. Doctors usually ask them to avoid stressful activities, like lifting heavy loads or climbing many stairs. Patients must also keep away from anything that may upset the pacemaker's generators, like cell phones, walk-through metal detectors and medical equipment like MRI scanners.

▼ *A pacemaker is a small device, about the size of a matchbox or smaller*

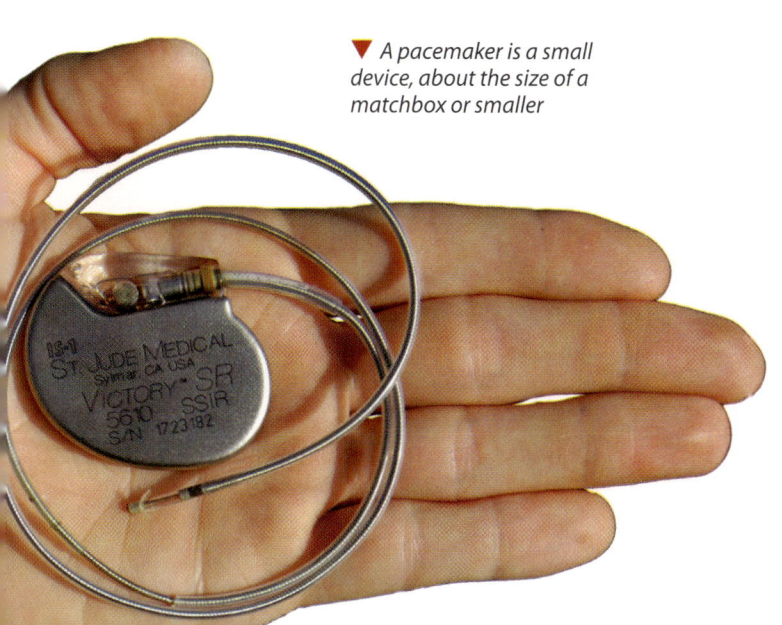

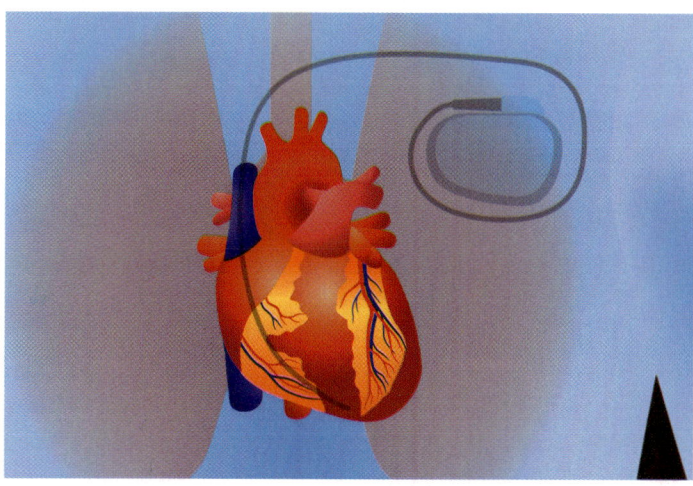

▲ *A pacemaker is surgically implanted in your chest*

Heart Troubles

By now, we know that the heart plays the most important role in the circulatory system. So, what happens when it takes ill? Like any other organ, the heart too has its share of diseases and disorders. But unlike other organs, diseases of the heart often cause death, because when it stops working, other organs are soon starved of blood and oxygen, they in turn, stop functioning. Let us try and understand the major problems the heart can have. Some are common and can affect anybody at any time, while others are rarer and affect only some kinds of people.

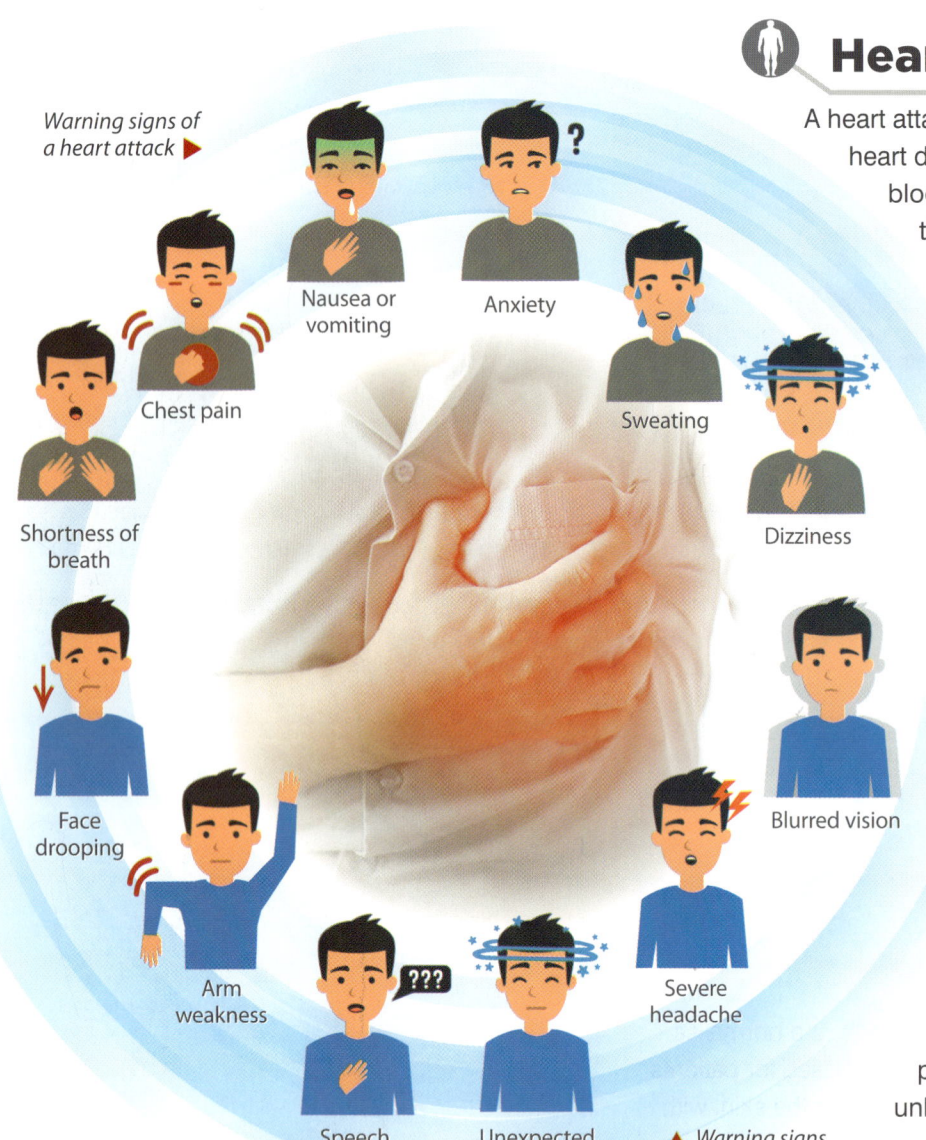

Warning signs of a heart attack ▶

Nausea or vomiting · Anxiety · Chest pain · Sweating · Shortness of breath · Dizziness · Face drooping · Blurred vision · Arm weakness · Severe headache · Speech difficulties · Unexpected dizziness

▲ *Warning signs of a stroke*

Heart Attack

A heart attack happens when the muscles of the heart do not get enough oxygen-rich blood. A blood clot in the artery of the heart reduces the amount of blood flowing and blocks the passage of blood cells. Some of the heart muscles stop getting oxygen and die. This causes paralysis of the heart, as it cannot pump anymore blood. This is called a '**heart attack**'.

Heart Failure

When the heart is not able to pump blood properly, it is called heart failure. A heart failure does not mean that the heart stops working completely. Instead, it does not have enough pumping force, and doctors see this as low blood pressure. This also means that your body is not getting all the oxygen it needs. You may also get heart failure if you are very excited or frightened, because the heart is pumping faster than normal, building up unbearable pressure.

Stroke

Blockage of the arteries of other organs can cause them to fail too. If the artery to the brain fails, the person gets a **stroke**. The brain's cells begin to die without oxygen, and the brain stops sending signals to the rest of the body, which means that they begin to die too. One of these signals is to the heart, telling it to keep beating. A patient with a stroke must get medical help immediately, before the heart also stops beating.

In Real Life

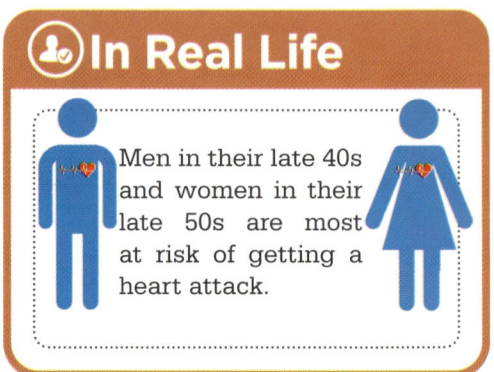

Men in their late 40s and women in their late 50s are most at risk of getting a heart attack.

Aneurysm

As our arteries carry oxygenated blood to our organs, keeping up the right pressure of blood is important. The organs' rate of taking up oxygen depends on that. If an artery swells up, or its walls become weak, then the person can get an aneurysm. If blood pressure rises due to any reason, the aneurysm may cause the artery to burst, which leads to internal bleeding. Blood flows into the patient's body cavity instead of the organs, its pressure drops sharply, and the patient is at death risk.

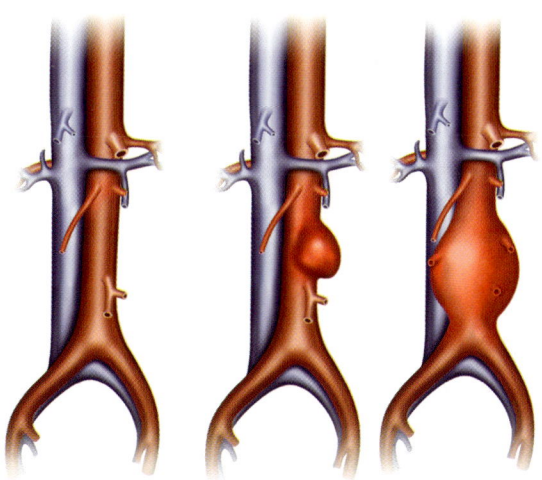

▲ *Aneurysm often occur in the aorta, brain, back of the knee, intestine, or spleen*

Sudden Cardiac Arrest

The heart keeps beating as it gets signals from the brain. A problem in the nerves may cause these signals to stop. The heart stops working, and doctors call this 'sudden cardiac arrest'. Patients stop breathing and lose consciousness. It causes instant death if not treated immediately.

Arrhythmia

Your heart beats at a regular rate or rhythm throughout life. Arrhythmia happens when the heart beats too fast, too slow, or skips beats. This disturbs blood pressure and can cause problems.

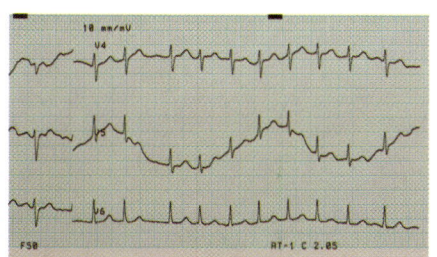

▲ *An ECG report shows the electrical activity of your heart at rest*

▶ *The heart rate is checked and noted on a chart with these kinds of lines called waves*

The Beat Regulator

There are many complications that a heart can face. Luckily, with the amazing advancements made in medicine and technology, there is relief even for heart patients, one of them being a heart pacemaker.

The Pacemaker

A pacemaker is a small device which helps to regulate and normalise the rhythm of the heartbeat. Pacemakers are primarily used for patients suffering from arrhythmia. It is placed in the chest, under the skin, with minor surgery.

The size of a pacemaker varies from about the size of a child's palm to as small as a capsule pill. It has two parts: the generator and the leads. The generator has the electrical circuit and the battery for the pacemaker while the leads are a couple of wires that carry an electrical message to the heart. Pacemakers can be placed for temporary or short-term use; or in the case of long-term heart problems, they can be permanent.

A pacemaker works on electrical pulses. It is an electrically charged medical device which prompts the heart to beat at a normal rate.

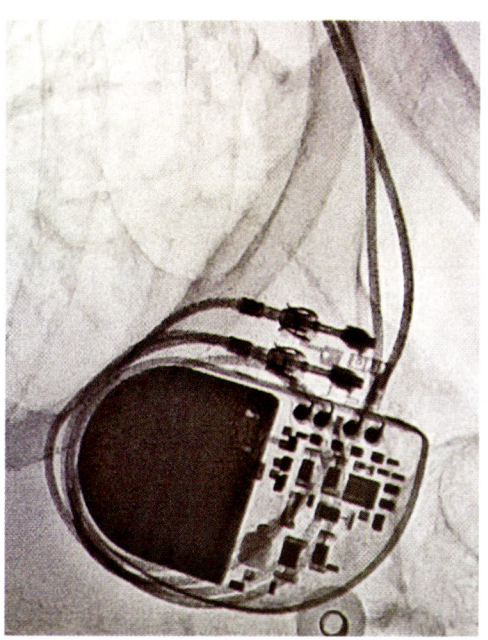

▲ *An x-ray showing a pacemaker fitted in a patient's chest*

Transplants
Gifting a New Life

People with a weak heart might not live for very long. But if treatment has not worked for them, do they have hope? Yes! They might receive a new heart which was 'donated' by a brain-dead person, i.e., a person whose brain has stopped working, but their heart can be transplanted to someone else. A **heart transplant** is a very difficult surgery. Prior to it, a surgeon checks the donor's compatibility with the patient and if the patient is healthy enough for the heart.

In Real Life

A heart must be transplanted within four hours of the heart being taken from the donor. Hence, once doctors have declared someone 'brain dead', everything must be kept ready for the patient to get the heart. Doctors, police and city authorities cooperate to create 'green corridors' between hospitals, so that an ambulance (sometimes even a plane or helicopter) can transport the heart.

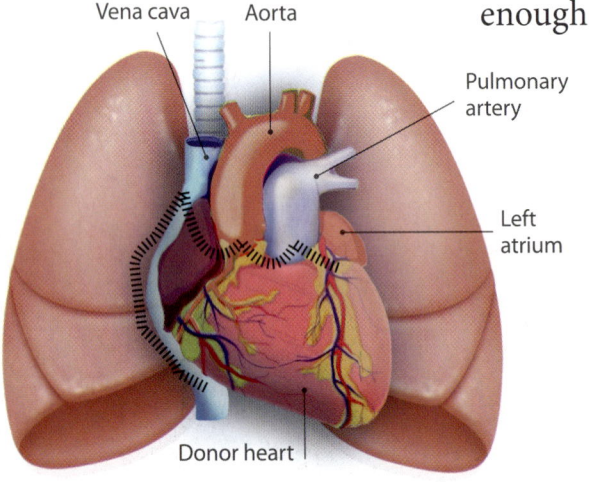

▲ The diagram shows a donor heart after heart transplant surgery

Who Gets a Heart Transplant?

A person gets a heart transplant at a time when their heart is at a very critical stage. It means that none of the medication, treatment, or surgery has worked in curing their heart problems. The heart is donated by a deceased person. This way, the person getting a heart transplant gets a new heart from another human being.

It is not easy to get a donor heart. Thus, it is important to make sure the person receiving the heart transplant really needs it. It is more important to make sure that the patient is healthy enough to survive after getting a new heart. A heart transplant must usually happen within four hours of the organ being removed from the deceased.

Incredible Individuals

Dr Christiaan Barnard (1922–2001) performed the first ever human-to-human heart transplant on 3 December 1967. The entire operation took nine hours and required a 30 person team.

▶ Dr Christiaan Barnard

Eligibility

Not everyone needs a heart transplant. Some that do need it, might still not get one because there are few donor hearts available. Here is how doctors decide on a transplant:

✓ The patient must meet health standards before going into the surgery and be healthy enough to heal after it.

✓ The patient should not have a serious disease like diabetes or cancer that could damage their heart.

✓ They must not have an infection at the time of surgery.

✓ They should be ready to change their lifestyle to keep their new heart healthy.

▼ During a heart transplant surgery, the patient is placed on a heart-lung machine

Know Your Blood

What is the composition of blood and why is it so important for our life? It is made of two parts—cells and plasma. Cells are born in the bone marrow and come in three types—red blood cells (RBCs), white blood cells (WBCs), and platelets. Plasma is made of water, salts, hormones, and many other biochemicals.

Blood makes up 7–8 per cent of your body's weight. Depending on your age, size, and gender, you may have between 4–6 liter of it in your body.

The cornea of the eye can get oxygen directly from the air. All other organs of your body must get their oxygen from the blood.

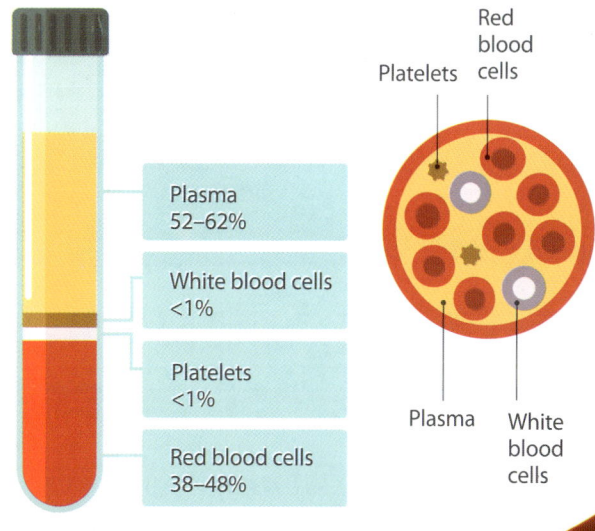

▲ *Components of the blood*

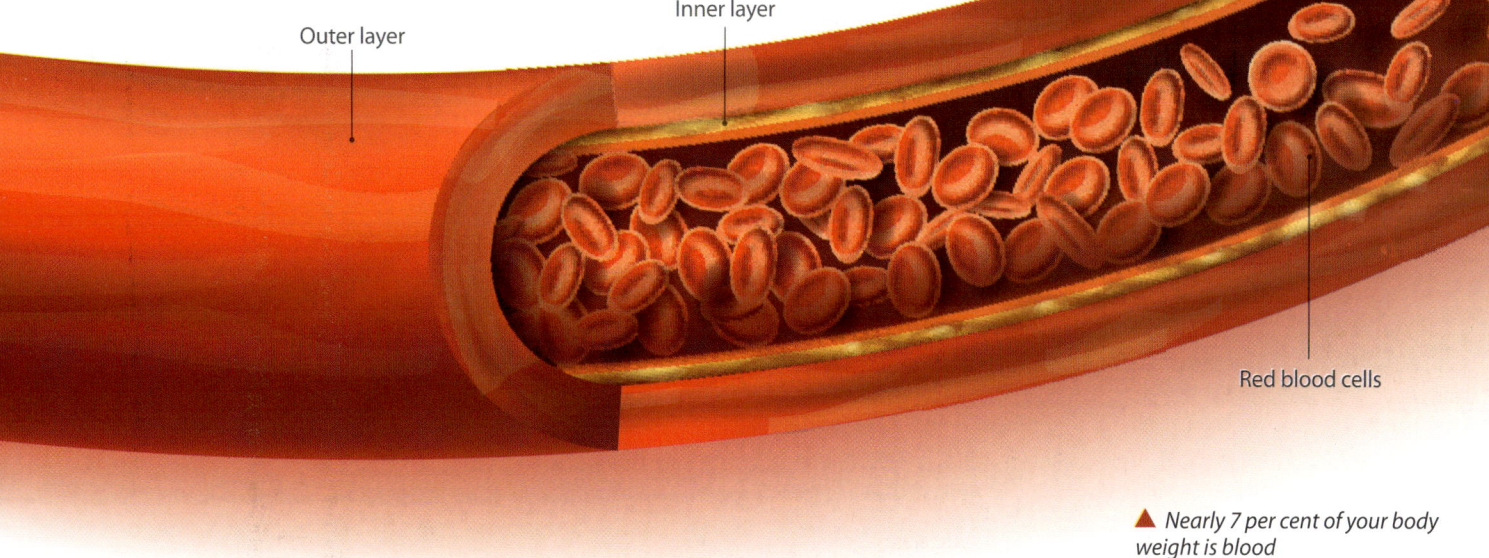

▲ *Nearly 7 per cent of your body weight is blood*

 ## Why is Blood Important?

The primary role of blood is to carry oxygen and nutrients to the cells of your body, and to carry out waste products to the lungs and kidneys. It heals wounds, by forming clots, it takes poisons to the liver for detoxification; and also helps your body maintain its temperature.

 ## Homeostasis

This is the word used by doctors and scientists to describe how your body maintains its temperature. When the weather is cold, the body can lose heat. The brain makes the superficial veins and arteries contract, so that less blood flows through them, and less heat is lost through the skin. In hot weather, the opposite happens. Blood vessels expand, more blood flows and more heat escapes through the skin.

Blood Cells
Little Transporters & Defenders

The cells of your blood do most of its important jobs. The most in number are the red blood cells (RBCs) or erythrocytes. They give your blood its colour and carry oxygen from the lungs to the organs through the heart. Each drop of blood has about 5 million RBCs. On the other hand, white blood cells (WBCs) or leucocytes make up less than 1 per cent of blood. They help defend the body by swallowing any germs that may enter your body. Both RBCs and WBCs are made in the bone marrow, and at the end of their lives, are destroyed in the spleen and liver.

▲ *You can donate RBCs to hospital blood banks. They specially freeze them, till somebody needs a blood transfusion. The RBCs can be stored for 42 days*

Isn't It Amazing!

If there are 5 million RBCs in one drop of blood, how many are there in the whole body? 30 trillion! Every fourth cell in your body is an RBC.

Red Blood Cells

Under a microscope, RBCs look like disc-shaped doughnuts (without the hole), making up 40 per cent of your blood. They travel through your body and live for about four months. Each second, 2 million new RBCs are born, and 2 million die. They are made from 'haematopoietic stem cells' in the bone marrow; it takes about a week for them to turn into 'mature' RBCs. Then they enter the blood through the 'sinusoid capillaries' to get on with their jobs!

What Do Red Blood Cells Do?

RBCs have a special protein in them called **haemoglobin**, which contains iron. The iron makes haemoglobin, RBCs and therefore your blood red in colour. In the lungs, each unit of haemoglobin binds four molecules of oxygen and carries them to the rest of the body's tissues, which are hungrily waiting for them. They also take carbon dioxide from the tissues to the lungs, where it is released into the air.

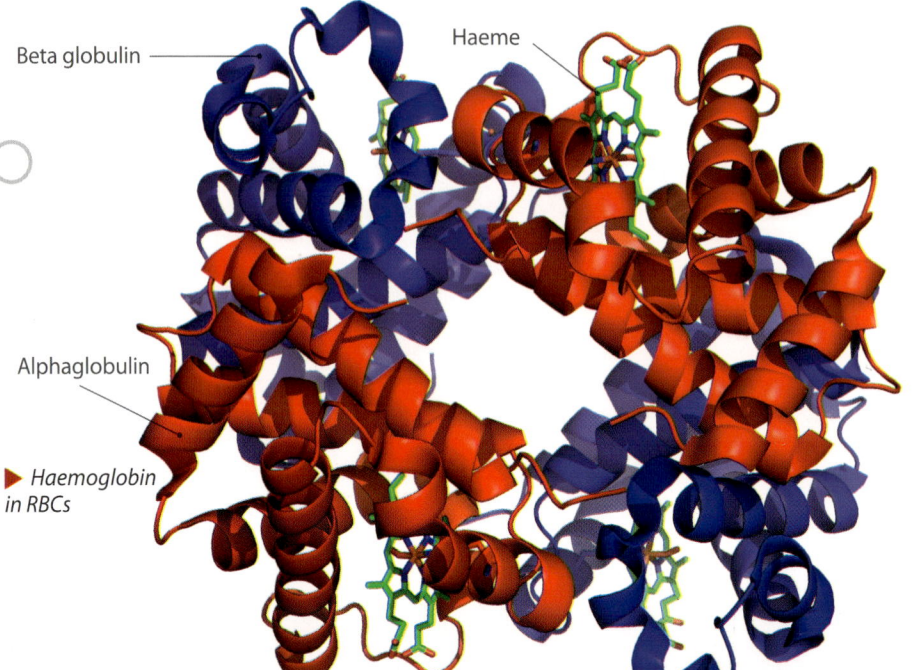

▶ *Haemoglobin in RBCs* — Beta globulin, Haeme, Alphaglobulin

White Blood Cells

WBCs come in many types—**macrophages**, lymphocytes, neutrophils, basophils, and eosinophils. Each type of cell does something different in protecting the body from germs and allergens. They are actually a part of the immune system, where the body fights diseases and infections. Some WBCs live only for a day, while some may live for several years, remembering a previous infection just in case you are infected again. WBCs cannot be donated easily as it is hard to extract them from blood.

How Do White Blood Cells Work?

The different types of WBCs work together like a team to fight germs. Along with the lymphatic system, they make up the immune system. Here is what each WBC does.

▲ White Blood Cell (WBC)

Types of White Blood Cell

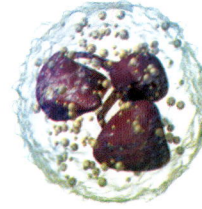

Neutrophil — Neutrophils go first. They attack most of the harmful germs and send signals to other WBCs to help.

Basophil — Basophils cause inflammation if an allergy-causing substance such as pollen or nuts enters the body.

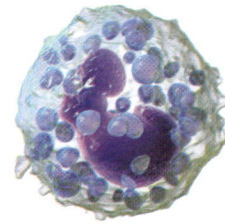

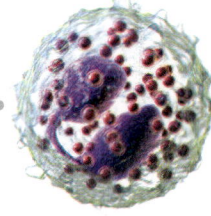

Eosinophil — Eosinophils kill bacteria and parasites in our body and clean up dead cells.

Lymphocyte — Lymphocytes come in many types themselves, making antibodies and other biochemical weapons.

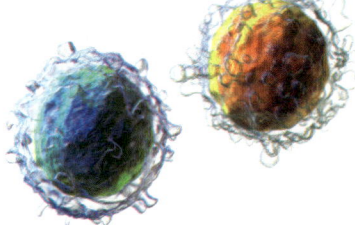

Monocyte — Monocytes transform into macrophages, which swallow infected cells, germs, and dead cells.

💡 Isn't It Amazing!

The cells also interact with each other through special biochemicals called cytokines and interleukins.

Platelets & Plasma

Platelet and plasma cells make up the rest of your blood. Plasma is the liquid part of blood and makes up more than half the volume of blood. It carries the nutrients your cells need, like the salts, glucose, amino acids, and lipoproteins dissolved in it. It is necessary for blood to flow smoothly, with an optimum blood pressure. On the other hand, platelets are the body's tiniest cells and make up a tiny fraction of blood but do important jobs like clotting of blood.

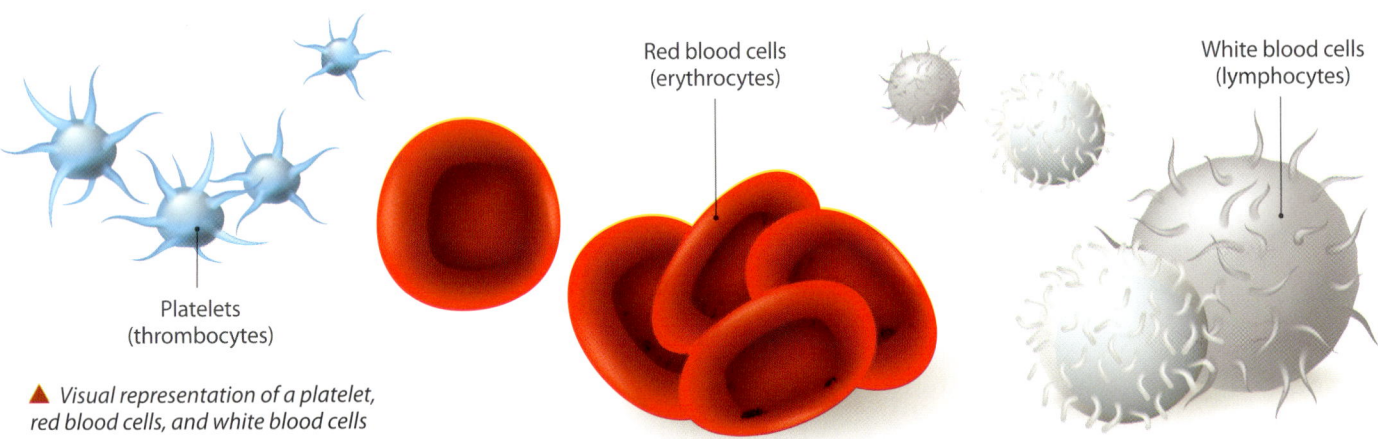

▲ *Visual representation of a platelet, red blood cells, and white blood cells*

 ## Platelets

Also called thrombocytes, these flat cells look like small plates under a microscope. When you get injured, platelets rush to the wound and clog it. Because they are sticky, they make a self-sealing bandage and do not let germs in. This is called clotting and stops you from bleeding. Without platelets, your body would be drained out of blood.

Unfortunately, you can only have a limited number of them. If your platelet count exceeds the limit, they can cause internal clots leading to aneurysms, heart attacks, and strokes.

Platelets can be taken out of the blood and donated separately. This is useful for people with dengue and other diseases where their natural platelet count drops sharply.

 ## Plasma

Plasma is 90 per cent water, so it is natural that it is the main tool of the body to maintain its electrolyte and fluid balance, called 'osmolarity'. Most people suffering from blood loss really just need plasma and/or platelets. The plasma is taken out and stored in blood bags. Unlike cells, plasma can be frozen and stored for a year.

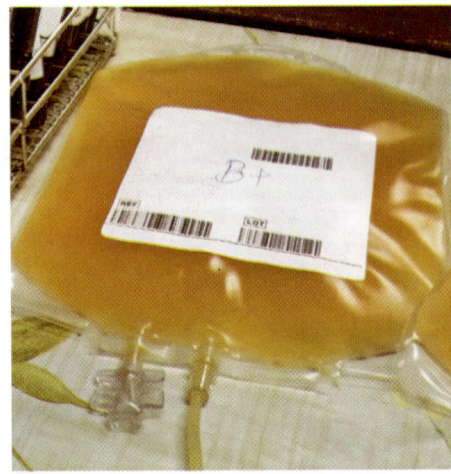

In Real Life

In a plasma donation centre, the blood you give is put into a machine called a centrifuge, which spins it very fast. The cells settle down at the bottom, and the yellowish plasma floats above.

▶ *Centrifuge uses centrifugal force to separate the blood components*

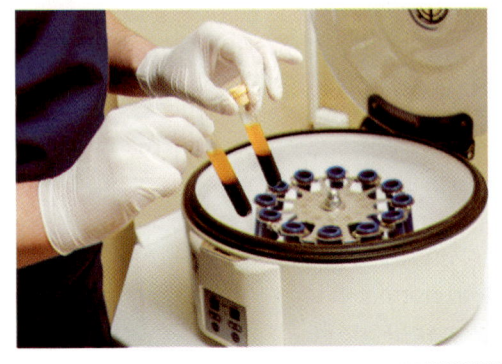

Donate Blood, Save a Life

Blood loss often happens if you have had a terrible accident or are undergoing major surgery. This can lead to oxygen starvation, as there is not enough blood to take oxygen from the lungs to the tissues. This, in turn, can cause organ failure and death.

Patients who need blood depend on people who donate it voluntarily (donors). When blood is injected into a patient, it is called a **blood transfusion**. In modern times, one does not always require all components of the blood to be transfused. Instead, doctors often call for individual blood components, usually plasma, platelets, or RBCs.

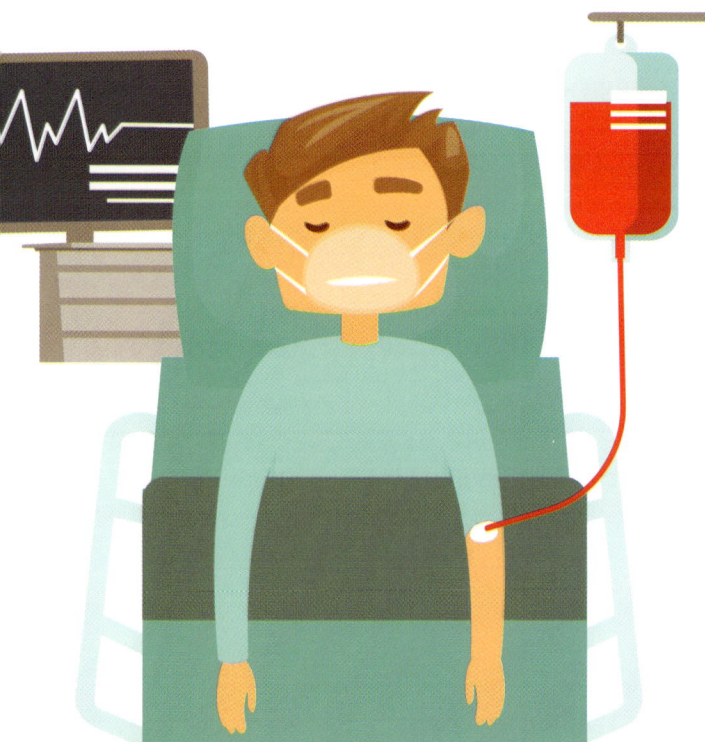

◀ It is not always easy to get a blood transfusion, but blood banks go a long way to help

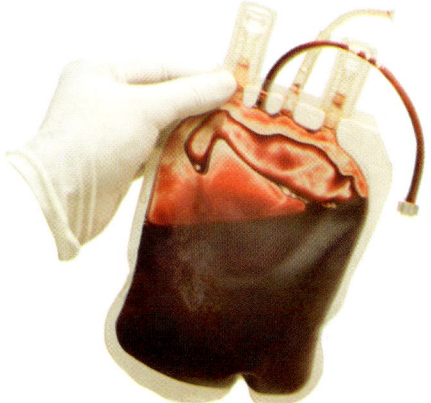

In Real Life

World Blood Donor Day falls on 14 June around the world. Celebrate it by donating blood at your nearest hospital.

Who can Donate Blood?

To donate blood, you have to be healthy and grown-up, between the age of 18 and 60 years old, and not weigh less than 45 kilograms (99 pounds). If you are anaemic or suffer from an infectious disease, or if you had hepatitis in the past year, you cannot donate blood.

How to Donate Blood?

A grown-up can donate up to 350 mililiters at a time, once every three months. Doctors will place a needle in the vein at the elbow and draw out blood into a specially made blood bag which stops it from clotting or getting infected. Some of the blood is used for testing and the rest is frozen till someone needs it. After a donation, you may feel dizzy or weak because of the loss of fluid. Drink some juice and eat something, and you will be back to normal in a few hours. Your body will replace the plasma within 48 hours and all the RBCs in 4–6 weeks.

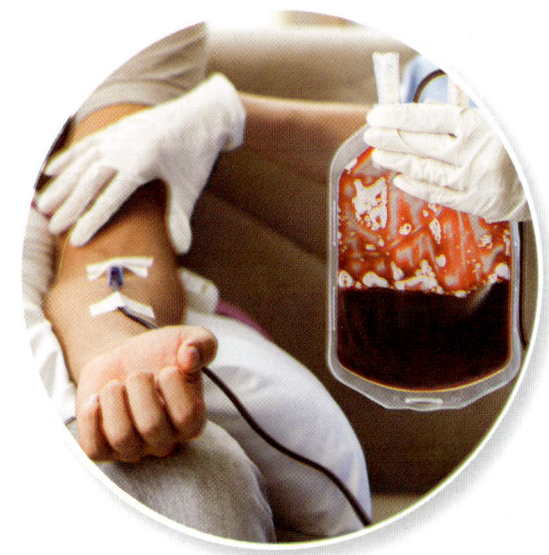

▲ Many neighbourhoods conduct blood drives where healthy people donate blood to local hospitals

Know Your Blood Type

There are two ways in which everybody's blood differs. The first is the ABO system, because of which we have A, B, AB, and O blood types. The other is the rhesus factor, usually written as Rh positive (+) or Rh negative (–). Thus, you get eight types: A+, A–, B+, B–, AB+, AB–, O+, and O–.

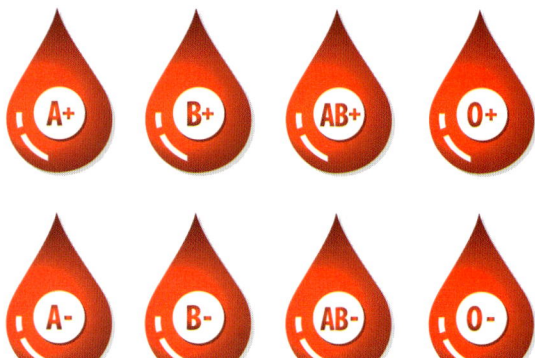

▲ The eight common blood types

Your immune system will only accept blood that matches the type that runs in your circulatory system and reject everything else. Therefore, you need to know your own blood type before you donate blood or receive any. Blood transfusion with mismatched blood leads to medical complications and sometimes even death.

 ## Why Must Blood Type be Matched?

The RBCs of your bodies have different 'antigens' on their surface. These are proteins that tell your WBCs that these cells belong to the same body. Some people have a type called A, so they belong to Blood Group A. Their bodies will make antibodies against the other types. People of Blood Group B have another type called B. Some people have both types that is, Blood Group AB, and some have none that is, Blood Group O. The rhesus factor works in the same way.

If you are of type B+ and receive blood of type A–, it may have antibodies that attack B+ RBCs, and kill them. Your body will develop antibodies against A–. These are called transfusion reactions and may cause severe problems in your body. That is why it is important to match. And that is why schools, universities, and workplaces put your blood group on your identity cards.

 ## Antigens and Antibodies in the Blood

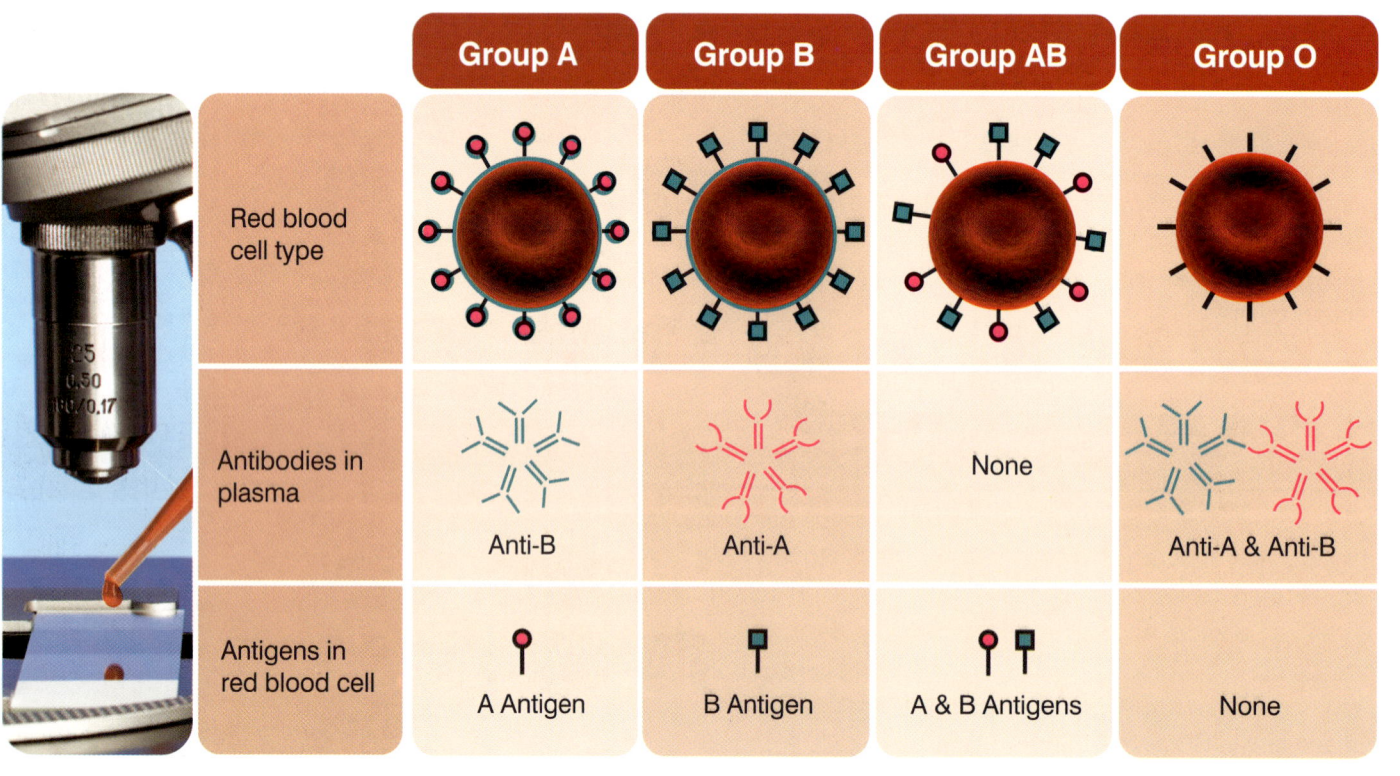

HUMAN BODY | HEART & CIRCULATORY SYSTEM

Incredible Individuals

Have you heard about the Man with the Golden Arm? His name is James Harrison, a Australian who donated blood for 60 years! Thanks to his donation, 2.4 million children suffering from a disease called Rhesus were saved. This became possible because his blood was found to have rare antibodies that could fight the disease.

▶ *James Harrison made his last donation in May 2018, since blood donation from those past the age of 81 is prohibited in Australia*

How Does One Match Blood Groups?

If you need blood, doctors will insist on an exact match at the blood bank. But if there isn't any of the right kind, and there is a real emergency, doctors use these rules:

- Give blood of type A to patients who have blood type A or AB.
- Give blood of type B to B or AB patients.
- Give blood of type AB only to AB recipients.
- Give blood of type O to patients of any blood type A, B, AB or O.
- Give Rh positive blood only to those who have Rh positive type, as long as their ABO is matched or follow the above rules.
- Give Rh negative blood to anyone whose ABO matches perfectly or follow the above rules.
- This makes those with O– blood capable of donating blood to anyone. So, people with this blood type are called universal blood donors.
- Those with AB+ type can take blood from anyone, so they are called universal blood recipients.

In Real Life

There are hundreds of rare blood types that people can have because there are more than 600 known antigens. It is the presence or absence of these antigens that determines the rare blood types.

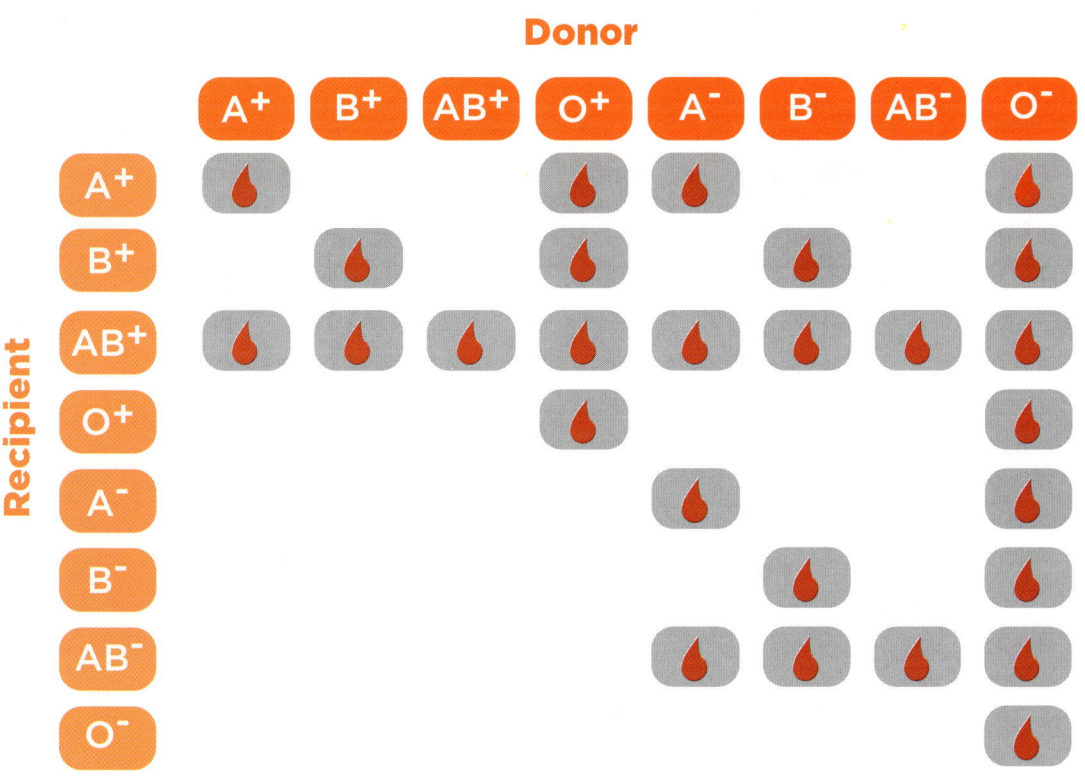

Blood Donation Compatibility Chart

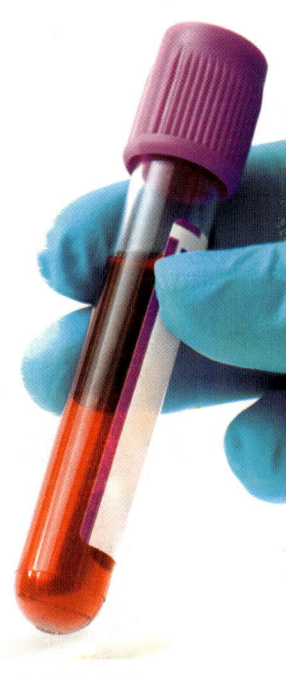

▲ *Anti-clotting vials to collect blood*

▼ *A finger-prick test is done to diagnose diabetes*

Know Your Blood Pressure

As our heart contracts and pumps blood into our arteries, it puts pressure on the walls of the blood vessels. This is called blood pressure (BP). BP goes up when the heart beats faster and goes down when you are sleeping or doing something quiet, like reading this book. There are actually two pressures, a lower one when the heart is waiting to be filled with blood (diastolic), and a higher one when it pumps out the blood (systolic). One act of pumping and one of relaxing makes up a single pulse.

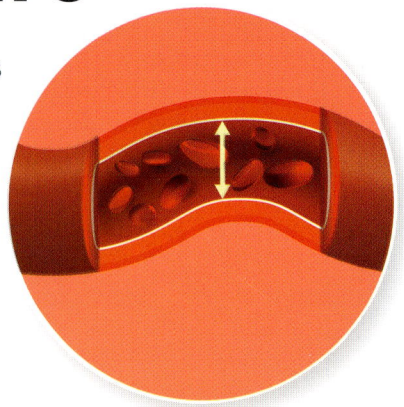

▲ Blood puts pressure on the walls of the blood vessels

How Do You Measure BP?

A **sphygmomanometer** measures BP. It has two parts. The doctor will first wrap a rubber cuff around your arm and inflate it to put pressure on the arteries in your arm. The second is the barometer.

Till a few years ago, this was a box that had a tube filled with mercury, like a thermometer. The height of the mercury in the tube (in millimetres/inches) gave the diastolic and systolic pressures. That is why BP is still written down as mmHg, that is millimetres of mercury in the tube. Now, digital sphygmomanometers tell you the BP electronically, along with the pulse. Blood pressure is written as systolic BP and diastolic BP separated by a forward slash, for example 120/80.

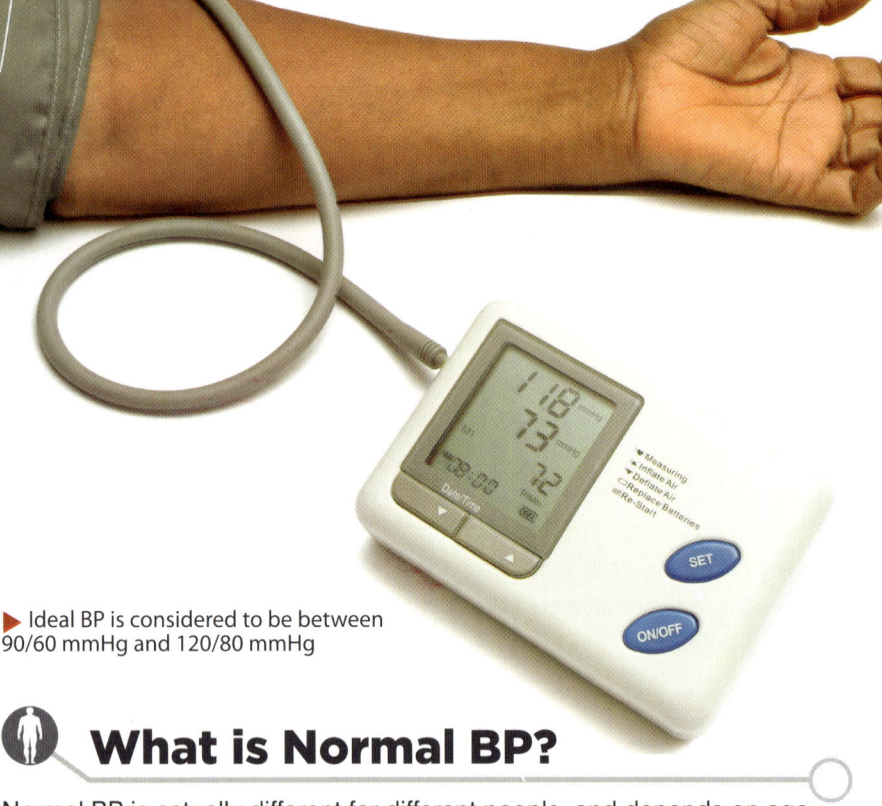

▶ Ideal BP is considered to be between 90/60 mmHg and 120/80 mmHg

What is Normal BP?

Normal BP is actually different for different people, and depends on age, sex, size, and health. But if your BP is regularly above 140/90 mmHg even if you are resting, you have high blood pressure, medically known as **hypertension**. You will feel normal, but you have a higher risk of heart attack, stroke, or kidney failure. On the other hand, if your BP is regularly below 90/60 mmHg, you have low blood pressure, or what doctors call **hypotension**. A sudden drop in BP may make you faint or feel dizzy and light-headed.

In Real Life

Some symptoms for issues with blood pressure are headaches, shortness of breath, nose bleeds, dizziness, nausea, difficulty in concentration, and blurry vision. While some are apparent, most will not show unless it gets serious. Always have regular health check-ups.

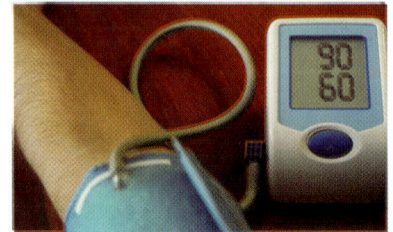

▲ The reading of 90/60 or lower indicates that the patient has low blood pressure

Lymph
The Second Circulatory System

Did you know that our body has a second circulatory system, with its own network of organs and vessels? This is the lymphatic system, and it gives the main system a helping hand. It also has WBCs in it, so it also helps the immune system. But more importantly, it helps drain the tissues of waste and excessive fluid, which would otherwise make them swell up and cause a condition called oedema. It recovers plasma that passes through capillaries. The fluid that travels through it is called **lymph**.

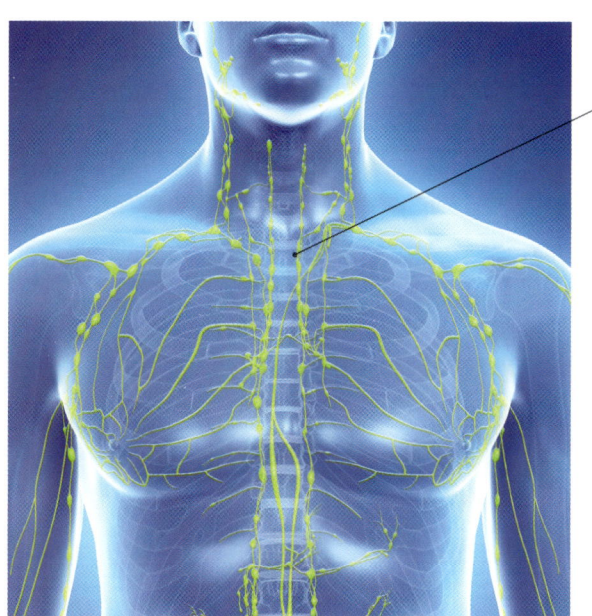

▲ An illustration of the lymphatic system

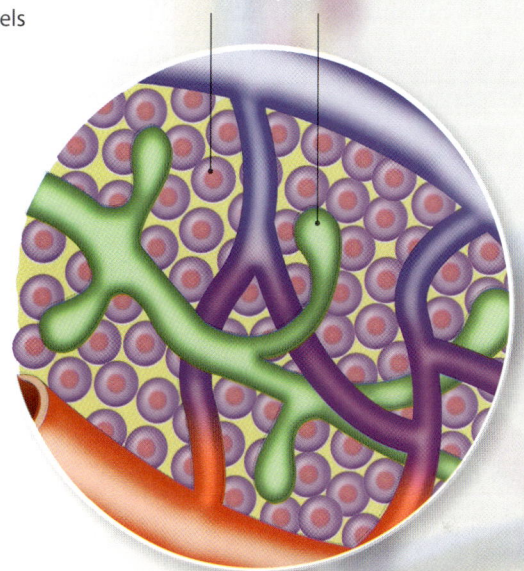

▲ The green lines in the diagram represent the lymph vessels

What is Lymph?

In insects, the lymphatic system is the main circulatory system, as there is a separate respiratory system. But in birds and mammals, blood takes over. But, the plasma from 'fenestrated' and 'sinusoid' capillaries oozes into the tissues, but not all of it comes back. Instead, the fluid becomes lymph, a colourless liquid, thinner than plasma. It flows into lymph vessels, which pump it through lymph nodes and finally pour it into the subclavian vein, near the shoulder bones. Thus, the extra fluid and the nutrients, salts, and hormones dissolved in it slowly come back to the circulatory system.

What Does the Lymphatic System Do?

The lymphatic system does not just carry away the extra fluid. It is also a way for nutrients to reach deep into tissues where blood cannot. From the capillaries, WBCs can crawl into tissues, where they can find and destroy germs. The destroyed germs are carried out through the lymph and finally cleaned up in the liver.

Lymphatic vessels also form part of a special tissue called **MALT**, in the lungs and intestines, where they act as a part of the immune system. Your body has 500 to 600 lymph nodes, which act as resting and training places for WBCs.

Lymphatic System
Your Body's Janitor

The lymphatic system is similar to the circulatory system, but lymph vessels are thinner than blood vessels. They end in two major lymph ducts, the right lymphatic duct and the much larger thoracic duct. Lymph only flows in one direction—back to the heart. For this, it has to travel upwards, as the two lymph ducts meet the circulatory system in the subclavian vein. The ducts have valves to stop lymph from flowing down.

As the lymphatic system does not have its own heart, it does not experience any pressure. Therefore, lymph flows slower than blood. This gives the WBCs in it more time to look for germs and fight them.

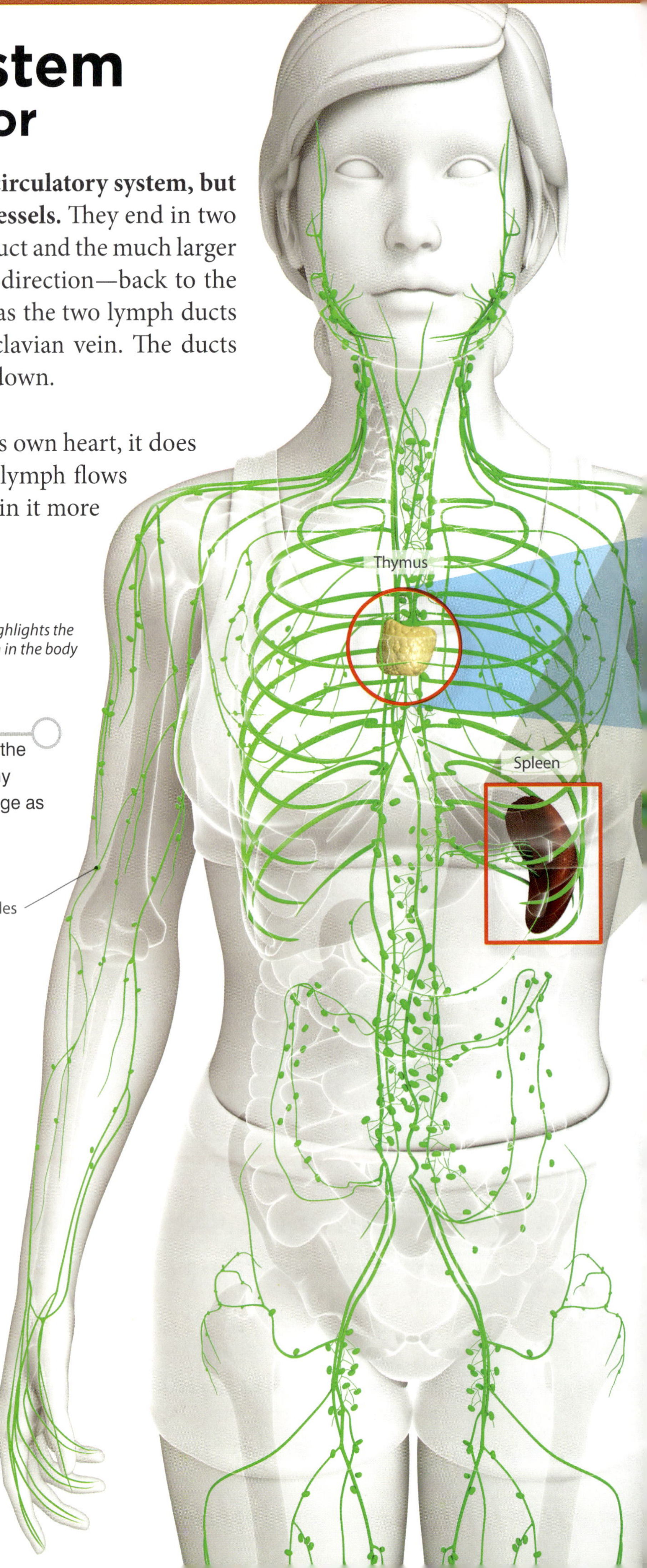

▶ The diagram highlights the thymus and spleen in the body

Lymph Nodes

Hundreds of lymph nodes do the work of pumping the lymph as it drains out of tissues. They come in many sizes: some as small as a pinhead, and others as large as kidney beans.

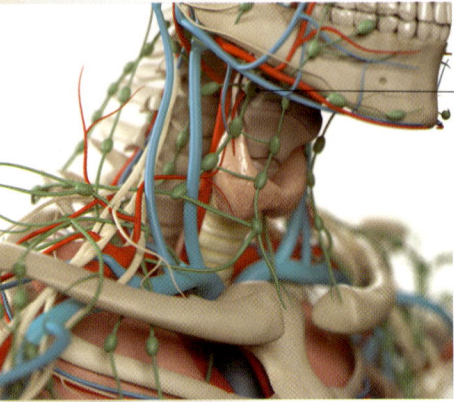

▲ A 3D representation of how lymph nodes appear in the body

They also act as filters as lymph carries all the rubbish from the tissues. The filtered rubbish is attacked by WBCs, thousands of which sit in the lymph nodes. This also helps WBCs to learn and identify germs which may be trying to hide. Hence, the lymphatic system has more WBCs than the circulatory system, which is why they are also called lymphocytes. Some parts of your body have more lymph nodes than the others, like the armpits, neck, throat, and groin.

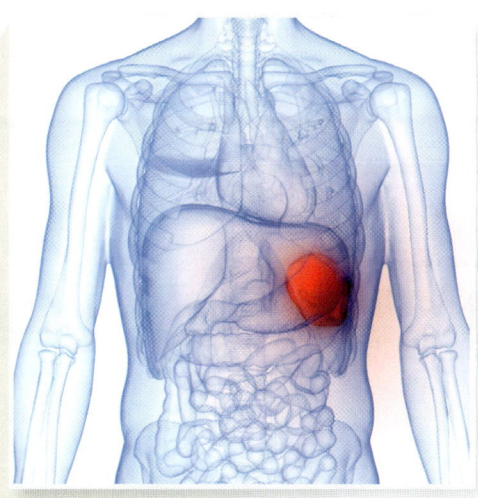

Spleen

Your spleen is a very important organ, which cleans your blood. You will find it on your left side, just behind and below the stomach. It looks like a very large lymph node, and it does many things a lymph node does. But it does something else too; it filters blood and removes old and worn out RBCs. It then destroys these and recovers the sugars, proteins, and lipids in them to be recycled by the body to make new cells.

The spleen also makes some WBCs and trains a whole lot more to detect germs. So it is a part of the immune system too.

◀ The diagram highlights the spleen in the body

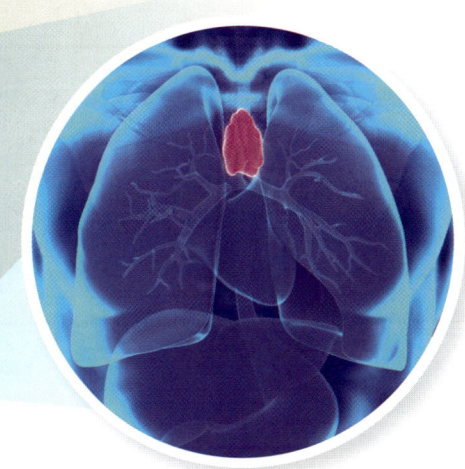

Thymus

The thymus is another organ of the lymphatic system. You find it between your breastbone and your heart. Like the spleen, it also filters blood and removes dead RBCs. But its main job is to make a kind of WBC called T-cells, which are born and trained here. T-cells are like commando cells. They do many things to fight germs especially as a part of the immune system, which protects our body.

◀ The diagram highlights the thymus in the body

Isn't It Amazing!

The thymus is the only organ in your body that actually decreases in size as you grow older, which means T-cells stop training after childhood.

▶ The thymus gets its name from its silhouette, that looks like a thyme leaf"

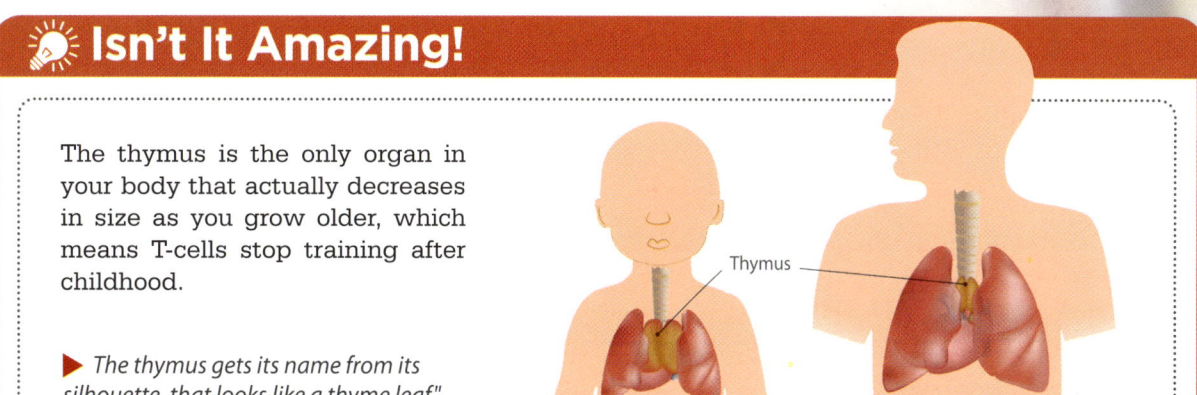

Mucosa-Associated Lymphatic Tissue (MALT)

Some parts of your body need more help from the immune system than the rest, because they are exposed to germs from the air, like your mouth, throat and lungs; or from food, like your intestines. Therefore, the body has evolved to put more lymph nodes and lymph ducts in these places, close to the lining of the lungs and intestines called the mucosa. This complicated tissue is called Mucosa-Associated Lymphatic Tissue (MALT). It is full of lymphocytes which are ready to fight any germs that try to enter the body.

Blood Disorders

You may have a healthy heart but unhealthy blood. The reverse can also be true too. Some of these illnesses are caused by genetics—you may have inherited them from your parents or grandparents. But others are because of poor nutrition when you were a baby, not eating well while growing up, and poor hygiene too. Let us look at a few of these and find out what can stop them.

Anaemia

Anaemia, a common blood disorder, is described by doctors as a condition of the blood when it has fewer RBCs than it should, or when the RBCs are misshaped and cannot function properly. An anaemic person's tissues do not get enough oxygen, so one cannot turn enough food into energy.

Such a person feels weak and tired all the time, and will have pale-looking skin. Most people have mild anaemia, but a few may have very severe anaemia, and need to be given RBCs.

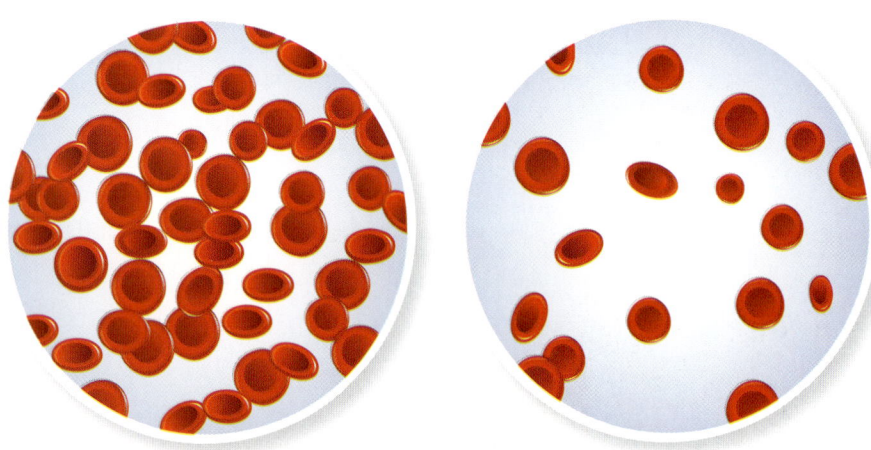

▲ The difference between the number of red blood cells present in the blood of a healthy person and an anaemic person

Sickle Cell Disease

The Sickle Cell Disease is a condition in which the RBCs, which are otherwise donut-shaped, lose their shape and become shaped like a sickle (a curved knife used to cut plants). This happens because the haemoglobin in them is not made right. The sickle-shaped cells cannot flow smoothly in the blood vessels but get tangled with each other, forming clumps. These clumps can lead to **aneurysms**, starving organs of oxygen. Patients often feel pain where these clumps are near nerves. The Sickle Cell Disease is a kind of anaemia that can be inherited from one's parents.

A sickle-shaped RBC

◂ A diagram representing sickle cell anaemia

Thalassemia Major

Thalassemia Major is also an inherited blood disorder, in which the bone marrow is unable to make haemoglobin properly. Therefore, it cannot make enough well functioning RBCs. There are many types of thalassemia, that can range from mild to life threatening. People with thalassemia are anaemic and look pale, feel weak and fatigued, get diseases easily, and many die young. The only real cure is a bone marrow transplant from a healthy, closely-related donor.

In Real Life

Thalassemia Minor is a genetic condition in which a patient has a faulty gene for haemoglobin but is otherwise healthy. An offspring of both Thalassemia Minor parents is likely to have Thalassemia Major.

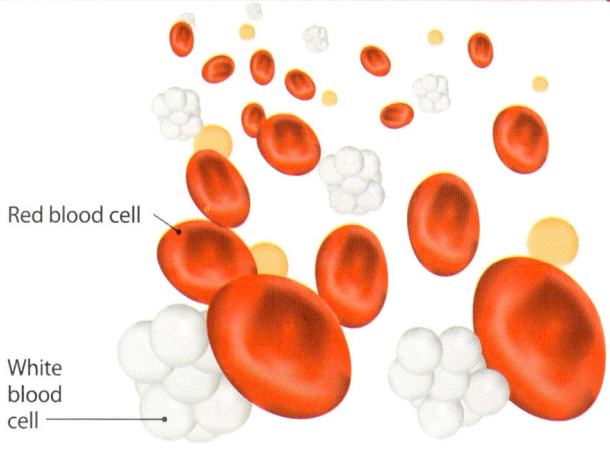
▲ The blood cells in the body of a healthy person

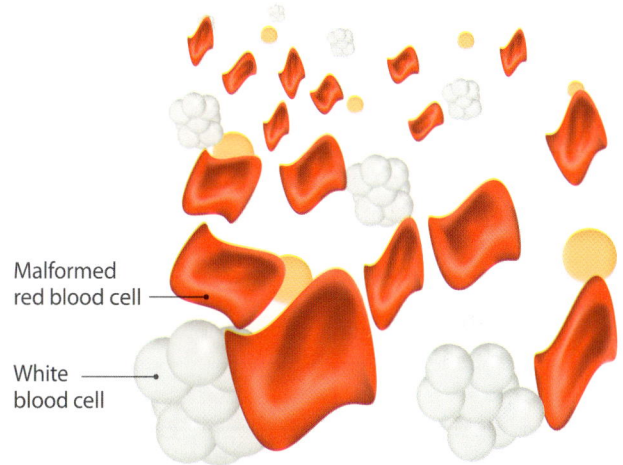
▲ Blood cells in the body of a thalassemic person

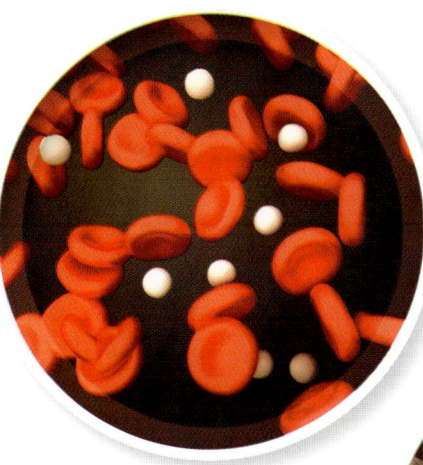

◀ The diagram shows the difference between blood cells in a healthy person and a person suffering from leukaemia

A healthy patient's blood

The blood of a leukaemia patient

Leukaemia

Leukaemia literally means white blood. This is a cancer which affects your WBCs. Your WBCs stop listening to signals from the rest of the body and stop fighting germs. Instead, they start multiplying at an unprecedented pace. It is these huge numbers that make the blood look whitish. They can interfere with the bone marrow's ability to make RBCs and platelets. Leukaemia is treated like other cancers, with radiation and chemotherapy.

A typical ITP dot

Immune Thrombocytopenia

Immune thrombocytopenia (ITP) is a disorder in which patients bleed a lot if they get injured, because they do not have enough platelets to form a clot. They must be careful not to get hurt, or even bruise themselves because they can get internal bleeding. They often bleed from the nose and mouth and show reddish-purple dots on the skin.

▶ *Immune thrombocytopenia causes purplish dots because of minor bleeding in the skin*

Life Cycle of Blood Cells

All your blood cells are born in the bone marrow. Many bones, like the vertebrae, ribs, sternum, hip bones, and bones of the arm and leg are hollow inside and filled with either yellow or red marrow. Yellow marrow is a jelly-like tissue made of fat-storing cells. Red marrow is more complex and is made of stem cells that make the cells of blood. These are the red blood cells (RBCs), white blood cells (WBCs), and platelets.

Until the age of seven, almost all of your marrow is red. After that, most marrow becomes yellow and stops making blood. But if there has been a bad injury or fever with lots of blood loss, yellow marrow can become red again.

The Birth of an RBC

To make a healthy red blood cell, your body needs iron. It also needs the compound haeme and the protein globin to make haemoglobin. Vitamin B12 and folic acid are important to make sure this happens correctly. The hormone erythropoietin tells the marrow when to start production and when to stop. After a few days, the RBC is ready in the marrow and can go into the blood to do its job. It will live for about 120 days, travelling thousands of miles inside your body, before it becomes too worn out to work anymore.

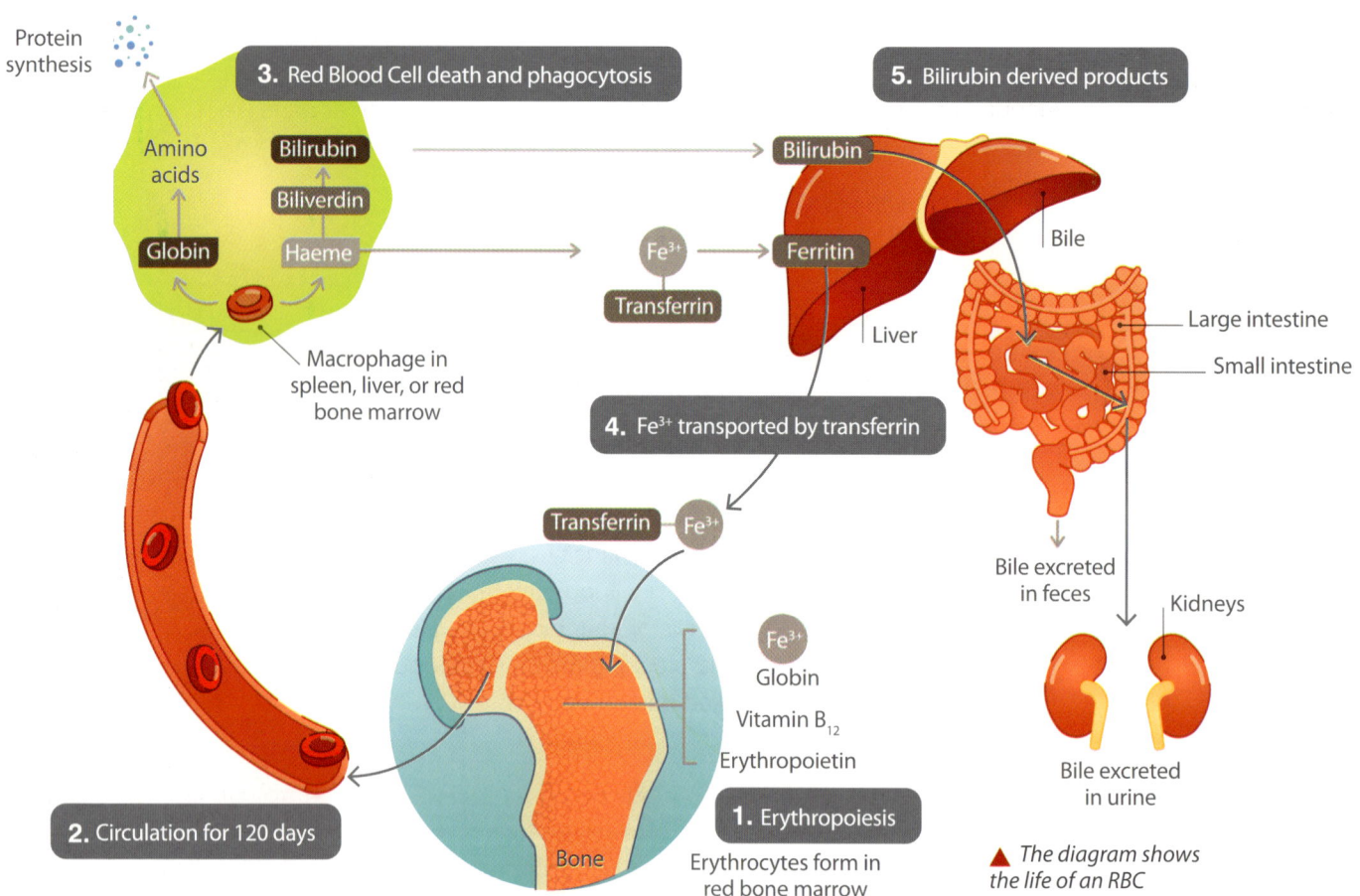

▲ The diagram shows the life of an RBC

The Death of an RBC

The worn-out RBCs are picked up by the spleen and liver. The membrane of the cell is broken, killing it. The globin is then broken up into amino acids, which are re-used by the body to make other proteins. The haeme of haemoglobin is turned into bile and goes into your digestive system, where it does a second job of digesting fat. The iron is tacked on to a protein called transferrin. This takes it to the bone marrow, where new RBCs are waiting hungrily for it.

Watch Out!

A healthy heart keeps a body healthy, for if the heart has any problems, it affects the whole body. An unhealthy heart cannot pump enough blood to get oxygen to all tissues so they too are sickly all the time. Adults who smoke cigarettes have a higher risk of cardiac disease. Alcoholism also affects the heart. Working too much without enough rest or sleep causes stress. This makes the heart beat faster and increases your BP. This increases the risk of sudden cardiac arrest.

▲ A small exercise routine daily, goes a long way in keeping your heart healthy

Smoking and Alcohol

Substances in cigarette smoke go into the lungs and cover them, stopping oxygen from passing into the blood. This leads to anaemia. Nicotine in cigarette smoke also makes the smoker excitable, increasing their heart rate and BP. Smoking throughout life makes the chance of a heart attack very high.

Alcohol makes people eat more fatty food, increasing their BP and the risk of an aneurysm or stroke. It also causes cardiac arrhythmia and the possibility of a sudden cardiac arrest.

◀ Drinking alcohol and smoking cigarettes are injurious to health

Cholesterol

You may hear doctors and parents talk of good **cholesterol** and bad cholesterol. What does that mean?

Cholesterol is mopped up from tissues by a molecule called lipoprotein and taken to the liver to be turned into bile, which then goes out of your body through the digestive system. Some of the cholesterol is mopped by a variety called high density lipoprotein (HDL), and some of it is taken up by another kind called low-density lipoprotein (LDL). LDL tends to stick to the walls of your blood vessels, forming 'plaques' and blocking the flow of blood, so it is called bad cholesterol. This can lead to aneurysms.

Eating foods rich in fats, such as cheesy food, deep-fried food, chocolates, etc. increases LDLs in blood, while smoking and lack of exercise reduce HDLs. Over time, this builds up the risk of heart disease.

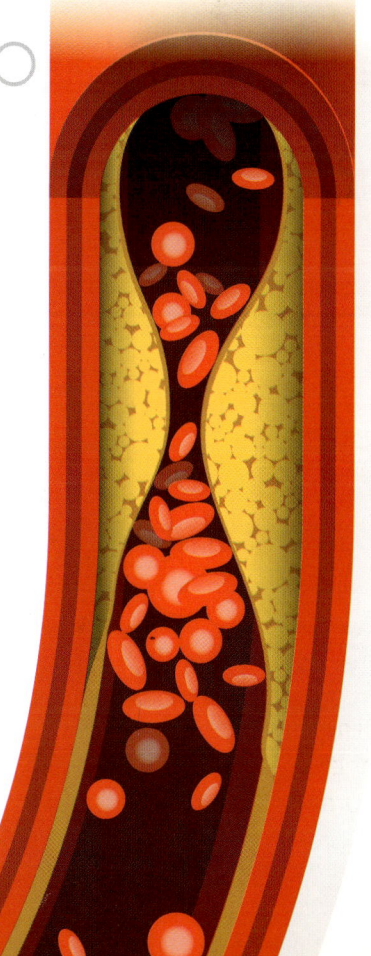

◀ Too much saturated fat is harmful for your heart ▶ Plaques make your arteries narrow and reduce blood flow

A Healthy Heart Makes a Healthy Body

If you listen to grandma talk about her childhood, she might tell you of times when she walked to school and back, played games in the playground, and ate fruits from trees. This may sound like how all old people talk, but they got some things right, which we do not in our times. Many of us now live 'sedentary' lives, working in offices, travelling by cars, and spending leisure time playing video games. All these make your heart work harder and weaker. Keeping the heart healthy is simple, but it needs a lot of discipline on our part.

Eat Right at Right Times

Eating a balanced diet will keep all your organs healthy, not just your heart. But, the heart specifically needs a few nutrients. For instance, fruits and vegetables are rich in vitamins and minerals which keep the muscles of the heart healthy. Doctors suggest you have four or five servings every day. Fish, seeds, and nuts are rich in omega-3 and omega-6 fatty acids, which are good for your nerves, which in turn keep the regular rhythm of the heart. Not skipping meals and drinking lots of water help maintain good blood pressure and water balance.

Foods high in cholesterol and salt add to the risk of artery blockages and high BP. You usually get these in packaged foods and drinks. But, now, many companies make low-salt and low-cholesterol foods, so choose wisely.

 Supplements can never replace real foods completely

Be Active and Get Enough Rest Too

Playing outdoors and exercising a lot keeps the heart healthy too. It helps burn up a lot of the food you eat and gives your heart the energy to keep going. It also stops cholesterol from clogging up your arteries. Some people like dancing or aerobics or 'cardio-exercises', which are good too. An active life helps you keep your BP and water balance normal.

But getting enough rest is important too. Too much work or exercise builds up stress, and makes your heart beat faster, and raises your BP. Getting a good night's sleep and warming up before exercise or playing helps reduce stress on your heart.

▼ *Stay active by playing different sports*

HUMAN BODY — HEART & CIRCULATORY SYSTEM

The Right Weight for Your Heart's Health

The more tissue there is in the body, the harder the heart has to pump so that blood reaches everywhere. So, it follows that obese people's hearts work harder. If they do not exercise enough, they face a very high risk of developing aneurysms. Obesity has multiple causes, including an unhealthy lifestyle, genetics, and environmental factors.

But people who are too thin can also get heart problems among other diseases. Eating right and exercising enough makes sure that your weight is optimally maintained.

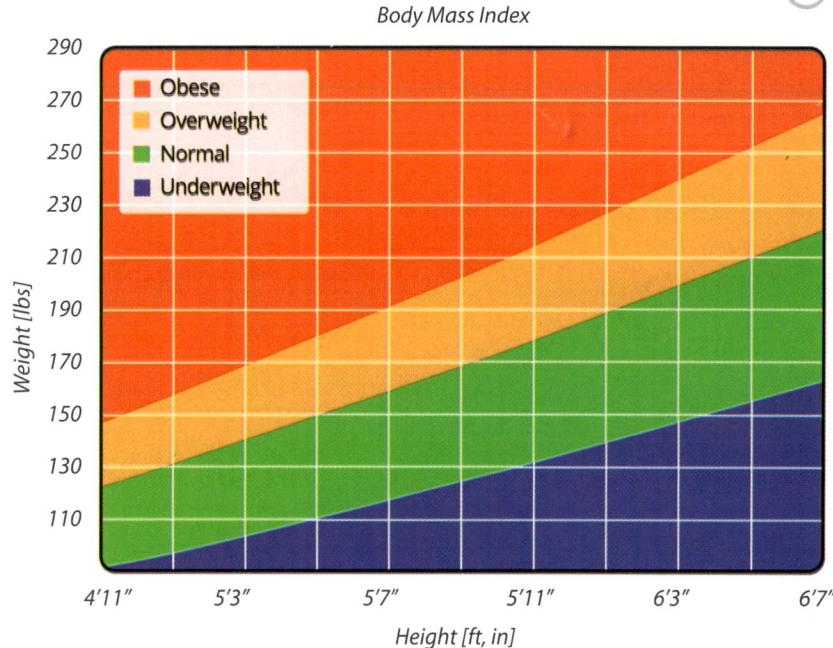

▶ Doctors have a chart of the right weight for you, depending on how tall you are

In Real Life

Love video games? Doctors suggest that you may be at the risk of becoming obese because of lack of exercise. But now a few game-making companies are coming up with games that help you exercise as you play and eat healthier too. These are called exergames.

▶ Playing active video games helps you stay fit

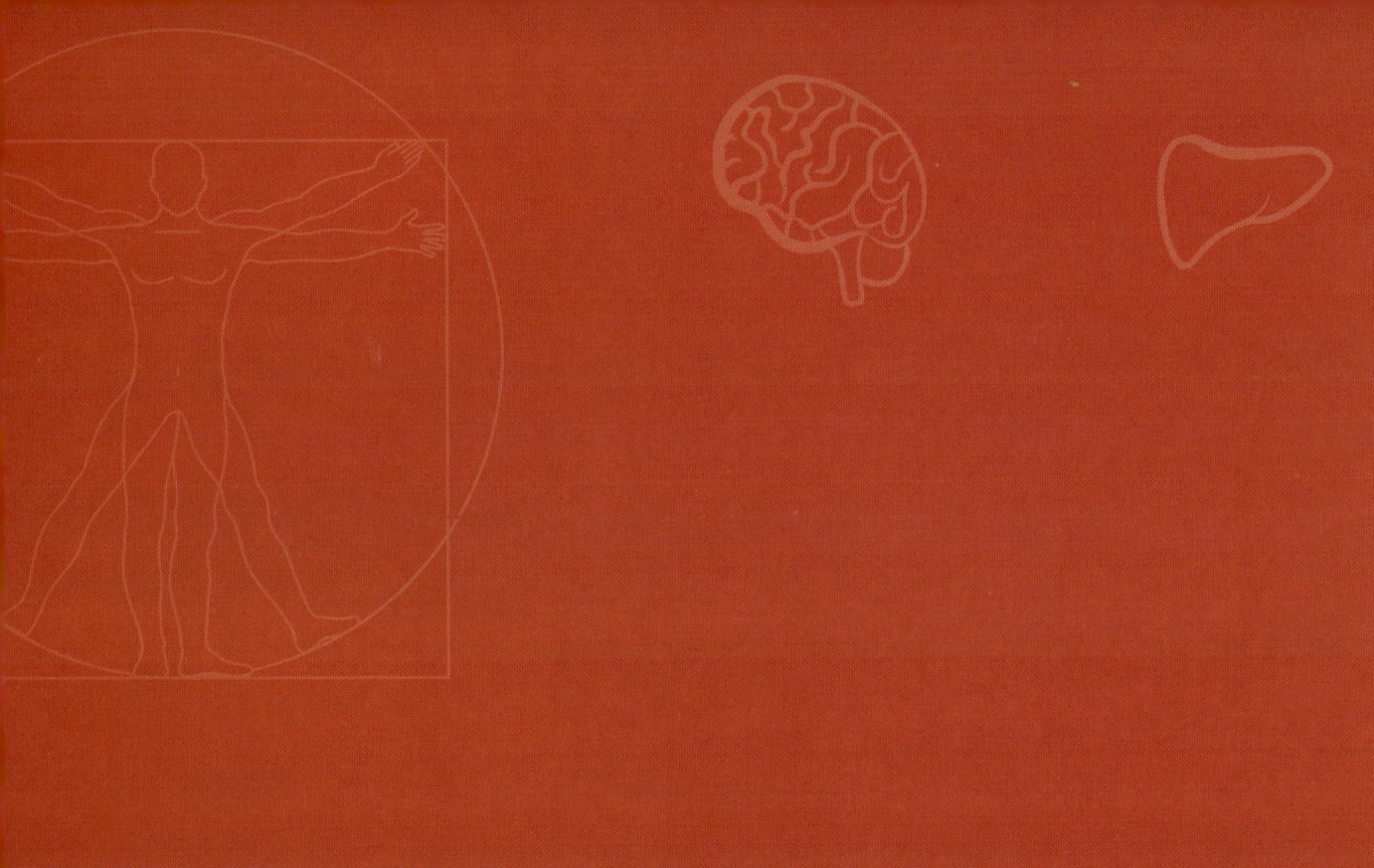

IMMUNE SYSTEM AND COMMON DISEASES

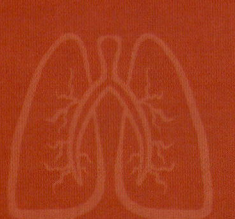

OUR BODY'S ARMY

If there were no bacteria, fungi, or viruses in the world, we would not need an immune system. But since they exist and some even cause diseases in our body, we need ways to defend ourselves. Just as a country has police officers to safeguard against troublemakers, the body too has a complex immune system that protects it from diseases.

Our immune system is always working to keep us safe, whether we are pricked by a needle or affected by a very serious disease. The skin is the first line of defence in our body. Our body has organs where immune cells are 'trained' to fight, just like the police is trained to fight. The immune system is a master of disguise—it has special cells that work like spies, travelling around the body looking for any invaders.

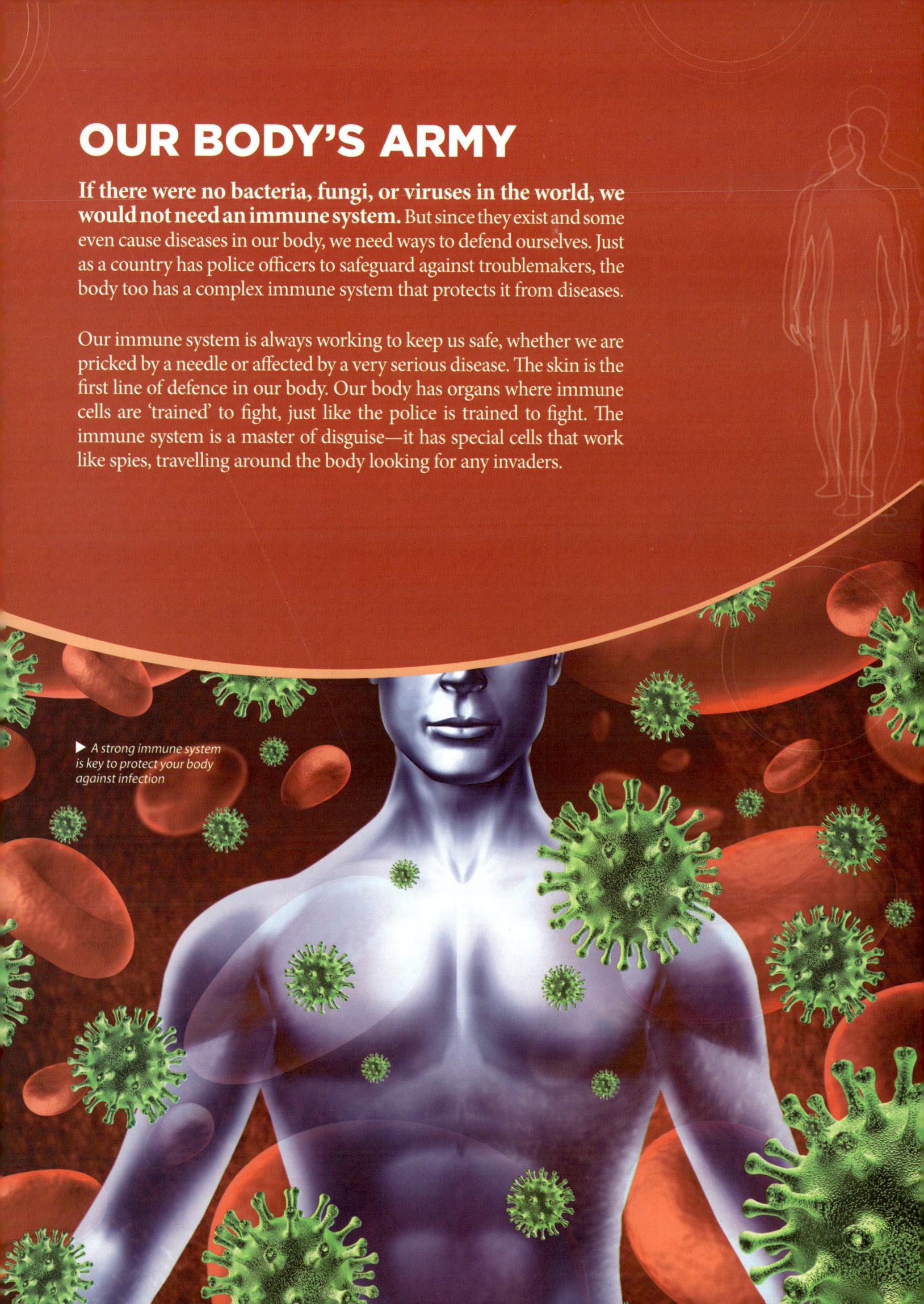

▶ A strong immune system is key to protect your body against infection

Our Skin: The Great Wall

Did you know that the largest organ of the body is your skin? It is the first defender of the body. It keeps the internal organs safe from the ill effects of varying temperatures. Our skin is made up of three layers—while the outermost layer is called the **epidermis**, the inner layer found beneath the epidermis is known as **dermis**. It contains hair follicles, sweat glands, and connective tissue. The third layer is the **hypodermis**, a deeper subcutaneous tissue which is made up of fat and connective tissue. Your skin is also your body's first immune warning system. It can stretch and also squeeze itself. That is why your skin does not tear when you extend your arms during a stretch or crouch down to find a coin that has rolled under the bed. Your skin has nerves that tell you when something has touched it.

▼ Children have thinner skin than adults. So, children absorb harmful agents faster than adults

Keratinocytes

Keratinocytes are wonder cells found in the epidermis that perform many functions. These cells are tightly packed together to keep out foreign particles like bacteria and viruses. They make vitamin D in the presence of sunlight. Keratinocytes also make **keratin**, which is needed to form the hair and nails. They detect wounds and heal the skin. They alert the immune system to destroy particles that might have entered from an open wound.

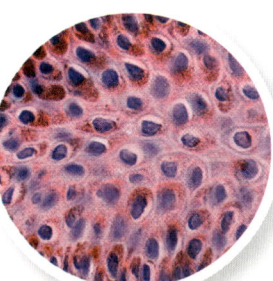

▲ 90 per cent of epidermal skin cells are Keratinocytes

Hair Follicles

Hair follicles are the organs from where hair grows. Hair is made of fibres of a protein called keratin. The hair on our bodies is of two types. First is the long wiry hair which we can see and which keeps us warm by forming a carpet that traps air. When we are cold or scared, the hair stands straight, causing 'goose bumps'. It then traps a thicker carpet of hair.

Our skin also has shorter, white hair, which cannot be seen without a magnifying lens. These act like burglar alarms. If a louse, bedbug, fly, or mosquito brushes against them, they trigger the nervous system, which in turn makes us 'itch'. We instinctively scratch and scrape off the creature.

In Real Life

The darker a person's skin is, the safer they are from ultraviolet (UV) rays of the Sun. That is because the darker a person is, the more melanin they have in their skin. Melanin stops 99 per cent of UV rays from entering the body and causing harmful diseases like cancer. However, no matter what the colour of your skin is, you need to apply sunscreen and protect your skin from sunburn or other damage.

The Inner Skin

In the dermis, there are **sweat glands**, oil glands, hair follicles, and blood vessels. Sweat glands keep the body cool by releasing water along with some salt and amino acids onto the skin, which evaporate and cool it. However, if we sweat too much, a lot of amino acids collect on the skin. Bacteria can grow on these and, in turn, produce smelly chemicals. That is why when sweat collects in the armpits, it begins to smell bad.

The oil glands are also called **sebaceous glands**, because they produce an oil called **sebum**. It spreads all over our skin and makes sure that water does not enter skin from outside. That is why, even if you get wet in the rain, you do not soak up water like a sponge. Sebum also prevents bacteria and fungi from growing on the skin.

▲ Sweat prevents your body from overheating

▼ Your skin is like an onion. The outermost layers of the skin are made of dead cells, while the inner layers are made of living cells

Isn't It Amazing!

Birds do not have oil glands throughout their skin, but a single 'preen gland' near their tail. They use their beaks and claws to spread this oil all over the bodies and feathers to keep themselves clean and healthy. We call this preening.

▲ Birds need their preen glands to stay dry, because without them their feathers would get wet and prevent them from flying

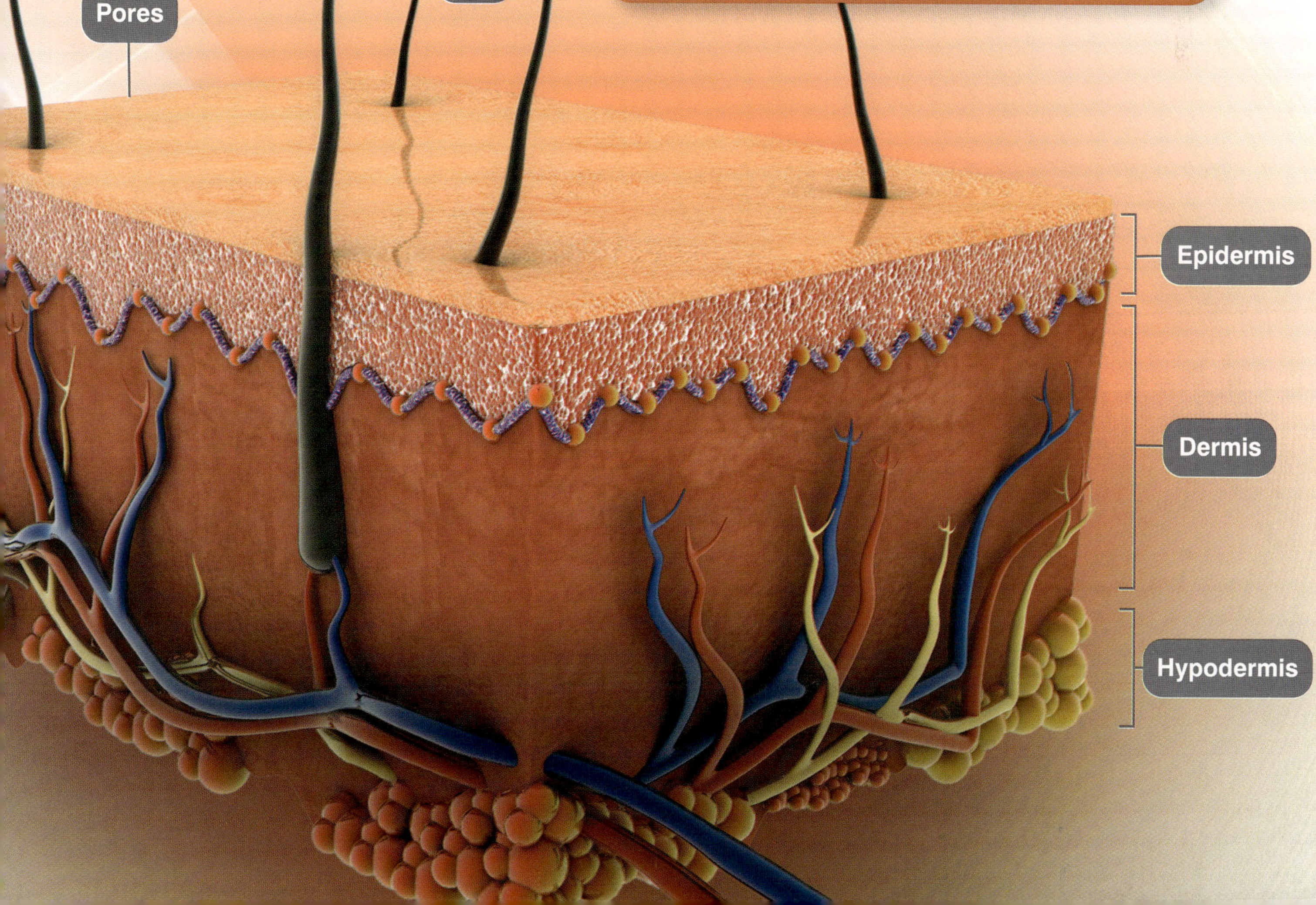

- Pores
- Hair
- Epidermis
- Dermis
- Hypodermis

The Organs that Protect Us

Did you know that our immune system is not really a system? As the name suggests, the immune system performs the critical role of defending the body against foreign bodies, viruses, bacteria, etc. However, it may surprise you to know that the immune system is unlike the other organ systems that make up the human body. That is because it has no organs dedicated solely to itself, as is the case with the digestive system or nervous system. Instead, the organs of the immune system work towards safeguarding the body in addition to performing other functions.

Lymph Nodes

The **lymph nodes** are part of the **lymphatic system**, which is our body's second circulatory system. This system carries lymph, a colourless liquid that contains the waste products of our tissues. Lymph passes through a number of tiny organs called lymph nodes, which filter the lymph and remove **pathogens**, broken cells, etc. which are then destroyed by the immune system, similar to how the spleen works.

The lymph nodes also act as resting camps for B- and T-cells. In an active infection, a lymph node will swell up, so that more soldier cells can come in and fight the pathogens as they are brought in by the lymph.

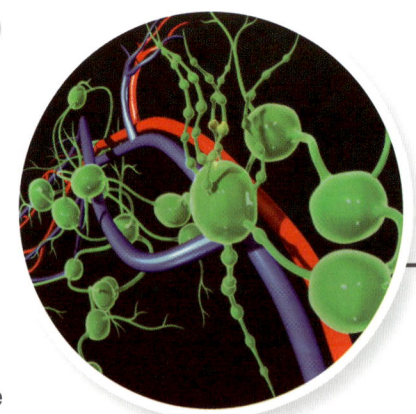
▲ Lymph nodes filter germs and train WBCs

Bone Marrow

This is the fleshy tissue inside most large bones. This is where red blood cells (RBCs), white blood cells (WBCs), and platelets are made. Their parent cells are called stem cells. The bone marrow also trains some of the WBCs to fight disease-causing germs called pathogens. These trained cells are called B-cells.

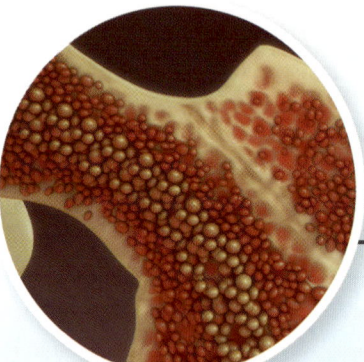
▲ Bone marrow makes up 4 per cent of an adult human's weight

In Real Life

Some people have a defective bone marrow, and therefore they get a disease called thalassaemia. This disease causes both anaemia, that is the lack of RBCs and low immunity which is the lack of WBCs. Patients feel tired all the time, have paler skin than normal, slow growth of the body, and have excess iron in their bodies. The only known cure is a transplant of bone marrow from a close relative.

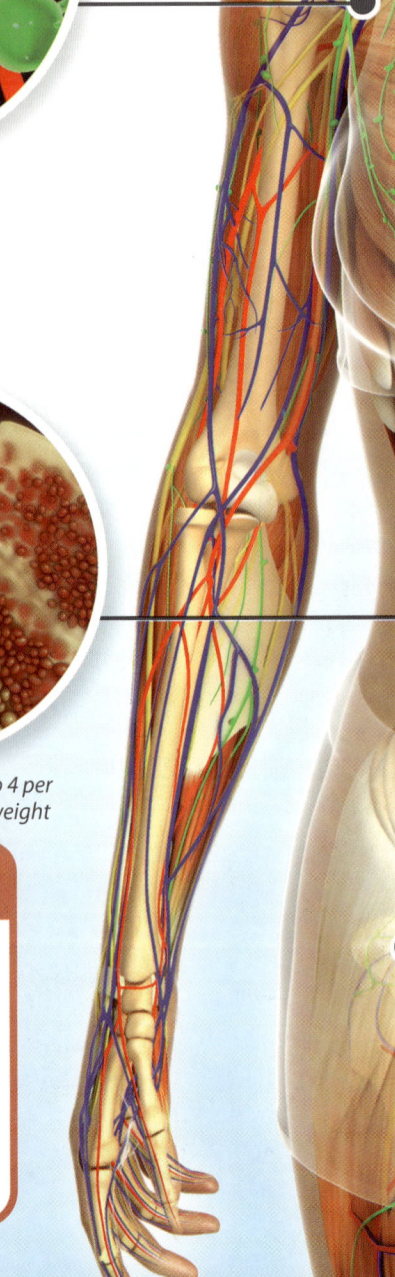

HUMAN BODY — IMMUNE SYSTEM & COMMON DISEASES

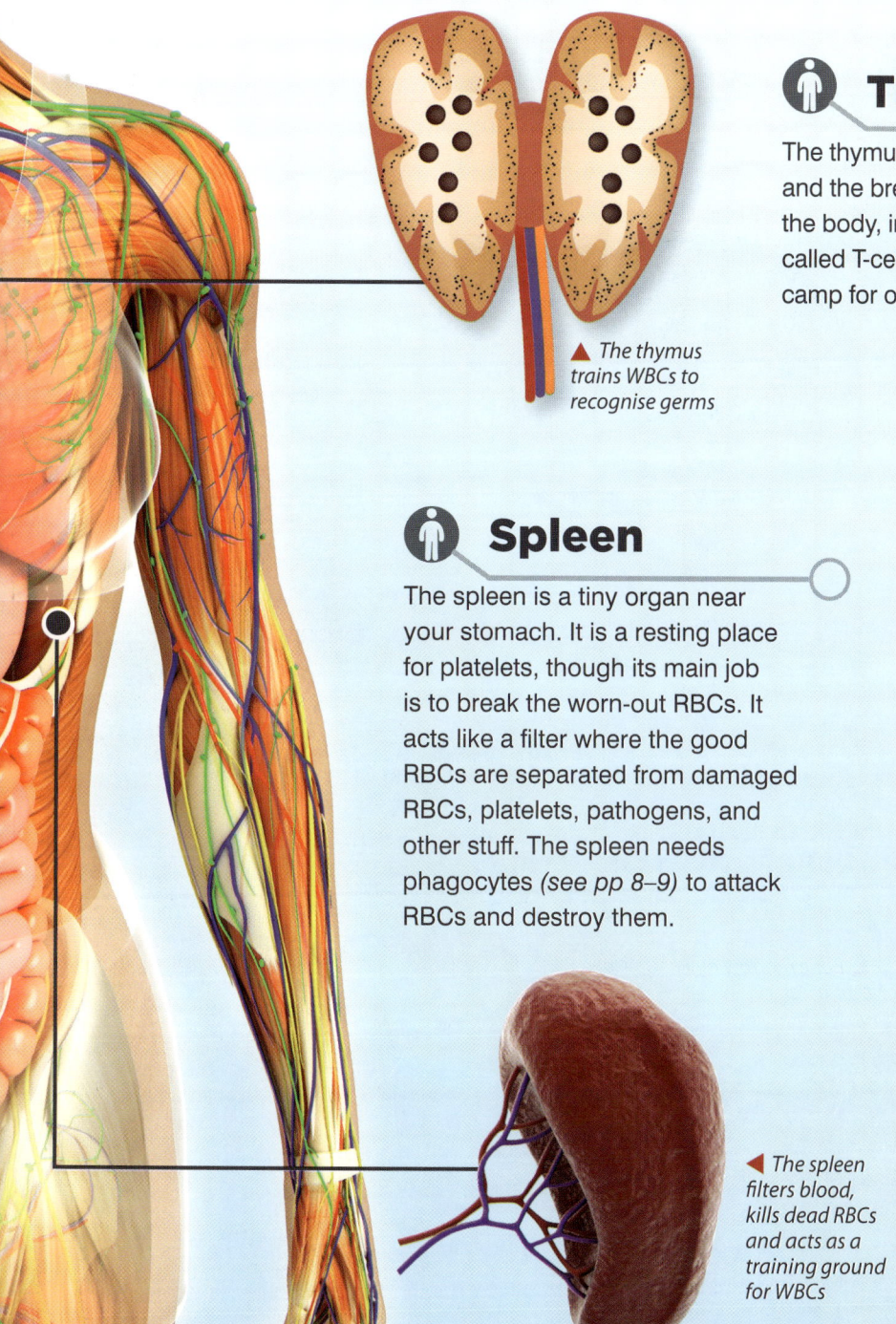

◀ The main organs of the immune system

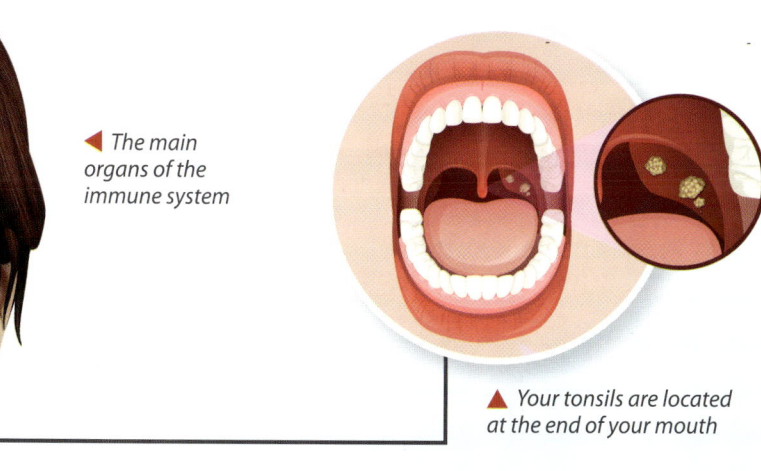

▲ Your tonsils are located at the end of your mouth

Tonsils and Adenoids

These are the lymph nodes that are present in our throat. They help to fight off infections in our mouth and nose respectively. Like all lymph nodes, they swell up during an infection. Unfortunately, this swelling can cause problems with swallowing food or breathing. That is why you have to have your tonsils removed if they swell up too much. The swelling of adenoids causes snoring, because they block the passage of air from the nose into the lungs.

Thymus

The thymus is a tiny organ between the heart and the breastbone. It performs many tasks for the body, including training a class of WBCs called T-cells. The thymus also acts as a resting camp for other kinds of WBCs.

▲ The thymus trains WBCs to recognise germs

Spleen

The spleen is a tiny organ near your stomach. It is a resting place for platelets, though its main job is to break the worn-out RBCs. It acts like a filter where the good RBCs are separated from damaged RBCs, platelets, pathogens, and other stuff. The spleen needs phagocytes (see pp 8–9) to attack RBCs and destroy them.

◀ The spleen filters blood, kills dead RBCs and acts as a training ground for WBCs

Isn't It Amazing!

Ever squashed a mosquito and seen a splat of red blood? That is your own blood! Insects have no blood, they only have lymph, which drains out the body's waste and has cells to fight pathogens. Blood only developed with the evolution of vertebrates, which includes human beings, as well as other mammals, fish, amphibians, reptiles, and birds.

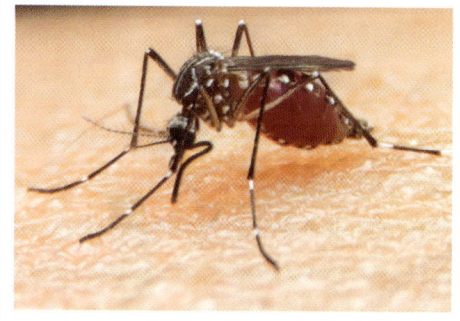

▲ A mosquito sucking blood. Note that its abdomen is full of human blood.

Immune Cells: Tireless Soldiers

An army has many kinds of soldiers. There are snipers, tank-drivers, fighter pilots, drone operators, and engineers. In the same way, our body has immune cells and the cells of the immune system come in many types. These cells are together called white blood cells. They can travel quickly to any part of the body through the blood and lymph, where they detect and destroy pathogens.

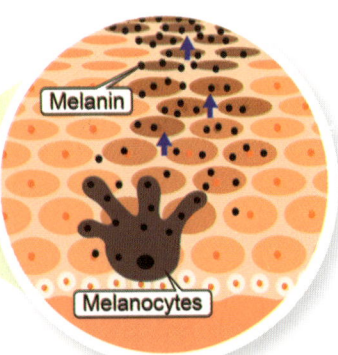

▲ Melanocytes are present in the skin, hairs, iris of the eye, and some places inside the body

 ## Melanocytes

These cells live in the skin and produce **melanin**, which gives our skin its colour. They also act as scouts, looking for bacteria, viruses, fungi, pollen, dust, smoke particles, etc., that might have entered our body.

Monocytes

Monocytes are WBCs that circulate in the blood. When they get the signal that an invasion has happened, they swell up and turn themselves into lethal **macrophages**. At this stage, they leave the blood to enter the body tissues, with a mission to eliminate the harmful particles.

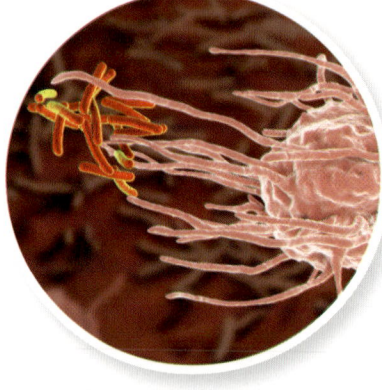

▲ A monocyte that has become a macrophage

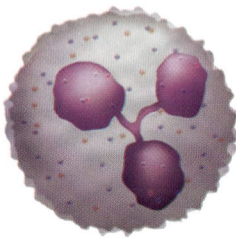

▲ 55–70 per cent of your WBCs are Neutrophils

 ## Neutrophils

Neutrophils are WBCs that are filled with chemicals called **histamines**. They attack the pathogen and release the histamine all over it. Histamines bring macrophages to the site, which finish the job.

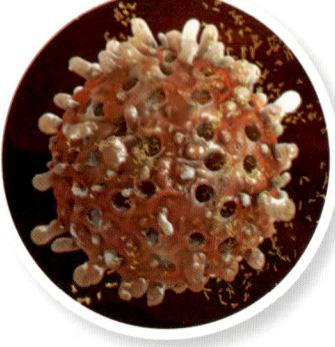

▲ B-cells are chemical weapons specialists

 ## B-cells

B-cells make a special type of protein called **antibodies**. The B-cells make thousands of different antibodies, each of which can identify a different bacteria or virus. Antibodies stick all over the pathogens and then the immune system's T-cells come and kill them.

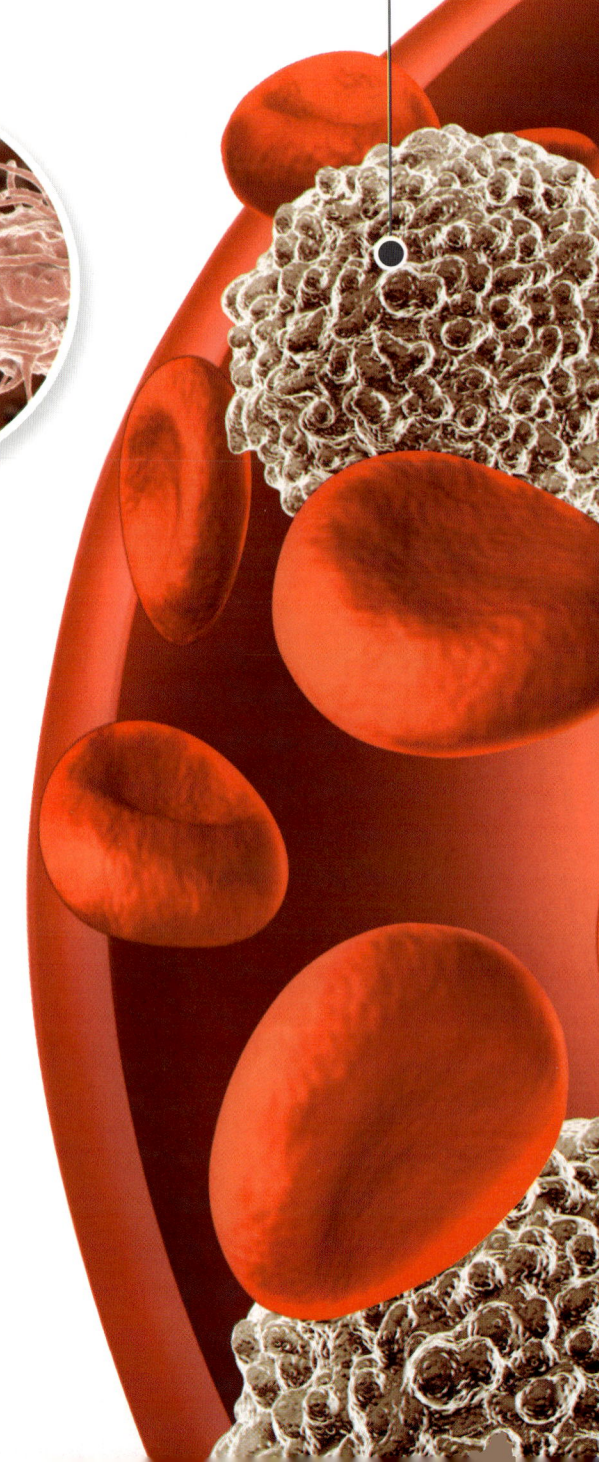

▼ White blood cells (WBCs) travel through blood to reach the infected tissue, following chemical signals

Basophils

There are not too many of these WBCs when we are healthy. But when we get bitten by an insect, or have a worm infection, they flock to the site and multiply quickly. They cause redness at the place where you got bitten, known as **inflammation**.

◄ *Basophils carry histamines and enzymes that smash up the pathogen and call for more help*

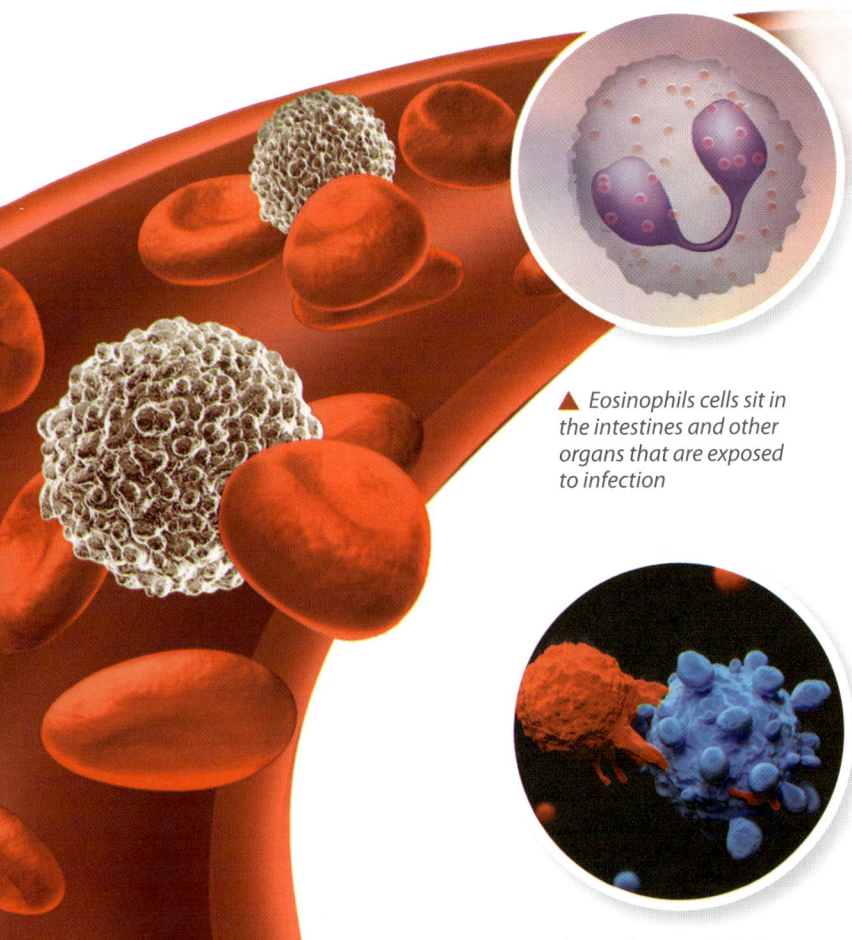

Eosinophils

Eosinophils work a lot like basophils, but they are also involved in allergies and asthma.

▲ *Eosinophils cells sit in the intestines and other organs that are exposed to infection*

T-cells

T-cells are like the special forces or naval SEALs. They perform many functions such as targeting and killing cancer cells and pathogens. You will read more about these cells in detail later.

▲ *T-cells are called when the others are unable to do the job on their own*

Phagocytes

Unlike the other immune cells, phagocytes are present throughout the body. When they are present in the intestines, lungs, and the lining of the nose, they are called **dendritic cells**. In the skin, there is a special type of phagocyte called **Langerhans cells**, which detect and destroy foreign invaders.

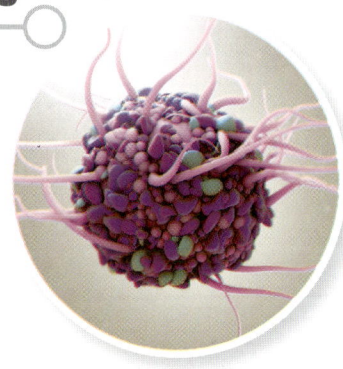

▲ *Phagocytes*

Natural Killer Cells

Natural killer cells perform the important function of finding and killing cancer cells before they take over the body.

Phagocytosis
Eating Our Body's Enemies

Phagocytosis is the body's way of getting rid of any other foreign agent that enters it. These could be bacteria, fungi, or viruses. Phagocytosis is also the main immune defence of all animals, from the most primitive sponges to insects and mammals alike. In our bodies, this job is done by immune cells.

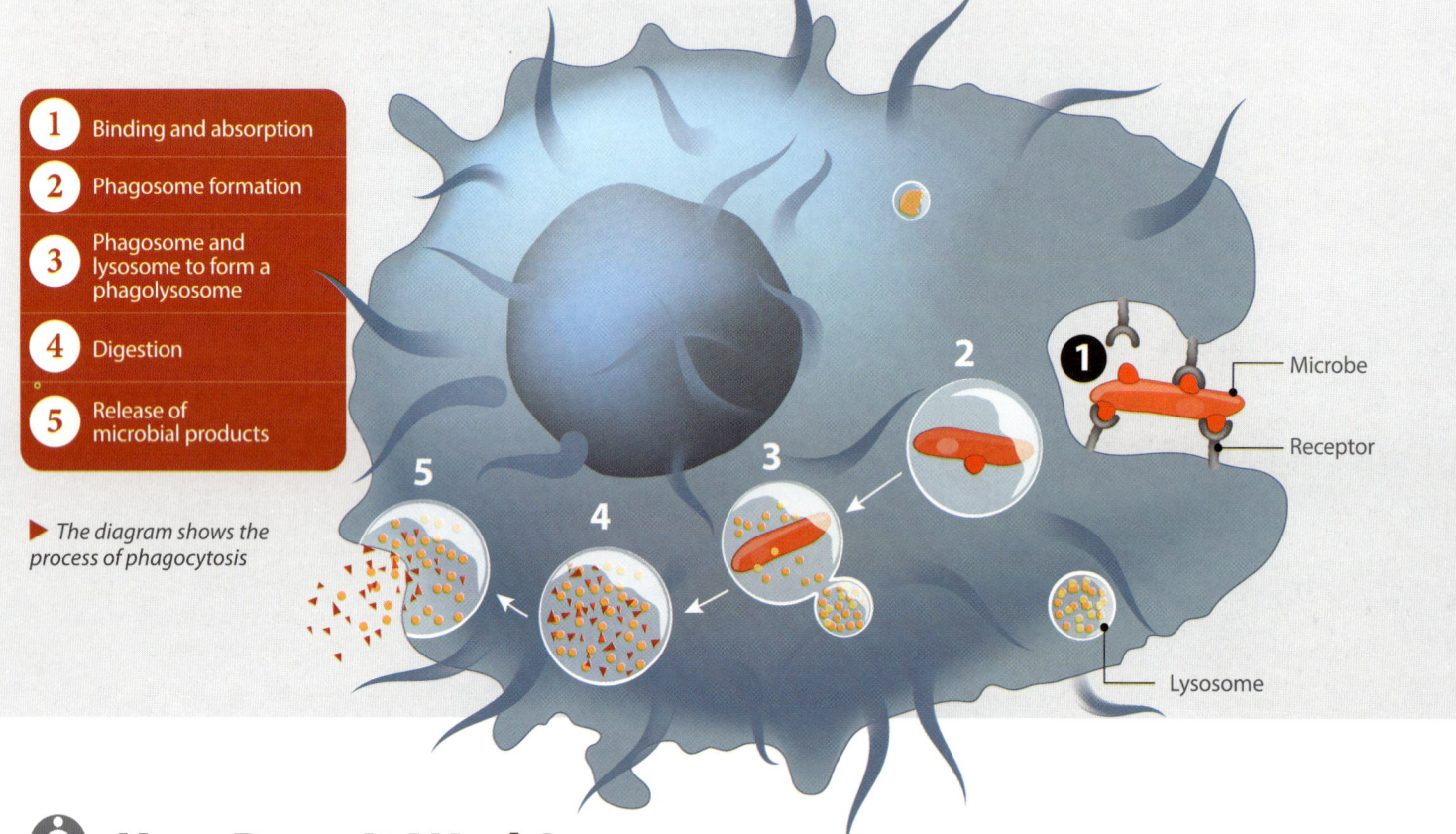

1. Binding and absorption
2. Phagosome formation
3. Phagosome and lysosome to form a phagolysosome
4. Digestion
5. Release of microbial products

▶ The diagram shows the process of phagocytosis

How Does It Work?

Phagocytosis works in a specific way. On the surface of the phagocyte cells, there are proteins called receptors that recognise a pathogen when they brush against it. Once they detect the pathogens, the cell membrane immediately surrounds the particle until it is completely enveloped or captured. It is like covering yourself with a blanket.

The ends of the membrane meet and the pathogen is captured in a **phagosome**. The cell has little bags called **lysosomes**, filled with enzymes. Lysosomes merge with phagosomes to form phagolysosomes. These phagolysosomes have enzymes that break up the proteins, carbohydrates, and fats from the foreign body. Most of the broken remains are removed from the cell, and then the body, through the lymph.

Many different kinds of cells can do this. Some of the immune cells also do another job—they pick up some of the remains of the things they ate up and display them on the surface of their cell membranes, like identity tags. Other cells of the immune system use these tags to identify the kind of infection we have. These immune cells can then respond in a way that ensures the infection is cleaned out completely.

Isn't It Amazing!

Ever seen an amoeba under a microscope? Our body's WBCs behave exactly like the amoeba. They can also move on their own, especially in the spaces between cells in tissues where the blood cannot reach.

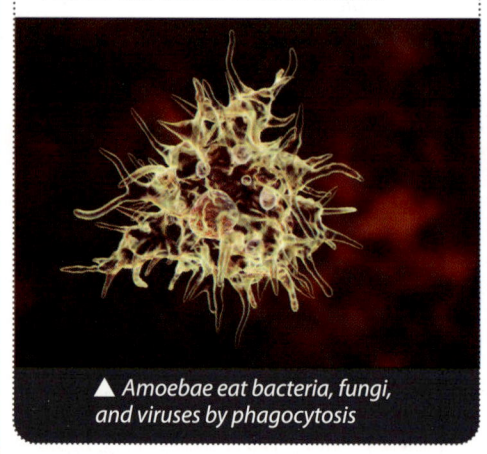

▲ Amoebae eat bacteria, fungi, and viruses by phagocytosis

Shoot at Sight

There is a special way by which our body fights diseases. Each pathogen (bacteria, viruses, fungi, plasmodia, and worms) is made of different kinds of proteins and other biochemicals.

Antigens

Bits of these pathogens are used by the immune system to recognise pathogens. These bits are called **antigens** and this job is done by the cells that perform phagocytosis. If the macrophages and dendritic cells are like the policemen showing the antigen, the T-cells and B-cells set out to look for the pathogen, just like a sniffer dog can identify a criminal based on a handkerchief that he has left behind at the scene of the crime. Both have different proteins on their surface that try to match the antigen. If the match is perfect, then the cells get to work.

▶ Parts of a pathogen displayed on the surface of a phagocyte that ate it. These parts are called antigens.

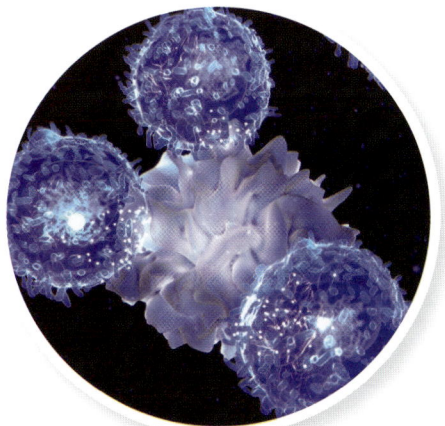
◀ A dendritic cell that presents antigens to the (blue) T-cells to see which one matches

In Real Life

Why do we have different blood groups? This is because the RBCs of our bodies have different antigens on their surfaces. Some people have a type called A, so they belong to Blood Group A. People of Blood Group B have another type called B. Some people have both types, that is Blood Group AB, while some have none, that is Blood Group O.

Antibodies

Antibodies are our body's chief biochemical weapons. Antibodies are proteins made by the B-cells and each one is made to match a particular antigen. For example, if the B-cell that makes anti-chicken pox antibodies meets a cell showing the chicken pox antigen, it will make these antibodies by the thousands. The antibodies are released into the blood, where they latch on to the chicken pox viruses. Other biochemicals in the blood called cytokines, recognise these covered viruses and get the commando cells to swallow them up.

▶ The illustration shows antibodies attacking a virus

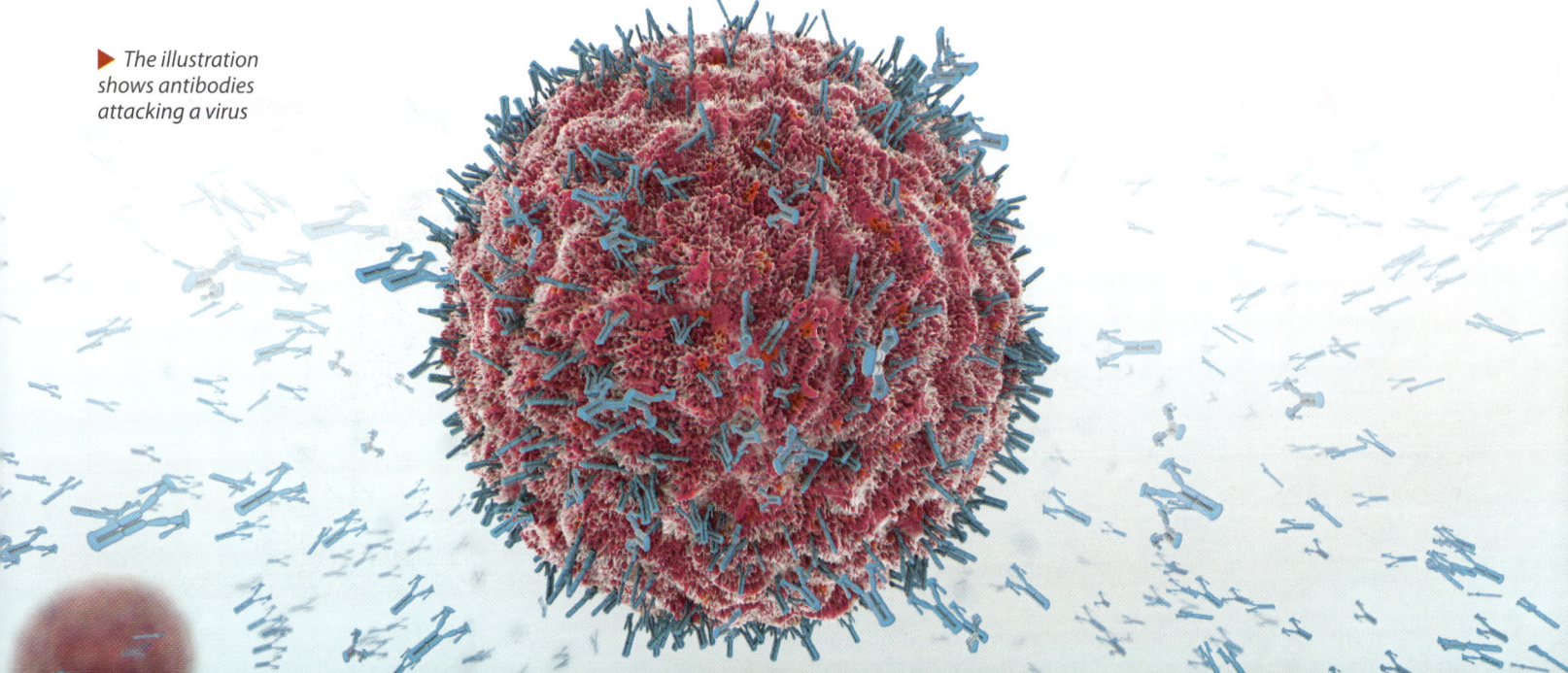

You are Under Attack

Our body can be attacked in two ways: either by pathogens or allergens. Pathogens are living creatures like viruses, bacteria, fungi, plasmodia, and worms. Allergens are often non-living things like dust and certain kinds of chemicals, as well as living things like pollen, nuts, vegetables, or meat products.

Pathogens

Many pathogens invade our body when our immune system is weak. Doctors call them facultative parasites. Some, like the one that causes malaria, however, must invade our bodies because they can only reproduce that way. These are called obligate parasites.

There are other microscopic creatures that are called commensals. They hang around on our skins or our intestines, living off sweat, and undigested food. Some of them help us fight off pathogens and make the vitamins we need.

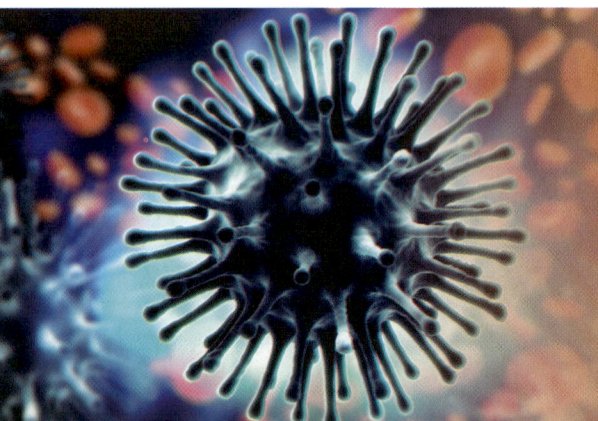

◀ Viruses hide inside human cells, so the immune system must catch them before they can get inside our cells

Viruses

Viruses are so tiny that you cannot see them with the naked eye. They are the tiniest living organisms. They are simple organisms made of DNA covered by a coat of protein—we are not even sure that they are really living things. Viruses are the cause of different kinds of diseases, from common cold to swine flu, dengue fever, and even HIV/AIDS.

Fungi

Most fungi are not infectious. But those that are, infect the tongue, causing diseases such as white-tongue or the skin, leading to infections like ringworm. Fungal infections often happen when the immune system is weak, especially when the body is recovering from a major illness.

▶ Cells of the pathogenic fungus Candida stick to the tongue, making it look white

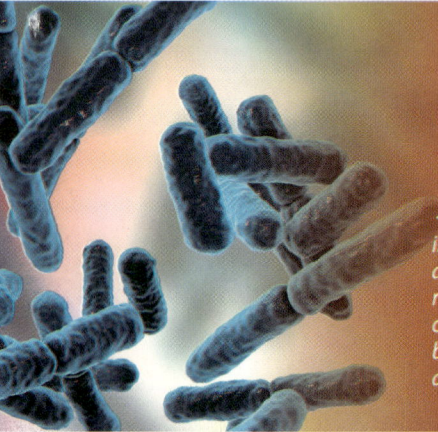

◀ Bacteria come in many sizes and shapes, from round, pea-like cocci to rod-like bacilli, spirals, and threads

Bacteria

Bacteria cause most of the diseases known to us, from pimples on the skin to fevers and cholera, leprosy, and tuberculosis. Bacteria are the simplest kinds of cells, with just cytoplasm covered by a cell wall.

There are two types of bacteria—those that need oxygen, called obligate aerobes and those that do not need oxygen, called obligate anaerobes. *Bacillus subtilis* is an obligate aerobe. *Clostridium* is a rod-shaped bacterium found in the intestinal tracts of animals and human beings. It does not need oxygen to grow.

HUMAN BODY | IMMUNE SYSTEM & COMMON DISEASES

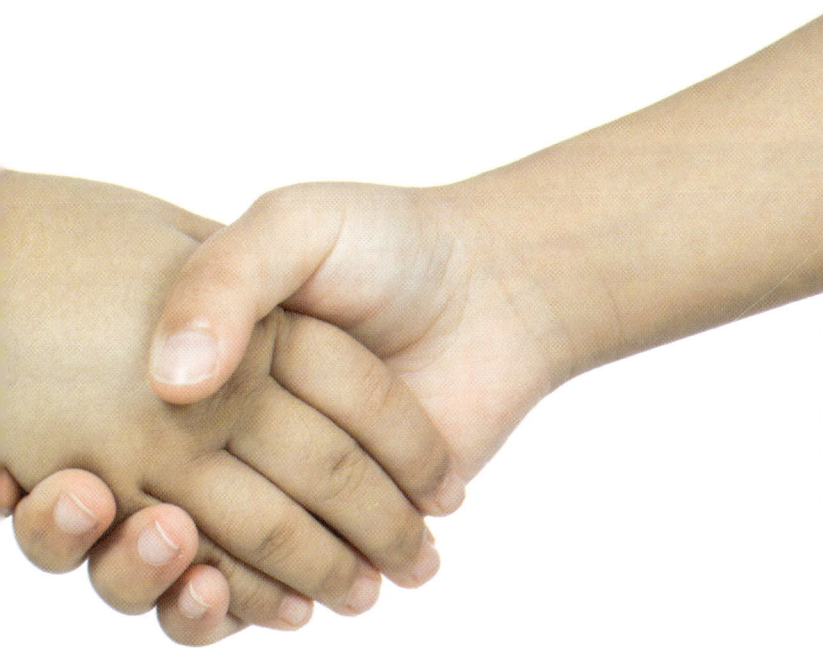

💡 Isn't It Amazing!

Viruses need to get inside another living thing to make more viruses. Therefore, they will infect any living thing, from human beings and other animals to plants and one-celled creatures like amoebae and even bacteria. These last ones are called **bacteriophages**.

▲ Scientists are using bacteriophages to treat some infections caused by bacteria

Worms

If you think of worms, you are probably thinking of earthworms. But the worms that attack human beings are really tiny, and are called **helminths**. The most common of these are tapeworms, which infect the intestine.

▶ Some helminths drill through the intestine wall and get into the blood, through which they can reach the brain

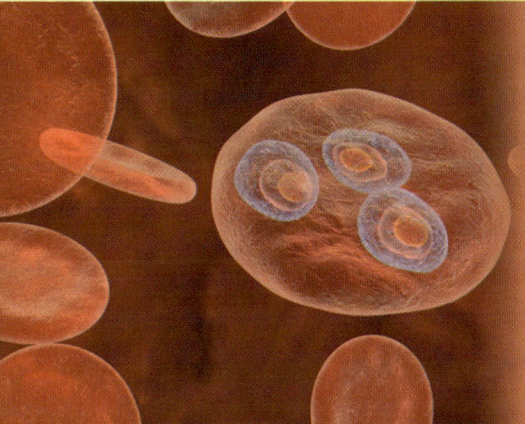

◀ Plasmodium hides inside red blood cells to escape the immune system, so it is a hard enemy to fight

Protozoa

Protozoa are one-celled animals that have a nucleus and no cell wall. Protozoa cause some of the deadliest diseases known to us, like malaria, caused by Plasmodium, and sleeping sickness caused by Trypanosoma. Protozoa cannot infect us without an agent, which is often an insect that bites us, like the mosquito. The insects live in tropical regions, so the diseases they cause are called tropical diseases.

Pollen

While pollen do not really attack or invade our bodies, when they are in the air, they get into our systems through breath and food. Pollen triggers the allergic response of the immune system in some people, which causes hay fever, leading to sneezing, reddening of the eyes, and itching in the throat.

▶ Allergic patients need to be careful in spring, as repeated exposure to pollen can cause death

Clotting and Wound Healing

Did you ever get bruised or cut, and bleed? Did you wonder why it stopped after a while? That is because blood clots. An injury, like a cut or prick, makes a hole in your artery or vein, but the body plugs it very fast, so that you do not lose any more blood. If your blood did not clot, it would keep bleeding. Major blood loss can be fatal.

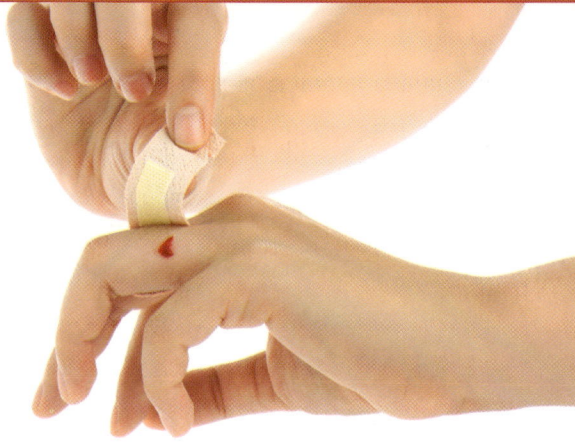

▶ Putting a band-aid on a wound helps to keep it sterile and prevents you from losing blood

Brain Clots

Blood is supplied to the brain by tiny arteries that have very thick walls. In case a clot forms in one of these, it is very hard to dislodge it.

The blood stops flowing to that part of the brain and it soon becomes starved of oxygen. If this continues for too long, the patient can get a stroke. In serious cases, it can lead to paralysis and even death. It is called a haemorrhage or an apoplexy.

Incredible Individuals

Some people suffer from a disease called haemophilia, in which their blood does not clot. They have to be extremely careful, as they can bleed to death even if they have a small wound. It is called the Royal Disease, because Queen Victoria's children and grandchildren suffered from it. Her son, Prince Leopold, died in 1884 at the age of 30 after he slipped and fell. He had no visible injuries but had bled internally.

◀ Prince Leopold was the eighth child and the youngest son of Queen Victoria and Prince Albert

◀ A clot is made of a big mass of fibrin threads and cells

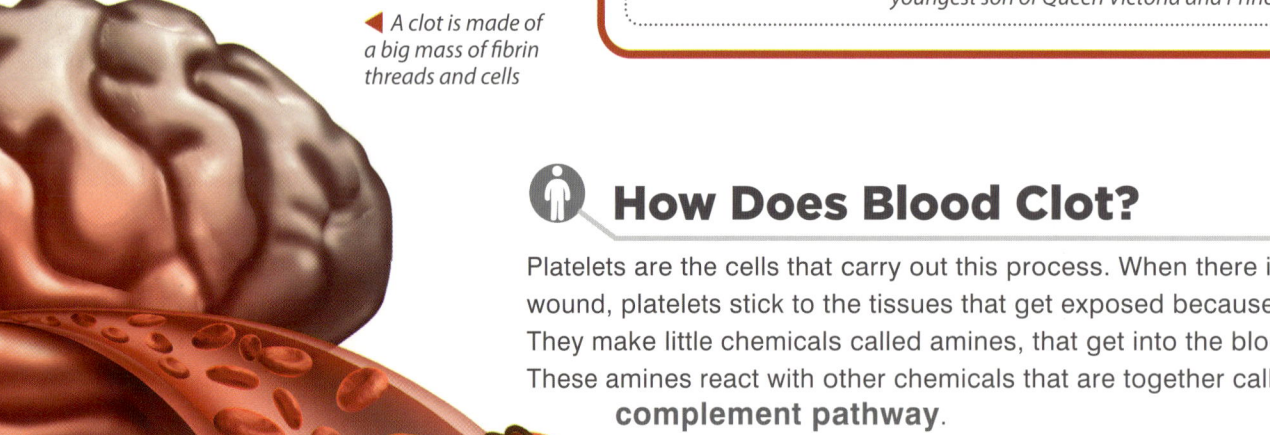

How Does Blood Clot?

Platelets are the cells that carry out this process. When there is a wound, platelets stick to the tissues that get exposed because of it. They make little chemicals called amines, that get into the blood. These amines react with other chemicals that are together called the **complement pathway**.

The plasma of your blood has a protein in it called fibrinogen. When the complement pathway meets it, the fibrinogen turns into **fibrin**. Fibrin is a thread-like material that gets deposited on the cut. It soon forms a net that traps the red blood cells flowing out of the wound. After this, the skin's self-healing mechanism takes over and closes the wound over the next few days. You get a black, scaly covering called scar tissue, that often itches. Once skin has grown back, scar tissue falls away. Vitamin K is very important for clotting.

HUMAN BODY | IMMUNE SYSTEM & COMMON DISEASES

Platelets: Our Body's Repairmen

Did you know that platelets are the cells with the shortest life span in our blood? They are born in the bone marrow and live for less than 10 days, after which they die in the spleen and liver.

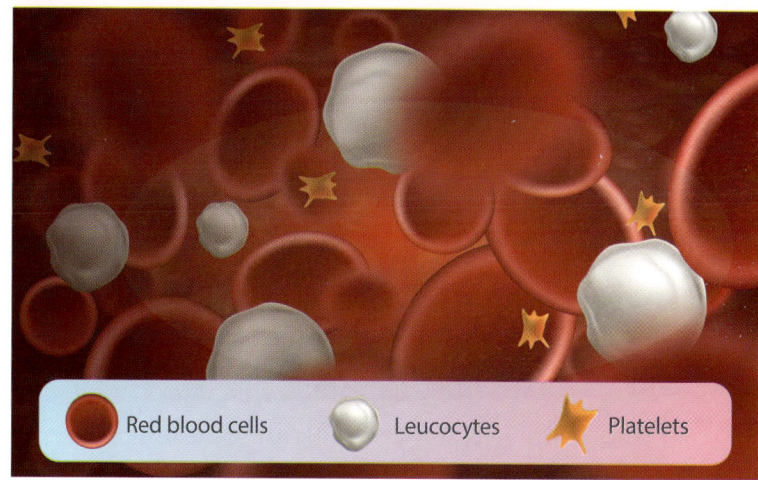
Red blood cells • Leucocytes • Platelets

What Do Platelets Do?

In their short lives, platelets do a lot of things. In the blood, they circulate at the edges, near the walls of the veins and arteries. Because of this, they are the first to spot a hole or tear in the vein and can act immediately. The inner wall of a blood vessel is non-sticky, like a Teflon-coated pan, but the rest of our tissues are not. So, in a wound, platelets stick immediately and start the clotting process.

Platelets have other jobs too, in healing inflammation, in the development of the lymph vessels, and in helping the liver regenerate.

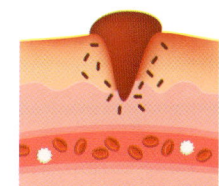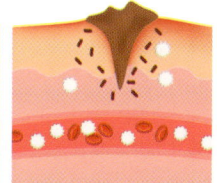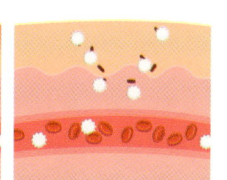

▲ After platelets seal a wound, new cells begin to grow

▲ The body stores extra platelets in the spleen, to prepare for emergencies

Platelet Count

A litre of blood in a normal person should have between 150 and 400 billion platelets. Diseases such as dengue and thalassemia cause extreme reduction of platelets. If a person's platelet count is below 50,000 platelets per microlitre, one is at a serious risk of bleeding to death. On the other hand, if one has more than 400,000 platelets per microlitre, one may get thrombosis.

Thrombosis

Thrombosis is a disorder of the body that causes clot formation inside the arteries and veins, because of which the flow of blood to organs gets choked, leading to organ failure. Thrombosis may be caused due to damage to the spleen, anaemia or cancer, or because the body makes too many platelets or too much fibrinogen.

In Real Life

Unlike a blood donation where your blood type has to be matched to the receiver's, you can donate platelets to anybody. Today, doctors realise that in most cases where a blood transfusion is required, giving only plasma or platelets is sufficient.

▼ A thrombus stops blood flow, starving the tissue around it of oxygen

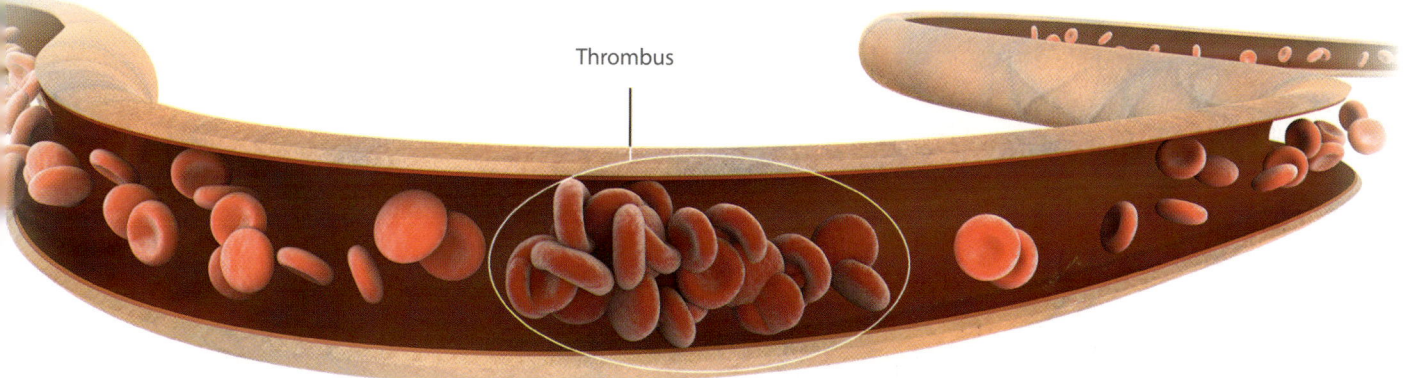
Thrombus

Keeping Our Gut and Lungs Safe

After our skin, our digestive system and respiratory system interact with the environment the most. Our immune system works in special ways to keep the organ systems safe. Just as countries have a border that is often fenced, the intestines and lungs have a strong wall called the epithelium. It also performs other functions like absorbing digested food in the intestine and oxygen in the lungs that need it to be really thin. Yet it must also form a stiff wall that prevents pathogens from the food or air from getting into the body. It does this by making sure that there are no gaps between cells.

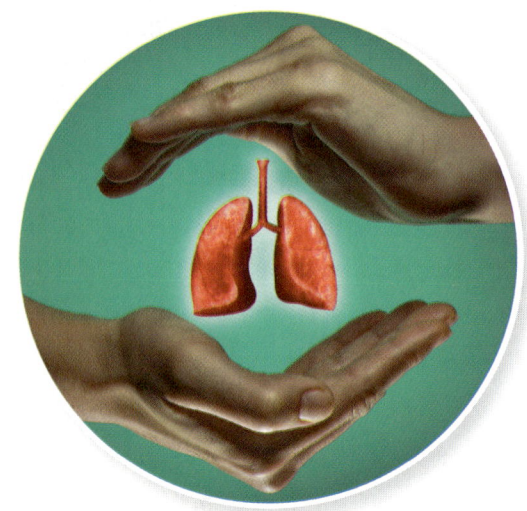

▶ The immune system keeps our lungs healthy and free of sickness

Epithelial Defences

The epithelium of the lungs has tiny hair called cilia that detect pathogens and alert the immune system. In the intestine, epithelia can sense harmful chemicals made by germs. In turn, they make cytokines and interleukins that alert the fighting cells of the immune system. *(see pp 20-21)*

The epithelia are also coated with a sticky, jelly-like material called **mucus**. This traps all foreign things and moves them to the digestive system. When you have a cold, your body makes a lot more mucus, which ends up coming out of your nose!

◀ Sneezing throws out germs that irritate the nose's epithelium

Isn't It Amazing!
Lysozyme is found in egg whites and keeps the growing chick safe.

▶ The term 'lysozyme' was coined by Alexander Fleming, the discoverer of penicillin

Lysozyme: The Killer Enzyme

Lysozyme is the enzyme that is found in our saliva and in our tears. This very powerful enzyme digests the cell wall of bacteria, which makes them die. When mum kisses you, the lysozyme in her saliva attacks the bacteria on your skin. Perhaps that is why they say 'a mother's kiss heals sores'.

The Forts of Our Body

Just as armies built forts near borders inhabited by soldiers who would be ready to face an attack, our body too has tiny forts called MALTs associated with the lungs, nose, intestines, and stomach. These have a large number of immune cells—T-cells, B-cells, and macrophages—waiting in them, ready to attack any pathogen.

Our Microscopic Allies

Did you know that we have more bacteria living inside our intestines than we have cells in our entire body? These bacteria get into our system just after birth and live inside us throughout our lives. But do not worry, they are our friends, not our enemies. Together, they are called **gut microflora**.

Gut Microflora

Gut microflora perform a lot of different functions for the body. They crowd the surface of the intestines, so disease-causing bacteria do not get a place to settle down and start the infection. They make biochemical weapons that kill other bacteria. These are called bacterio-toxins. They also help to train the immune system when we are young, to recognise antigens and to distinguish good bacteria from the bad ones. They prevent us from contracting illnesses like diarrhoea.

Other bacteria live on our skin. They help themselves to the nutrients they can get from dandruff and sweat and make sure pathogens do not get to them. They turn them into smelly things, which is what gives you bad body odour, especially under the armpits. Showering regularly keeps them away.

▲ Bacteria in the large intestine finish up digestion

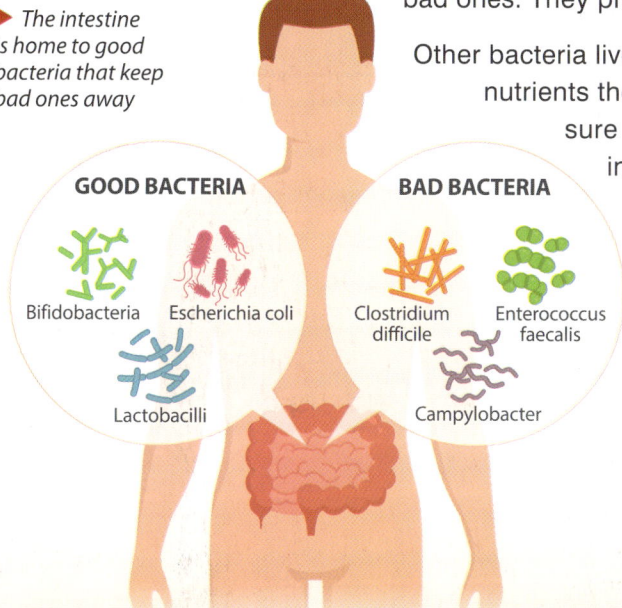
▶ The intestine is home to good bacteria that keep bad ones away

Incredible Individuals

Many cultures throughout the world eat a lot of fermented food. Fermentation is a process by which bacteria break down hard-to-digest things in food. Elie Metchnikoff was a famous scientist who noted that Bulgarian villagers ate a lot of yoghurt and lived long lives. He soon found out that it was because yoghurt had a bacteria called Lactobacilli in abundance. When you eat yoghurt, these bacteria start living in your intestine, and prevent harmful bacteria from getting a chance to grow.

▲ Bulgaria played a vital role in introducing yoghurt to the West

Isn't It Amazing!

Did you know that a bacterium called Escherichia coli or E. coli inhabits the human intestine? This bacterium has helped us study how diseases are caused, how the body fights diseases, and how we can develop drugs against them. It also makes Vitamin K12, which our body cannot produce for itself.

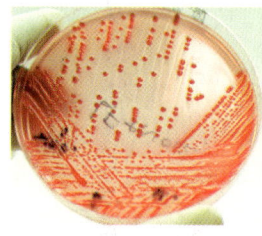
▲ Escherichia coli being grown in a petri dish

How Our Immune System Fights

If you have seen an action movie, it usually ends with the heroes saving the day. In the same way, in the case of a severe infection, the B-cells and T-cells take over the fight. Otherwise, they rest quietly in the lymph nodes. T-cells come in three types, based on how they act, while B-cells come in two.

Why Our Body Needs Immune Cells

Animals, like insects, did not have immune cells. But as mammals evolved to live longer lives, they needed a stronger immune system, so as to identify germs and be able to get rid of them.

B-cells and T-cells act much faster than phagocytic cells and neutralise the enemy. That is, they make it impossible for germs to hide or defend themselves, so that the phagocytes can then eat them up. They also have a lot of chemical weapons that other cells do not have.

Killer T-cells

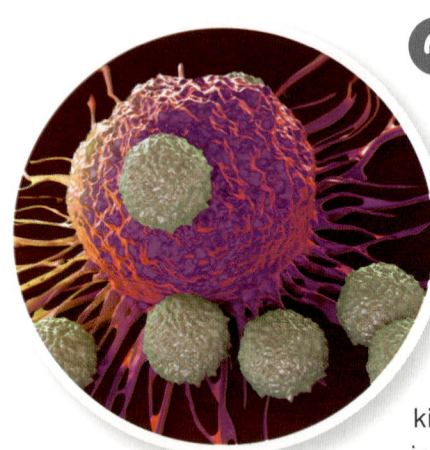
▲ *T-cell activation by phagocytes*

Killer T-cells are like assassins. Once they know their enemy, they kill it directly. Killer T-cells are called by phagocytic cells that have swallowed infected cells. If the antigen presented by the cell matches the record that the T-cell has, it immediately launches chemical weapons called granzymes and perforins. These invade the infected cell, killing both the cells and the pathogens that infected it. This is the immune system's way of getting at viruses that hide inside cells.

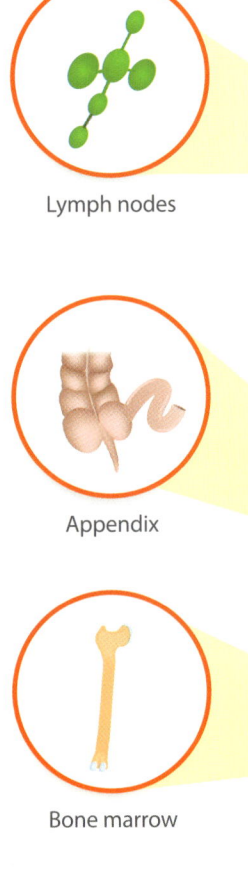

Lymph nodes

Appendix

Bone marrow

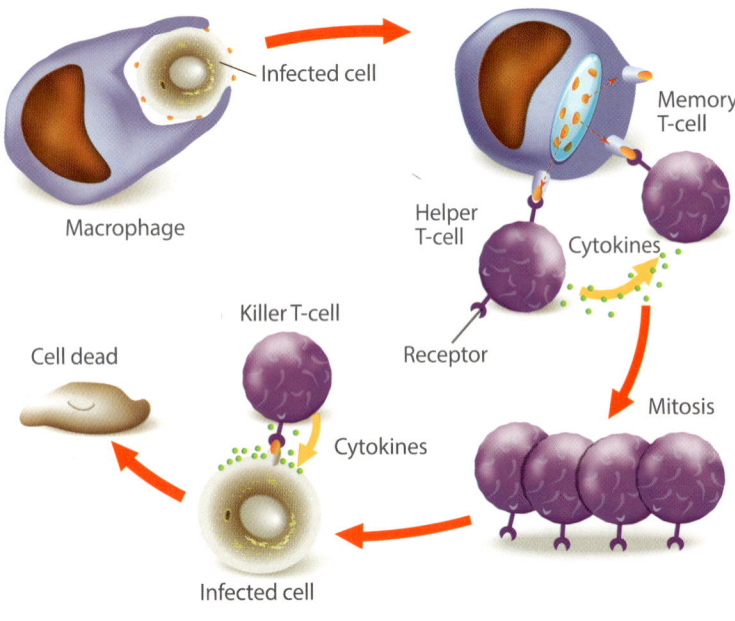
▲ *The diagram shows how the helper T-cells work*

Helper T-cells

Helper T-cells are cells that do not fight but call other T-cells and B-cells to fight. Once a T-cell has recognised the antigen, it puts out a kind of biochemical called an interleukin, that acts like a military siren. When a killer T-cell or a B-cell finds these interleukins, it knows that it needs to get ready for the fight. Such a cell is called an activated cell.

HUMAN BODY | IMMUNE SYSTEM & COMMON DISEASES | 83

 # Memory T-cells

When a killer T-cell or helper T-cell is activated, it makes many copies or clones of itself. While most of the clones go off to fight harmful cells, some of them stay behind. They hang around in the lymph nodes and spleen until the pathogen comes back a second time. This time they become the helpers and start the fight all over again and also make copies for the next time.

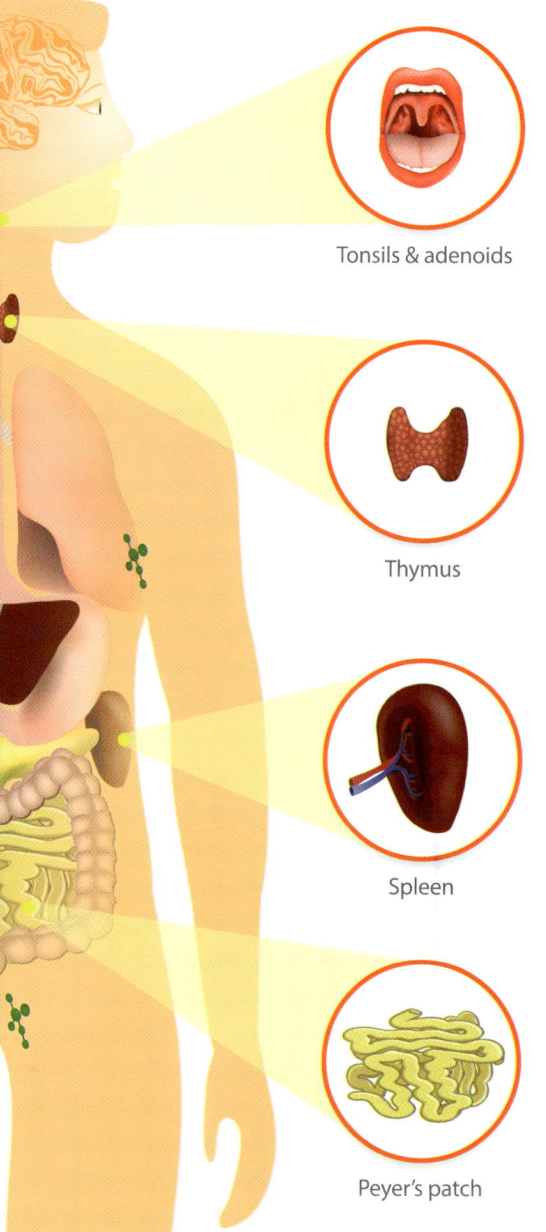

Isn't It Amazing!

WBCs travel in two ways. For most of the journey, they go with the pressure of flowing blood. Nearing the target, they stick to the blood vessel's walls. Here they put out pseudopodia, just like amoeba and squeeze through the wall lining to enter the target tissue. The same pseudopodia also grab the bacteria.

▲ A sample of amoeba proteus

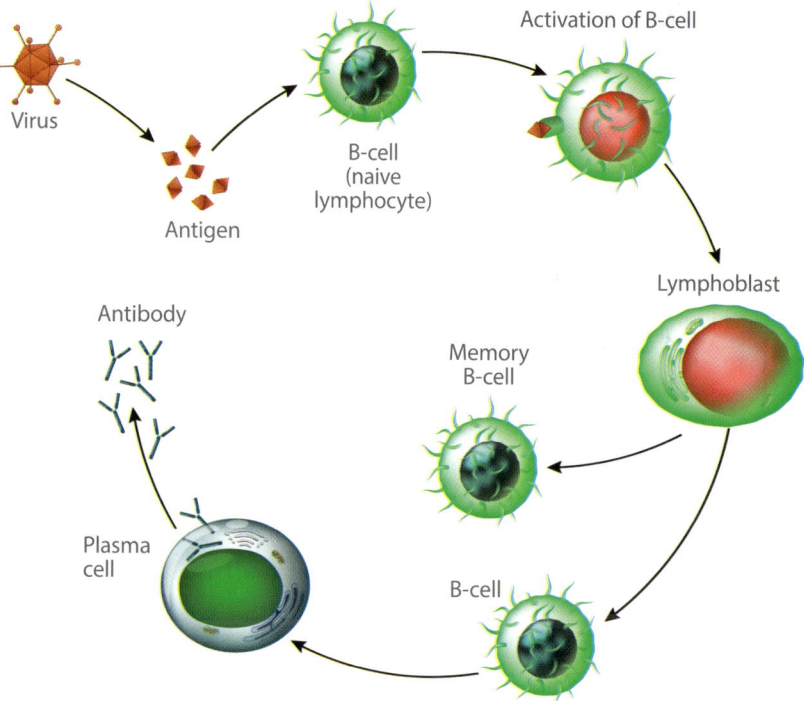

▲ Different kinds of B-cells make different kinds of antibodies

 # B-cells

B-cells are activated just like T-cells. But they do not kill by direct contact. Instead, they make antibodies and release them into the blood and lymph. When the antibodies find their matching antigen, they stick all over the pathogen and call the T-cells and macrophages to do the killing.

Like memory T-cells, there are memory B-cells too. Other activated B-cells turn into plasma cells, which are simply huge factories for making lots of antibodies. It is only the B-cells that match with antigens that turn into plasma cells. The other B-cells that do not match, stay quiet. This makes sure that only the infecting pathogen is destroyed, and the remaining tissue is safe.

Places where there are repeated infections, like the lungs and intestines, have a lot of B-cells and T-cells lying in wait for germs, in the tissues called MALT.

Our Immune System's Chemical Toolkit

Modern armies have all sorts of weapons and gadgets to fight a war—including radars and satellite phones, night vision goggles, and jammers to block enemy communications. Our immune system also has gadgets like these, except that they are all chemical in nature. These tools give the immune system an incredible ability to act against any infection. They help the immune system increase or decrease the strength with which it responds, and also helps coordinate among all its cells. This is known as adaptability. They also prepare the rest of the body for fevers, and warn cells to look out for pathogens. During organ transplants, doctors give patients drugs called immuno-suppressants. These make sure that these immune chemicals do not attack the new organ.

Hormones

Many immune cells work in the same way as hormones in the body. Hormones are chemical messengers that travel through the blood and bind to proteins called receptors on the surface of the cells.

If a receptor is a lock, a hormone is like the key that opens it. The same key can open different locks, i.e. the same hormone can interact with different kinds of receptors on different cells. Some hormones signal immune cells to stop being active, while other hormones make inactive immune cells active.

Granzymes & Perforins

These are enzymes that are stored in the granules of natural killer cells and also killer T-cells. When released, the perforins drill holes in the membranes of infected cells, causing them to burst. The granzymes (enzymes from the granules) then digest all the material.

Cytokines & Interleukins

These are the messenger proteins of the immune system. There are many kinds of cytokines, each of which carries a special message. Cytokines can be made by most cells of the body, but only when infected. They are also made by macrophages that have found antigens. Both are received by T-cells and B-cells that then get activated to do their jobs.

Interleukins are cytokines that are used by B-cells and T-cells to talk to each other. They are mostly made by helper T-cells.

▲ Model of serotonin, the key hormone that stabilizes our mood and feelings of well-being

▶ Perforin, a protein responsible for pore formation in the membranes of target cells

▶ A cytokine molecule attaching to a receptor on the surface of a cell

HUMAN BODY — IMMUNE SYSTEM & COMMON DISEASES

Interferons

These are messenger proteins that are released by the cells infected by viruses. They communicate to the immune system to send killer cells and macrophages to finish them off.

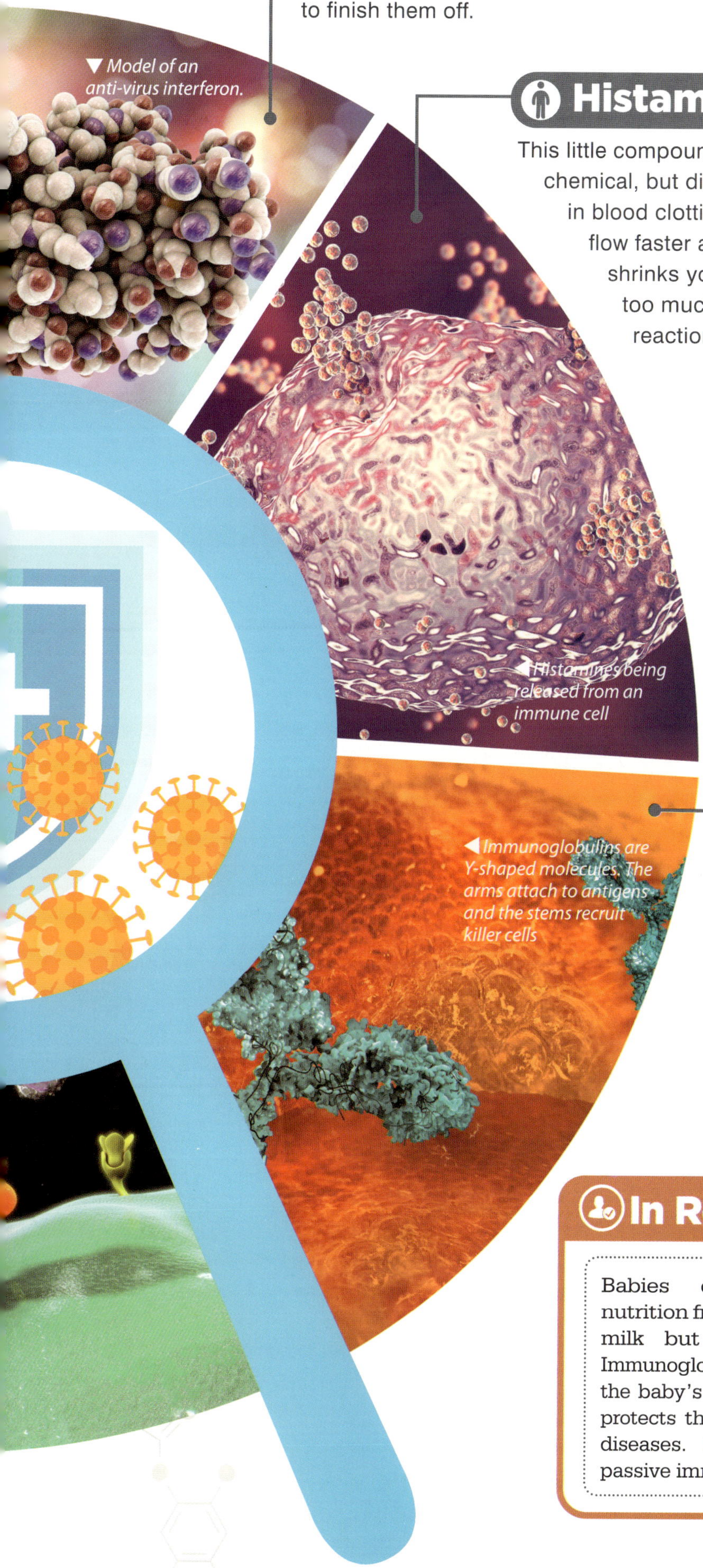

▼ Model of an anti-virus interferon.

Histamine

This little compound is the body's alarm siren. It is a messenger chemical, but different cells read it differently. Histamine is involved in blood clotting, increasing heartbeat and making the blood flow faster and widening blood vessels among other things. It shrinks your lungs so you do not breathe in germs. However, too much histamine in the body causes an excessive allergic reaction, called anaphylactic shock.

◀ Histamines being released from an immune cell

Antibodies

These are the body's wonder weapons, and are made by activated B-cells. The medical term for them is **immunoglobulin**, Ig for short. They can be of the following five types:

IgM: This is the antibody made by B-cells the first time there is an infection.

IgG: These antibodies are made the second time there is an infection. IgG is made in huge quantities and can finish off the pathogen very quickly.

IgA: This antibody is released into the mucus of the intestines, lungs, and tears to fight pathogens before they enter the body.

IgE: This antibody is meant for allergic reactions, though it works like the rest.

IgD: This antibody stays on the surface of the cell to help it recognise the pathogen.

◀ Immunoglobulins are Y-shaped molecules. The arms attach to antigens and the stems recruit killer cells

In Real Life

Babies don't just get nutrition from their mother's milk but also immunity. Immunoglobulin A travels into the baby's stomach, where it protects the baby from many diseases. Doctors call this passive immunity.

▲ Feeding on mother's milk is important for immunity among all mammals

Our Soldiers Never Forget

Have you ever wondered why you never get some diseases twice? For example, if you got chicken pox once, you will never get it again. This is because the immune system remembers all the pathogens that it has encountered. How does it do so?

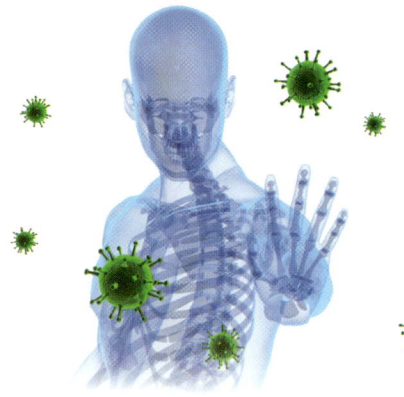

▶ A depiction of a healthy body fighting immune diseases

 ## Memory

Whenever there is an infection, you know that the phagocytic cells eat up the pathogen and display bits of it on their surface. Thousands of B-cells and T-cells try to see whether they can match this antigen. This is done using a protein called a receptor, and each cell has a different one. The cell whose receptor matches becomes activated.

This activated cell will make thousands of cells like itself. Most of these cells will join the fight against the pathogen. But a few will remain in the lymph nodes, doing nothing. This is because their job is to wait for the next time the same pathogen attacks. Scientists call them memory cells.

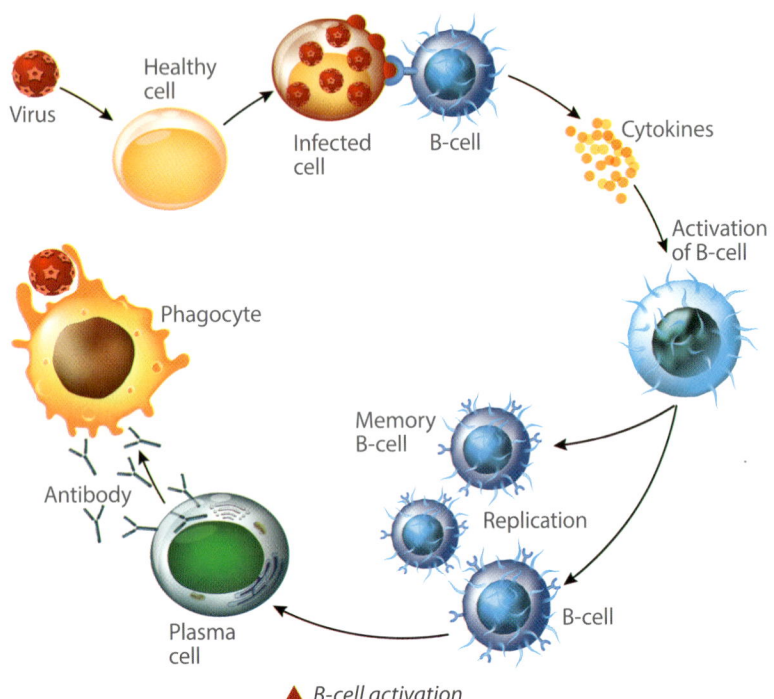

▲ B-cell activation

In Real Life

Our body's soldiers remember their targets through their receptors, so they have a large database of these receptors. Something like this is used by police everywhere—when they collect the fingerprints of people.

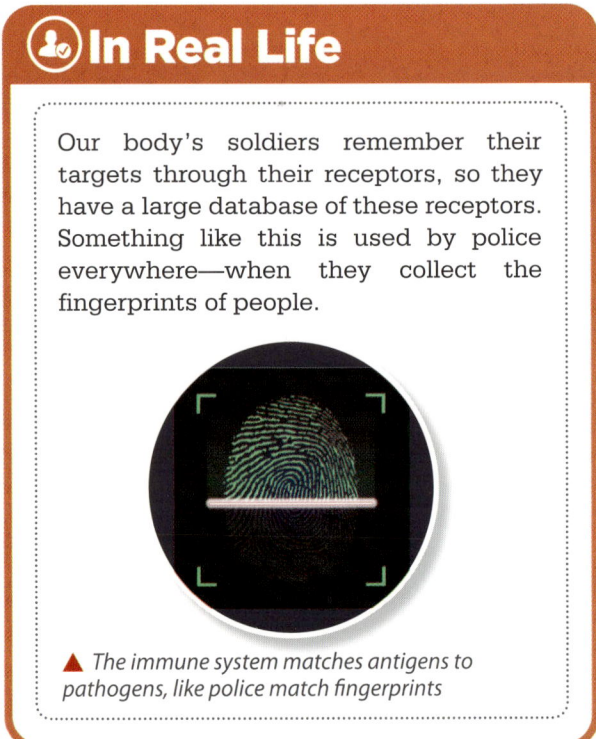

▲ The immune system matches antigens to pathogens, like police match fingerprints

 ## Making up Memories

The immune system responds best to a second infection by the same pathogen—but what if the first infection was fatal? When scientists found this out, they thought to themselves, what if there was a way to teach the memory of some common diseases to the immune system, so that when a real infection happened, it was fully ready for the fight. This is the thought behind vaccination. When you receive a vaccine, your body is administered with the disease-causing pathogen in a weakened state. This trains the body's immune system to fight a disease that it has not previously suffered. Thus, vaccines are meant to prevent disease.

◀ In the past 60 years, vaccination has helped in eradicating one disease—smallpox

Why We Get Allergies

Some of us are allergic to dust, some to pollen, and some even to nuts. All of these are called allergens. But why do we react to them, even though none of these cause any diseases?

▲ You can take tests to find out exactly what you are allergic to. People can be allergic to food additives, animal fur, pollen, etc

▼ Sneezing is a typical allergic reaction caused by inhaling allergens

Perceiving Danger

Our immune system really cannot make out whether anything foreign in our body causes diseases. However, it knows that all foreign bodies have to be destroyed. This trick works against germs. So, it makes antibodies against bacteria and interferons against viruses. But the body still does not know how it must deal with pollens and dust particles.

If You Can't Beat It, Overreact

Anything foreign in the body is seen off either by the phagocytes, or by the T-cells and B-cells. But many allergens still remain in the body, so the immune system mounts a bigger response. It produces a lot of histamine, making the blood flow faster. The site of allergy becomes red and swollen. It makes a lot of IgE antibodies, which call in more killer cells. This leads to tissue damage, causing the rashes and welts that you see. Sadly, there is no true cure for allergy, except finding out what causes the allergy, and then trying your best to avoid it throughout your life.

 Asthma

Asthma is a kind of allergy caused by dust, pollen, smog, and other things in the air. Whenever the allergen enters the lungs, it triggers the lymph nodes present there. A huge amount of histamine is produced and the lungs shrink, making it difficult to breathe.

▶ Asthma patients use an inhaler, which contains a drug that makes the lungs expand again

Fighting Cancer

Every cell in our body undergoes wear and tear, which is not unusual. However, sometimes it so happens that the cell stops responding to the body's signals. This could be due to various factors such as infection by a virus, exposure to chemicals or UV light, etc. Under such a situation, the cell starts to function like a pathogen, and begins to multiply on its own. This uncontrolled multiplication leads to cancer.

As we evolved to live longer lives, the risk that we would get cancer also increased. Some scientists suggest that our complex immune system also evolved alongside to find and destroy cancers. Today, as vaccines and medicines have reduced the trouble caused by germs, cancer is becoming the biggest cause of death due to disease.

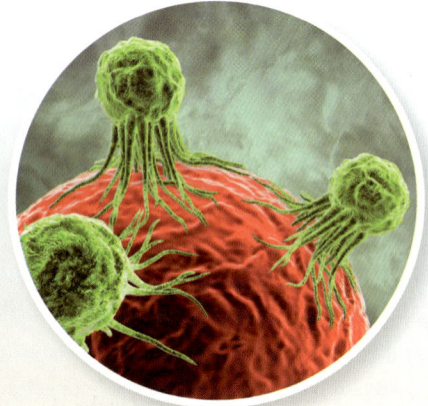

Spies Who Spot Cancer

All cells in our body have little proteins on their surface that act like ID cards. When we are inside our mother's womb, the immune system learns to not attack any cells that show these ID proteins. When a cell becomes cancerous, it stops showing these proteins. The neighbouring cells sense this and warn the immune system. The immune system in turn sends natural killer cells to the site to destroy the cancer cells.

◀ *When the immune system loses, cancer cells multiply*

◀ *Meditation and some quiet time are said to be helpful to fight diseases*

What Happens Next

Once a cell has become cancerous, it undergoes mutation of its DNA. It stops showing the self-ID antigens. The immune system then learns to recognise new antigens from the cancer cells and new T-cells and B-cells take up the fight. Some cancer cells die, but some mutate again, so that the antigens disappear. The immune system needs to start all over again. If the immune system wins, the cancer stops growing, and the lump of cells is called a benign tumour.

In Real Life

▲ *Healthy food not only prevents diseases, but also improves overall lifestyle*

Does eating healthy food like fruits and green vegetables help us fight cancer? Healthy food generally helps the immune system, however, scientists are still working on finding a direct link between a healthy diet and the suppression of cancer.

When Our Soldiers Turn Against Us

You have seen how the immune system reacts when it cannot overcome an allergy. Sometimes, this goes so far that the immune system loses its ability to differentiate the body's own cells from a foreign agent. Then the body begins to attack its own cells, causing an autoimmune disease.

This may happen because there is a genetic defect in the immune system, or because the antigens made from a pathogen may be very similar to one of the body's self-ID, which confuses the immune system. But in many autoimmune diseases, we do not know the cause. Luckily, very few of us will ever get them.

Major Autoimmune Diseases

Over 80 **autoimmune diseases** are known. Some of them are listed below.

Celiac disease
This happens when the immune system attacks the small intestine, causing indigestion, diarrhoea, and pain.

Autoimmune Diabetes
If the immune system attacks the pancreas, it kills the cells that make insulin.

Rheumatic Fever
An infection by Streptococcus bacteria may cause this fever, because the antigens made from bacteria are very similar to proteins found in the valves of the heart.

Lupus Erythematosus
This is a very serious disease affecting many tissues of the body as the immune system attacks many proteins.

Myasthenia Gravis
The immune system attacks the meeting points of nerves and muscles, causing terrible weakness.

Rheumatoid arthritis
The immune system attacks the bone joints, causing pain all the time.

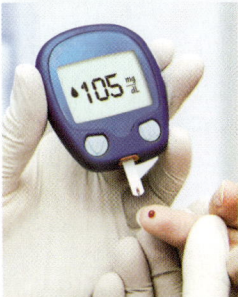
▲ Celiac disease causes poor digestion

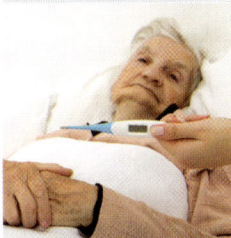
▲ Blood sugar testing machines are called Glucometers

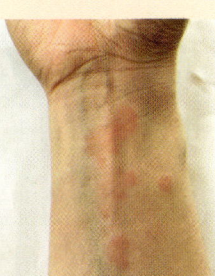
▲ Rheumatic fever causes frequent fever and pain

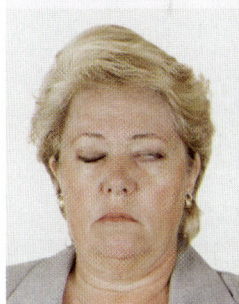
▲ Red blotches on the skin can point to lupus

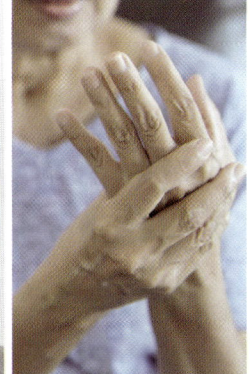
▲ Myasthenia gravis is a disease that leads to slow muscle loss

▲ Rheumatoid arthritis, leads to bent joints

Blood-Brain Barrier

The brain is not directly nourished by blood or lymph. Hence, the immune system has never seen brain cells, nor learned to see them as part of the self. If there is an injury to the brain, and blood enters it, your immune system will attack it like it's a foreign body.

In Real Life

If you are someone who catches too many infections, you have a good chance of not getting autoimmune diseases later in life. This may be because the immune system is better trained to make out self and foreign antigens after it suffers from many infections.

▲ Playing in the mud exposes you to many pathogens, and helps the immune system learn better

Fighting Common Diseases

Our body fights diseases in many ways which help slow down pathogens and speed up the immune system. Even though it is not a part of the immune system, the brain plays a very important role too, as it communicates to the other organs to perform their functions. A part of the brain called the hypothalamus, is responsible for controlling body temperature, blood pressure, hunger, and thirst. It works closely with the pituitary gland, which in turn controls other glands that make hormones. It is these hormones that communicate with other organs involved in performing functions such as storing iron, sneezing, making muscles contract, etc.

▼ *Fever causes fatigue, making the body want to rest so it can save the energy needed by the immune system*

Fever

Most infectious bacteria and viruses do best when the temperature of our body is 37° C (98.6 °F). The brain tells all our tissues to raise their temperature. Bacteria and viruses are then made to feel the heat, literally! Fever happens in many diseases like typhoid, malaria, dengue, anthrax, and Ebola among others.

Diarrhoea

Our body has ways to eliminate pathogens so that they don't harm us or make us ill. The intestine has many sensor cells that can find germs and toxins. These tell the autonomic nervous system and it immediately tells the intestine's muscles to contract and push the infected food out, along with the invaders. Doctors call this **diarrhoea**.

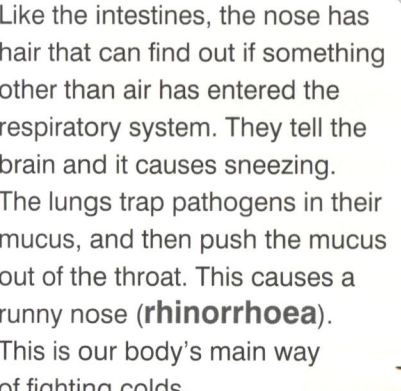

▲ *Diarrhoea cleans out the intestine, but without good hygiene, there can be other infections*

▼ *Drinking water and other fluids helps in treating a runny nose*

Rhinorrhoea

Like the intestines, the nose has hair that can find out if something other than air has entered the respiratory system. They tell the brain and it causes sneezing. The lungs trap pathogens in their mucus, and then push the mucus out of the throat. This causes a runny nose (**rhinorrhoea**). This is our body's main way of fighting colds.

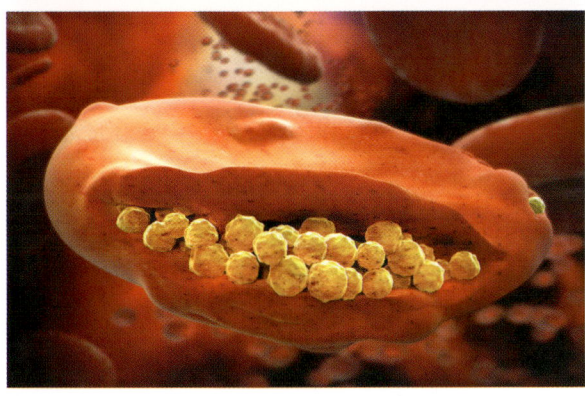
▲ Plasmodium cells bursting out of an infected RBC

Malaria

Malaria is a serious disease and the body has a tough time fighting it. This is because the pathogen that causes malaria, Plasmodium vivax, gets inside the RBCs, where it can escape the soldiers of the immune system. It multiplies there and breaks out of the RBCs, when the immune system catches it and triggers fever. But these pathogens get into other RBCs, this time causing chills in the body. Thus, you see many cycles of fever and chills in malaria.

Hiding Iron

Bacteria need iron to survive and multiply. So, our body makes sure that they do not get it. It pulls out iron from the blood and hides it in the tissues, till the bacteria have been eliminated.

In Real Life

An allergen makes your immune system respond faster and stronger each time. In some cases though, it may lead to shock called anaphylaxis.

HIV/AIDS

HIV is the Human Immuno-deficiency Virus. Like the malaria bug, it hides in cells. But it is deadly because the cells it hides in are the very cells of the immune system! During an HIV infection, the virus destroys many WBCs, dangerously weakening the immune system. If it becomes severe, it causes AIDS—Acquired Immuno-Deficiency Syndrome. Patients with AIDS easily get other dangerous diseases like tuberculosis and malaria.

Today, however, we have drugs that fight HIV, called antiretrovirals. But what we need most is to help patients with HIV and treat them like anyone else with a disease. Sadly, many people avoid them because of ignorance.

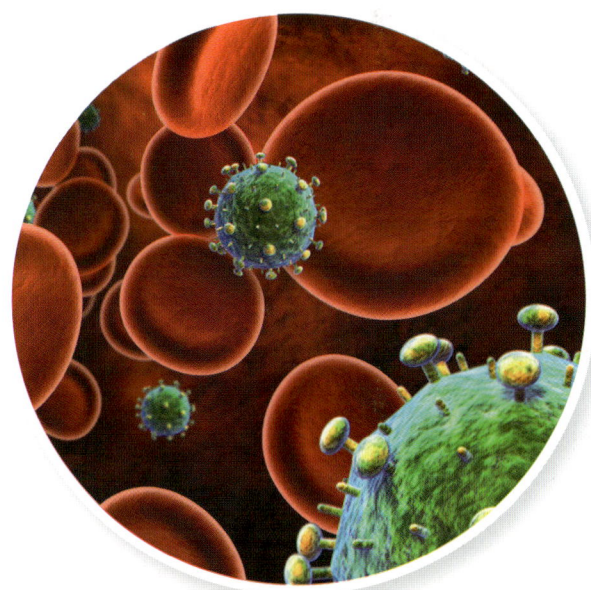

▲ HIV cells spread from one WBC to another through blood

Drink, Drink, Drink

You may have heard the old proverb—*starve the fever, feed the cold*. But is it true? Scientific studies have shown that you should in fact 'feed the fever, feed the cold'. For a cold this is fine—many colds run for over a week and your immune system needs the energy to keep fighting. During fever, which usually lasts for a day or two, you may not feel like eating. But your body's temperature rises, and you need more calories to keep up energy. Also, fever makes your body lose water and electrolytes, so you need to keep them up, even though you might not feel like it. The best thing to do, whether you have a cold, or a fever is to wrap up in warm clothes and drink lots of hot fluids—soup, milk, or hot chocolate.

Isn't It Amazing!

Different animals have different body temperatures. For example, your pet dog has a natural body temperature of 38.9°C (102°F), which can feel feverish to us.

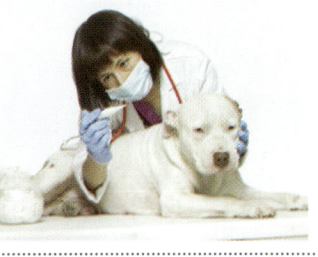

▶ Dogs are the most popular animals on a vet's table

Vaccines: A Jab of Safety

When the vaccine enters the body, macrophages catch hold of it, and eat it up by phagocytosis. Some of the bits of the dead germ are shown on the outside of the cell *(see pp 11)*, and T-cells and B-cells try to recognise the antigen. Those who do, become activated. Most will fight the vaccine, but a few will turn into memory cells *(see pp 22)*. So, your body is tricked into fighting a disease that wasn't there, but it left you vaccinated.

History of Vaccines

Edward Jenner (1749–1823) was the first to discover that having one disease can save you from others. He saw that the people in his village who got a disease called cowpox, did not get the much deadlier smallpox. Cowpox leaves little sores on the skin, and if you scrape material from these sores and inject them into healthy people, they too become immune. His discovery became a revolution in Europe; in 1811, Emperor Napoleon of France had his entire army vaccinated!

After Jenner, Louis Pasteur in France and Robert Koch in Germany continued to do a lot of research and developed a large number of vaccines against rabies, anthrax, cholera, and many other diseases. Pasteur's students Calmette and Guerin developed a vaccine against tuberculosis, and Dr Robert Salk of Canada developed one against polio. Most vaccines have to be injected, but polio is given as drops in the mouth. You get polio by drinking contaminated water, and your body fights it in the intestine's MALT *(see pp 16)*.

Today, the government and doctors recommend that we give several vaccines to our children in order to help them become immune to many diseases.

▲ *A child getting a polio vaccine in Brazil*

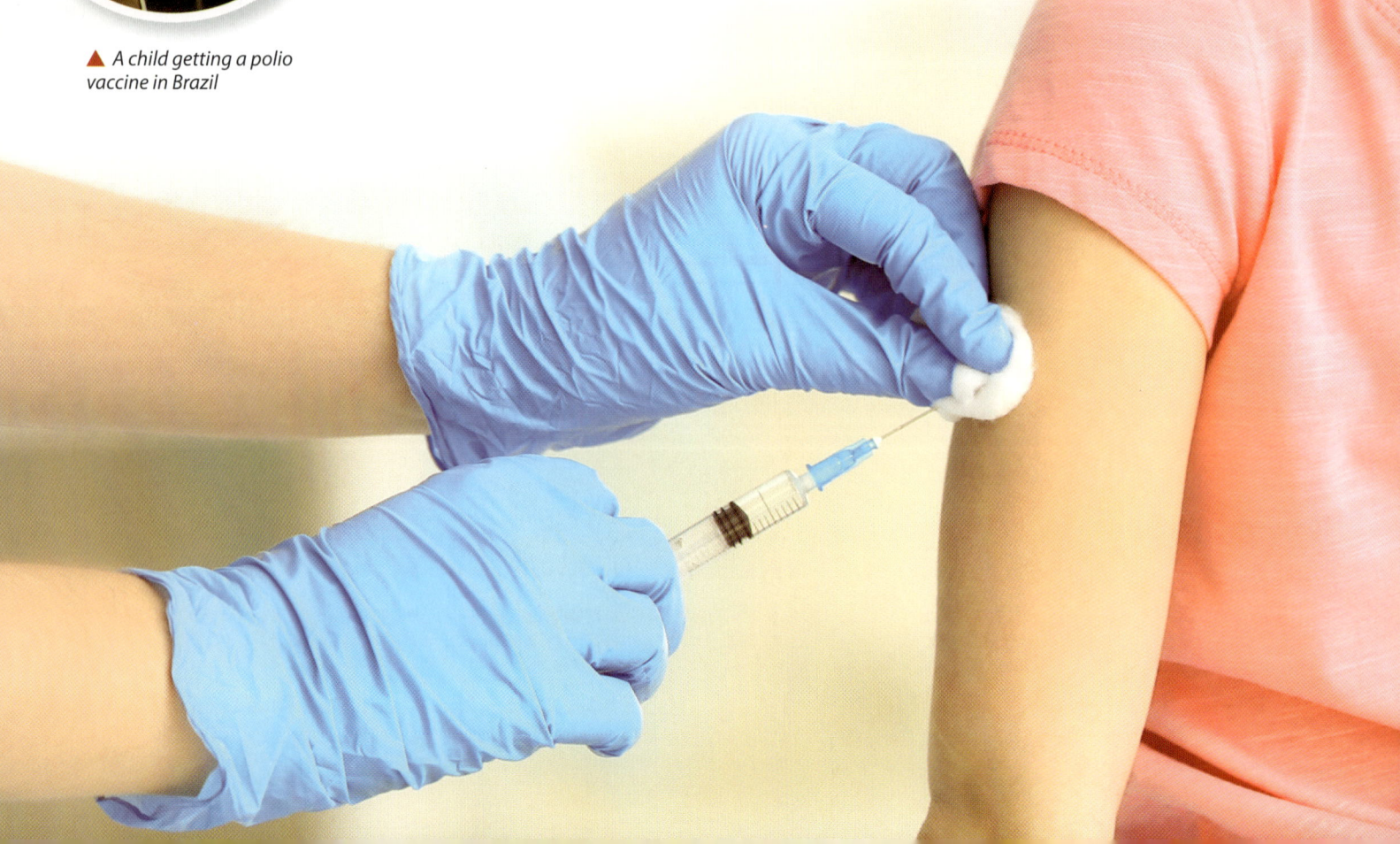

HUMAN BODY | IMMUNE SYSTEM & COMMON DISEASES

💡 Isn't It Amazing!

Since the first vaccine came from cowpox, a disease that people get from handling cows, Louis Pasteur named this way of protecting our body after cows. The term vacca in vaccination is Latin for cow.

Cancer Vaccines

Today, researchers have found that some kinds of cancers can be prevented by making antibodies against them. It means we can make vaccines against such cancers. For now, we have vaccines against cervical cancer, but many more are being tested.

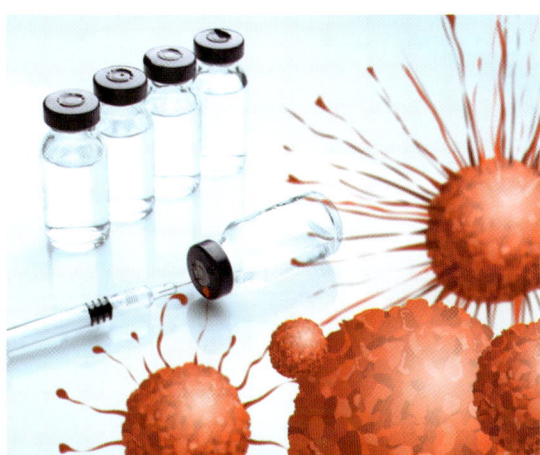

▶ Vaccines for many viral diseases are still being developed

▼ This is what an influenza virus looks like

Flu & HIV Vaccines

Influenza is a disease that is caused by many kinds of viruses. Because of this, we don't have a single vaccine that works against it. Also, many influenza viruses mutate, and their antigens keep changing, baffling the immune system. The HIV virus also does the same thing.

But there are different vaccines coming up against flu, and doctors say that you should get a jab every year to stay safe.

🏅 Incredible Individuals

Catherine the Great (1729–1796), the empress of Russia, wanted to introduce vaccination in her country. She made an offer: the first child who got vaccinated against smallpox, the Russian government would pay for their education and give them a pension. The first baby who got the vaccine was named Vaccinov!

▲ Catherine the Great, remains Russia's longest-ruling female leader till date

Immunity in Plants and Animals

Do plants and animals have immune cells? Learning about how other creatures defend themselves from diseases can help us make better medicines to fight diseases.

How Plants Keep Themselves Safe

Plants also get diseases from viruses, bacteria, and fungi. Plant cells respond to these like our immune system does: by finding out what antigens are self, and what are foreign, through 'receptors' on their cell surface. If a plant cell detects a foreign body, it initiates phagocytosis.

◀ A plant with fungal disease

How Insects Stay Safe

Insects don't have an immune system like mammals, but they do have a lymph, and phagocytes that float around in it. Insects also make proteins called defensins. These proteins stop bacteria from growing.

How Bacteria Stay Safe

We read earlier that viruses infect bacteria too. But bacteria have a unique way of fighting them off. When a virus infects a bacterium, or any living thing, it injects its own DNA into the cell. This DNA makes copies of itself using the infected cell's resources. Bacteria have enzymes called Restriction Endonucleases (REs). These enzymes chop up the viral DNA into little bits.

Incredible Individuals

The Irish Famine (1845–1852) happened when Ireland's main crop, potato, was infected by fungi. This disease, called late blight, kills the potato plant just as the potatoes are getting ready to eat. A million people died, and nearly 2 million emigrated to America. Among these were the Fitzgerald and Kennedy families, whose great-grandson John Fitzgerald Kennedy (1917–1963) became president of the USA in 1961.

How Fungi Keep Themselves (and Us) Safe

Fungi live in places where they are regularly attacked by bacteria. They protect themselves by releasing chemicals into the environment, which prevent bacteria from growing. The first of these was discovered by Alexander Fleming, when he saw that a fungus called Penicillium had killed his bacterial cultures. He soon found the chemical that was doing this and named it penicillin. Ever since then, scientists have discovered hundreds of such chemicals from fungi, which we together call **antibiotics**.

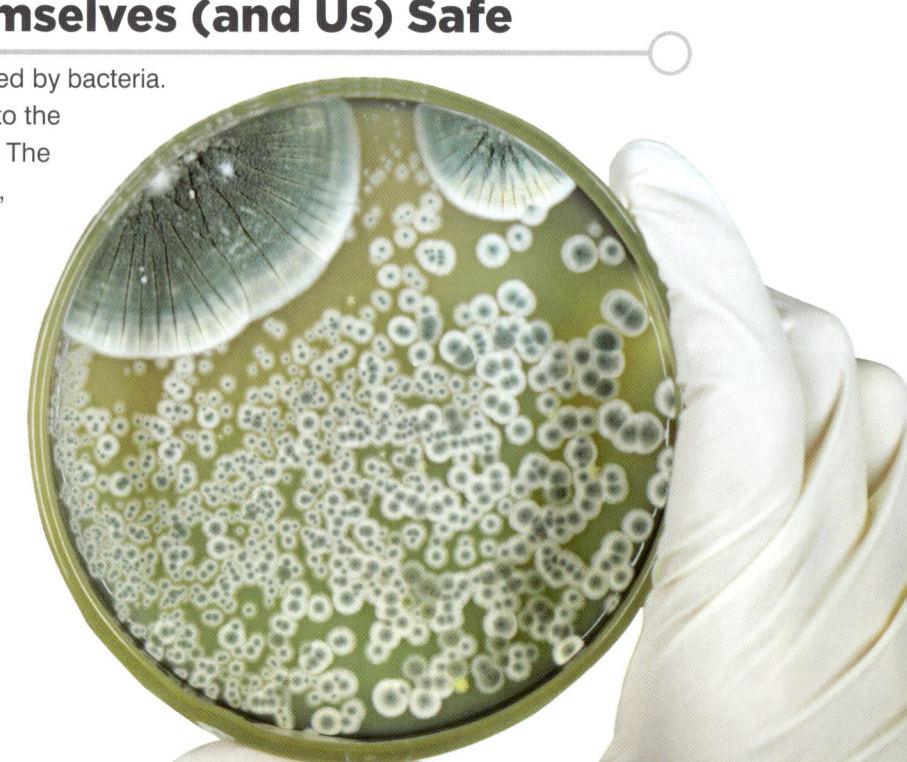

▶ Penicillium growing in a lab culture of bacteria

Helping Our Immune System

Healthy food keeps the body fit, and generally helps the immune system. Fighting illness needs a lot of energy. You also need a lot of vitamins and minerals to ensure that the immune system is healthy and is able to make enough of its messenger molecules and chemical weapons. But most of all, you need to exercise a lot to keep your heart and circulation system healthy, so there's enough oxygen for all cells of your body.

Eating Healthy

What the immune system needs most are Vitamins K and D. Vitamin K is a necessary part of the clotting process. Without it, blood would not clot, and you could bleed uncontrollably. Good sources of vitamin K are green leafy vegetables, cheese, and eggs.

Vitamin D is necessary for T-cells to function correctly. Vitamin D is made in our skin in the sunlight, but you can also get it from fish and mushrooms.

Sleep and Exercise

Lack of sleep or a lot of stress in the body causes the production of steroid hormones, which reduce immunity. On the other hand, exercise boosts immunity. It keeps up healthy circulation of blood so, WBCs can travel faster and also reduces stress hormones. So, go out and play a lot!

◀ Eating vegetables rich in vitamins is as necessary for your body as exercising

◀ Young children should be encouraged to play outside so that they can make friends and get some exercise

Hormones

Some steroid hormones in our bodies, like testosterone and cortisol, suppress the immune system. The body makes more of them when the immune system has fought off the pathogen, and energy is needed for other things. In autoimmune disease, these hormones are used to protect the body from being attacked by its own immune cells.

In Real Life

Did you know that listening to music for 50 minutes every day can help boost your immunity? This is because music helps calm the mind and reduces stress in the body. So, when you're studying or travelling, listening to music is a good idea!

◀ Listening to music is good for reducing stress

LUNGS & RESPIRATORY SYSTEM

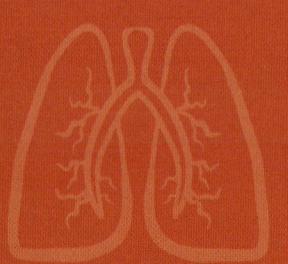

THE BREATH OF LIFE

As long as we are breathing, we are alive. Life starts with a baby's first breath when it comes into the world. The shock of life wakes babies up as air rushes in through the nose, past the windpipe, and into the lungs. From then onwards, the lungs work without rest, to filter oxygen from the air and pass it into the circulatory system. Here, oxygen is pumped by the heart and taken to every cell of the body. The waste, carbon dioxide, is taken to the lungs where it is forced out of the body by the contraction of the lungs.

But what if there is a problem? If the tissues sense a lack of oxygen, they send a signal to the brain, which then makes the lungs work harder. The intercostal muscles of the ribcage then pull together, shrinking and expanding the lungs harder to push out stale air and pull in fresh air. The nose and the sinuses work day and night to filter germs from the air before letting it into the body. The pharynx works to separate air from food, making sure you do not choke and also making sure no air enters your stomach—so you do not feel bloated. In between all of these processes, the larynx works quietly to make the sounds that make a language. Perhaps it is working right now, as you read this book out loud.

The day the system stops, because of a weak heart that cannot pump any more, or weak or polluted lungs that cannot breathe in fresh air any more, one dies.

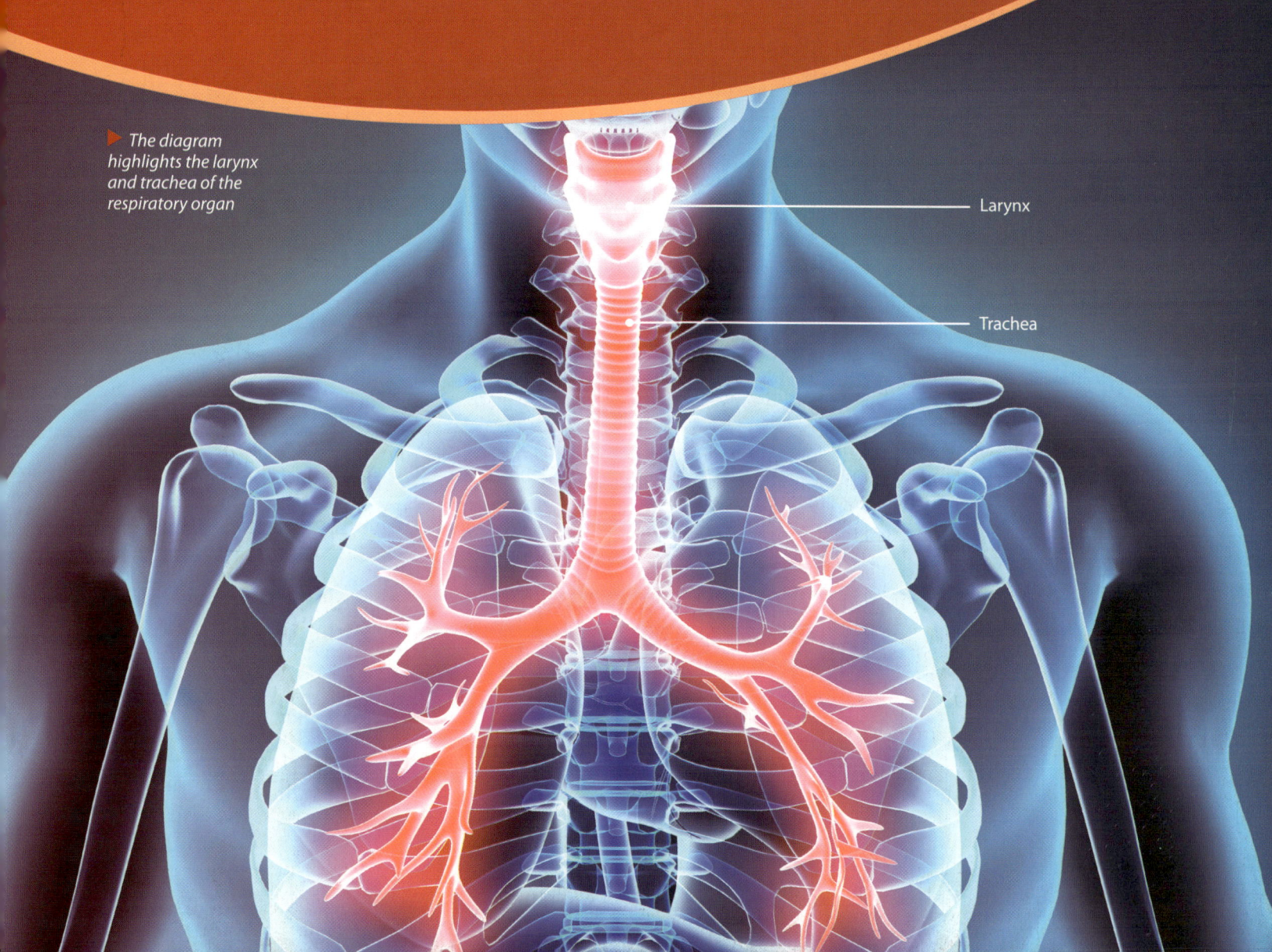

▶ The diagram highlights the larynx and trachea of the respiratory organ

Larynx

Trachea

Nose: Your Breath's Gatekeeper

Most of the respiratory system is made up of the vessels that bring fresh air right into the body. First there is the **conducting zone** which includes the nose, throat, windpipe, and air passages inside the lungs. The **respiratory zone** consists of the tissue where oxygen filters from the lungs into the blood.

The nose is the gateway of the conducting zone. There is more to it than the part that sticks out of your face, as it occupies a large space inside your head, between the brain and the mouth. It cleans the air that comes in through the nostrils, removing dust and germs. That is why you normally breathe through your nose, else air enters the lungs from the mouth.

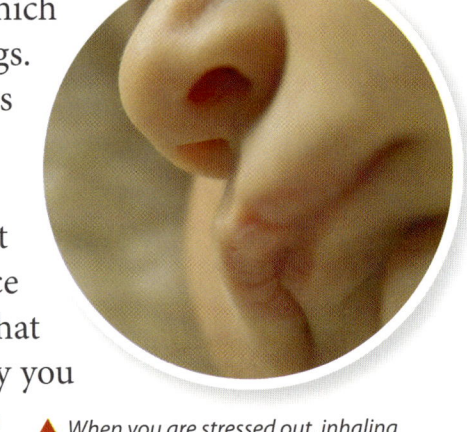

▲ *When you are stressed out, inhaling and exhaling makes you feel better*

Outer Nose

The part of the nose that you can touch and see is called the bridge of the nose. It is made of a cartilage septum and thickened skin that make the nostrils on either side. The inside of your nostrils is like the outer skin with sebum glands and hair. This hair keeps big dust particles, dirt, and even insects from entering the nose. Interestingly, human beings mostly breathe through one nostril at a time, even though we have two nostrils. If you want to check this fact, put your hands close to your nose and breathe in, then breathe out.

In Real Life

Some germs, like the flu virus, can infect your sinus. Your body's immune response causes them to swell. You can feel them on your face. This swelling causes facial pain, headache, and sometimes fever. It often follows a cold, leaving you with a runny nose.

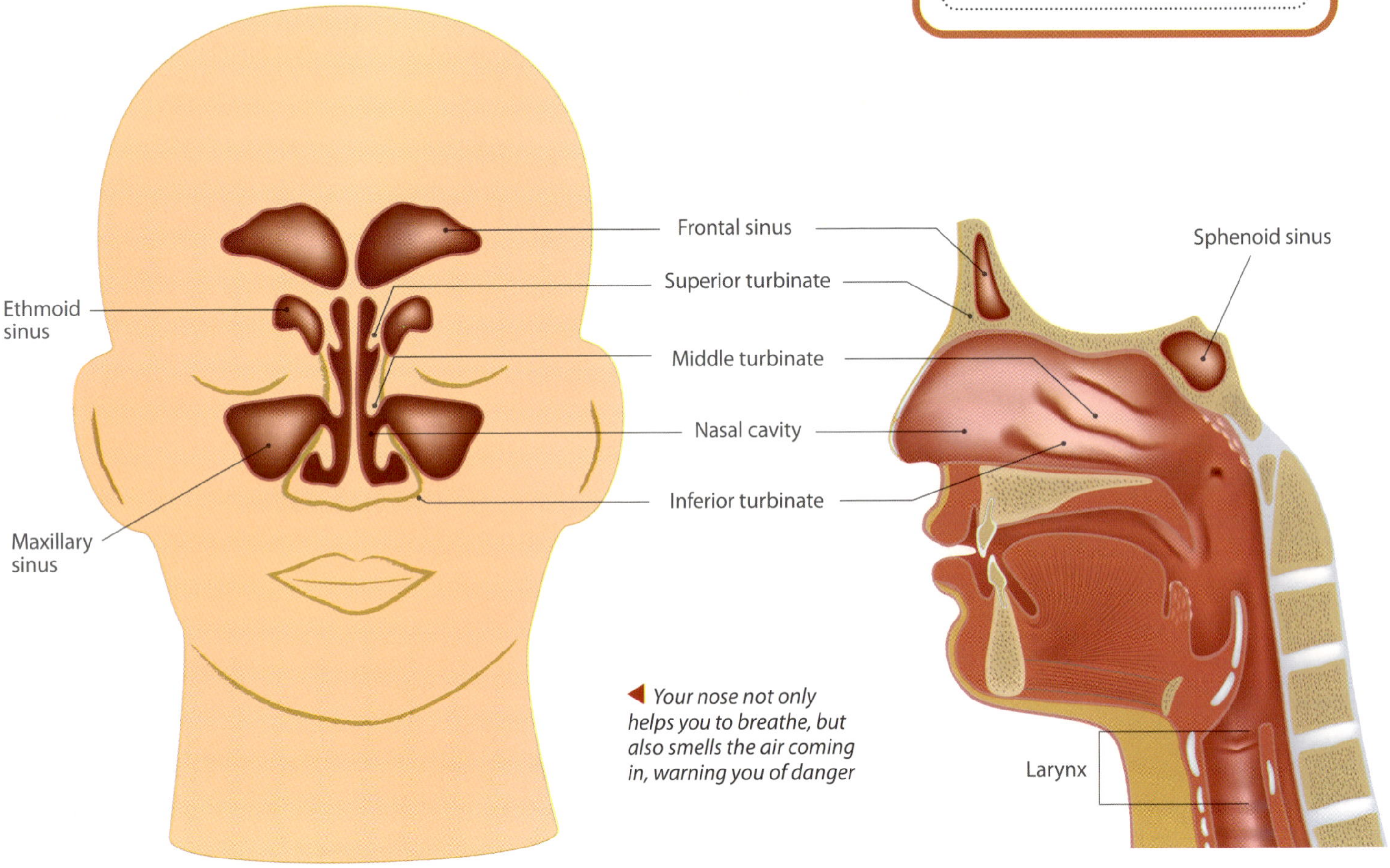

◀ *Your nose not only helps you to breathe, but also smells the air coming in, warning you of danger*

Inside the Nose

The septum extends right through the inner nose, where it is made of bone. It is not always in the centre, but often slightly shifted to the right or left. The bony walls of the nose are folded into **conchae** or turbinates. They swirl the air within the nose, like turbines in a generator. When you are breathing out, the nasal conchae trap water vapour so your nose does not become dry. The inner nose also has extensions called **paranasal sinuses**, of which there are four pairs—the frontal, maxillary, sphenoidal, and ethmoidal. The air in them makes the skull light.

Incredible Individuals

Cleopatra, the Queen of Egypt, had a long nose. Blaise Pascal once famously remarked, that had Cleopatra's nose been shorter, the whole face of the world would have been changed. He was referring rhetorically to the collapse of the Roman republic, as both, Julius Caesar and Marc Antony were spellbound by her beauty.

▲ A bust of Cleopatra

Sneezing

Sneezing is our body's way of keeping itself healthy by expelling germs, pollen, and other foreign objects with great force. The hair in your nostril are sensitive to the smallest things and trigger the nervous system to make the muscles of the chest and throat contract very fast. When you sneeze, air is forced out of your mouth and nose together. As the air moves out, it carries mucus droplets with it, in which the germs are trapped. If somebody happens to be near you when you sneeze, they might inhale your germs. That is why you should cover your nose and mouth while sneezing.

Cleaning the Air

The nose works like your body's own air purifier. The nasal cavity and sinuses are lined with a special **respiratory epithelium**. This is made of cells that have microscopic hair called cilia. Scattered in between are **goblet cells** that make mucus. As the air you breathe in swirls, the cilia and mucus catch all the dust and germs in it. Under the epithelium is a special tissue called **nasal-associated lymphoid tissue** (NALT), which has immune cells in it. These destroy the germs caught by the cilia. The mucus makes the air humid, while blood capillaries under the epithelium warm the air before it goes into the lungs. The 'purified air' is now ready for the lungs.

Isn't It Amazing!

Birds do not have external noses. Instead, they have a pair of openings called nares just above their beaks that open directly into their inner noses. The nares can be used to tell male from female. For instance, budgerigar males have blue ones, while females have brown ones.

▲ Some species' nares are covered with feathers

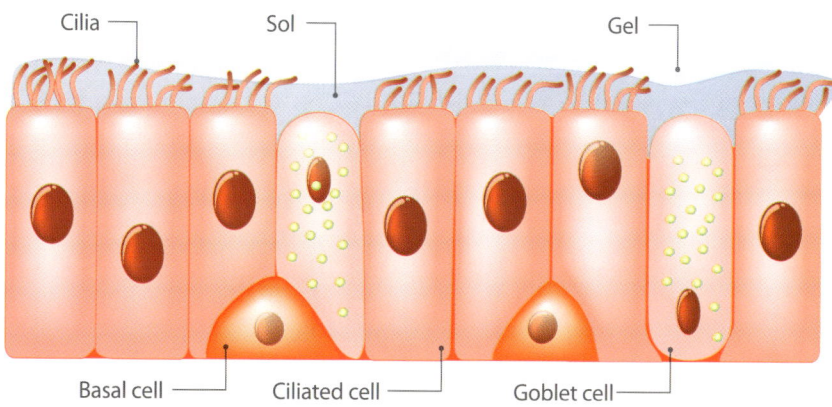
▲ The nasal mucus is made of a sticky, outer gel that traps germs and a fluid inner sol that moves it

Taking Air to the Lungs

Once air is cleaned and conditioned in the nose, it passes into the pharynx. This acts like a railway junction, where two tracks meet. The digestive system brings food from the mouth to the food pipe. The respiratory system brings air to the windpipe. So why do they meet at all?

If these two 'tracks' did not meet, we would not be able to speak at all *(see pp 27–29)*. During an emergency, when more air has to be sucked in or forced out, you breathe from the mouth. The flipside is that it creates the danger of choking, if food entered the windpipe. There is one simple way to prevent it—do not talk while eating!

▶ You should avoid talking while eating because the food might accidentally enter your trachea and interrupt your breathing

Parts of the Pharynx

The pharynx is made of muscle that helps you inhale air and also swallow food. It is made of three parts—the **nasopharynx**, the **oropharynx**, and the **laryngopharynx**. The nasopharynx is the extension of the inner nose to the throat. It has the **pharyngeal tonsils**, also called the adenoids, which hang down from the top like a fold of tissue. These are full of immune cells that kill any germs that have made it past the nose. It has another flap of tissue called the **uvula** that extends from the palate, which separates the nose from the mouth. While swallowing, the uvula stops food from entering the nose. The nasopharynx also has two canals called eustachian tubes that connect it to the ears so that they can be of the same pressure as the atmosphere, or else the ear drums will burst.

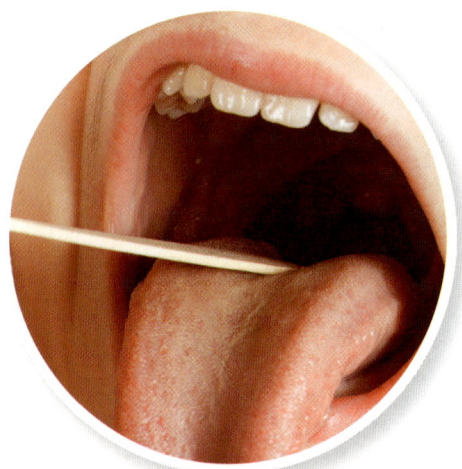

▶ A patient's tonsils being checked by a doctor

The oropharynx is the middle part, where the two tracks cross. The hyoid bone and the spinal column surround it. The hyoid attaches two sets of muscles. The first is the tongue inside the mouth. Under the tongue are the **lingual tonsils**. The other set are the muscles that control the **epiglottis**. The epiglottis shuts off the wind pipe while you swallow food. At the end of the palate, are the **palatine tonsils**. The final part is the laryngopharynx which leads the air into the trachea.

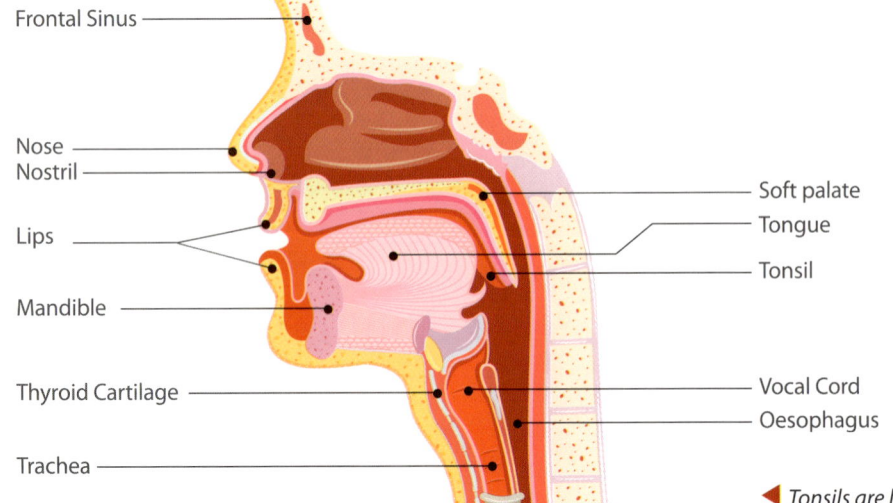

In Real Life

Infections acquired during childhood may leave you with enlarged tonsils. These cause obstructions to breathing that might need to be removed by surgery. If not, they can cause snoring or breathlessness.

◀ Tonsils are large in children but shrink as they grow older

HUMAN BODY | LUNGS & RESPIRATORY SYSTEM

 # Trachea

The trachea or wind pipe takes air from the pharynx to the lungs, sharing space in the neck with the oesophagus or food pipe. It is ringed by 20 C-shaped pieces of cartilage that make sure that it does not collapse onto itself like the oesophagus does. The **trachealis muscle** attaches to these pieces and, along with other tissues, makes the main air tube. It helps to stretch the trachea when breathing out, and shrink it when breathing in. In the lungs, the trachea branches into **bronchi**.

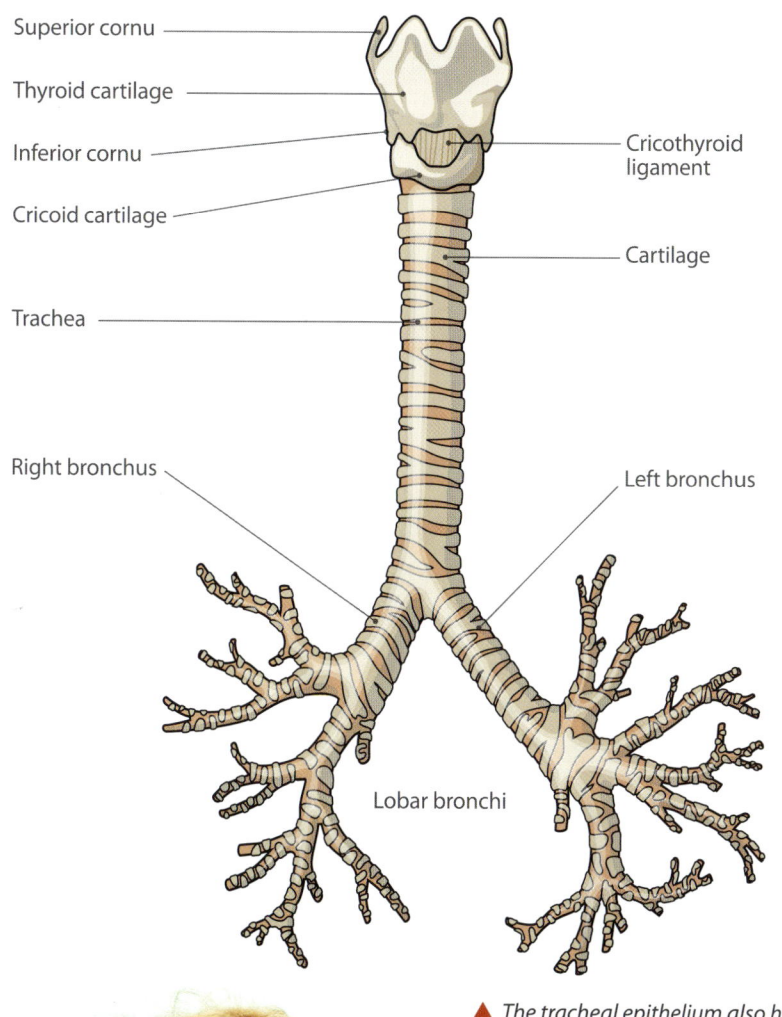

▲ The tracheal epithelium also has cilia, which brush germs towards the nose, so they can be sneezed out

Isn't It Amazing!

A giraffe's long neck also means they have really long pharynxes and tracheas. Their lung muscles have to work a lot to inhale and exhale air. The air passing through the trachea does not have enough force to vibrate the vocal cords, so giraffes are mostly silent animals.

▶ A giraffe's wind pipes take air down or up over 6 feet

▶ You might sneeze because you are exposed to dust particles, so keep your surroundings clean

The Last Mile: Bronchi and Bronchioles

When the trachea reaches a point in the chest just below the neck, called carina, it splits into two bronchi (singular: bronchus) to the left and right. The carina has nerves that can sense whether anything other than air has fallen down the trachea. If there is, it will trigger coughing. Muscles in the trachea will try to push the body out before it enters the lungs. Otherwise, it may cause you to faint or obstruct your breathing. (*see pp 26*)

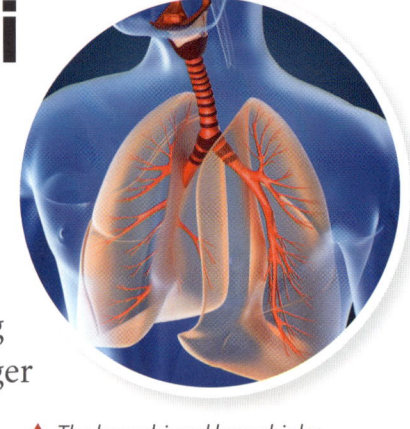

▲ *The bronchi and bronchioles make the bronchial tree*

The bronchi enter the lungs, where they further split into secondary and tertiary branches. All of these have cartilage rings to stop them from collapsing. Doctors call the point of entry the **hilum**, where the nerves, pulmonary arteries and lymph vessels enter and the pulmonary vein leaves.

Isn't It Amazing!

Insects have a very different respiratory system. They have tiny pores under the bodies called spiracles, which let air into large sacs. Tubes called trachea take air around the body, branching into tiny tracheoles, which deliver oxygen to cells.

▶ *Unlike vertebrates, an insect's respiratory and circulatory systems do not interact*

Bronchioles

Bronchioles branch off from tertiary bronchi and further branch into terminal bronchioles. These feed the units of gas exchange in the lungs called **alveoli**. They have no cartilage.

Cross-section

The bronchioles are lined inside by epithelial tissue which has cilia to brush germs and dust out of the lungs, and also goblet cells that make mucus to help them. Surrounding it is a piece of loose tissue called the **lamina propria**. Like the nose has NALT, the lamina propria has BALT or **bronchial-associated lymphatic tissue**. It has immune cells in it, which destroy germs. This is surrounded by smooth muscle that keeps the air moving.

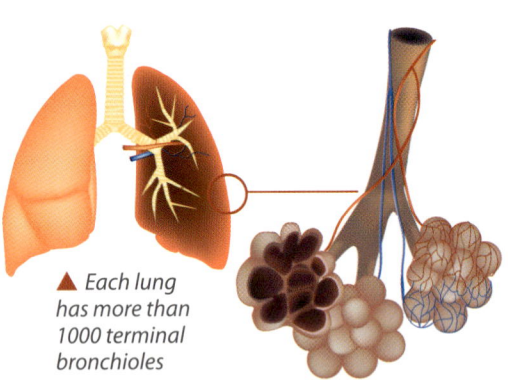

▲ *Each lung has more than 1000 terminal bronchioles*

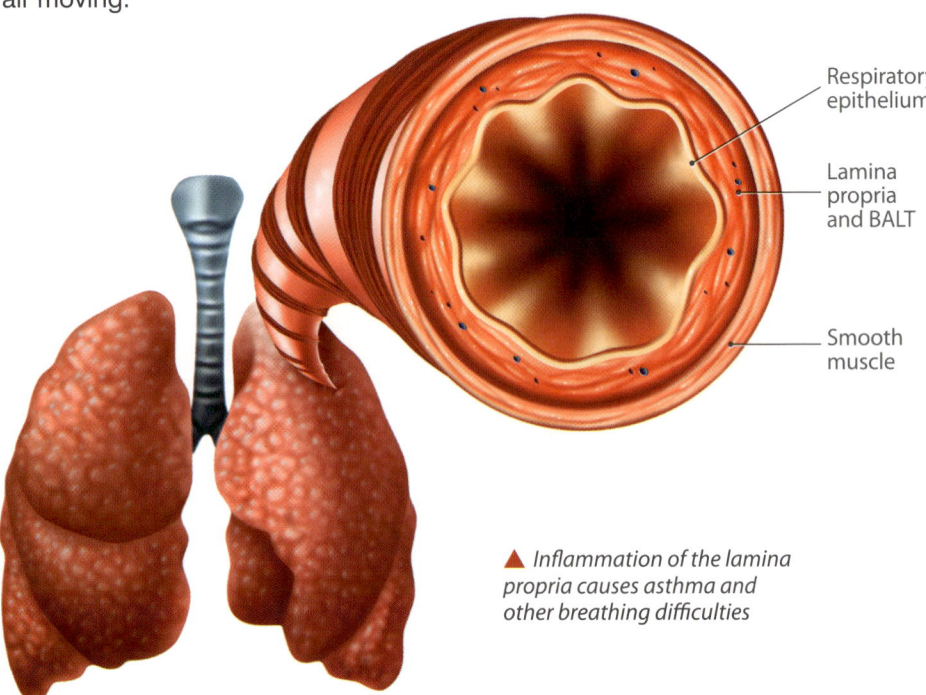

▲ *Inflammation of the lamina propria causes asthma and other breathing difficulties*

In Real Life

Infection in the bronchioles leads to a build-up of mucus, which causes congested lungs. This is called **bronchitis**.

Destination: Lungs

The air you take in from your nose has to reach the respiratory zone. Here, the respiratory system meets the circulation system. This zone has two parts, the **respiratory bronchioles** and the **pulmonary lobes**. It is made up of special tissue which allows oxygen to diffuse right into blood, where haemoglobin in the red blood cells is waiting for it. The blood has let go of carbon dioxide from deep within your body's tissues. It is now ready to make the journey to the outside world. Thus, you complete one breath.

Lung Structure

Your lungs come in a pair, with the left lung slightly smaller than the right one. Both lungs are encased in the ribcage and covered by a bag called the **pleura** (plural: pleurae), which cushions them against the ribs. The **diaphragm**, a giant, dome-shaped muscle, seals off the ribcage from below. The left lung has the cardiac notch, a space it makes for the heart to fit in.

Each lung is made up of lobes, which are served by secondary bronchi. The larger, right lung has three lobes, while the left lung has just two. Each lobe is divided into segments, which are then divided into lobules, each of which gets a bronchiole. Lobules are divided into alveolar sacs that, in turn, have alveoli within them. The alveoli are the functional units of the lungs.

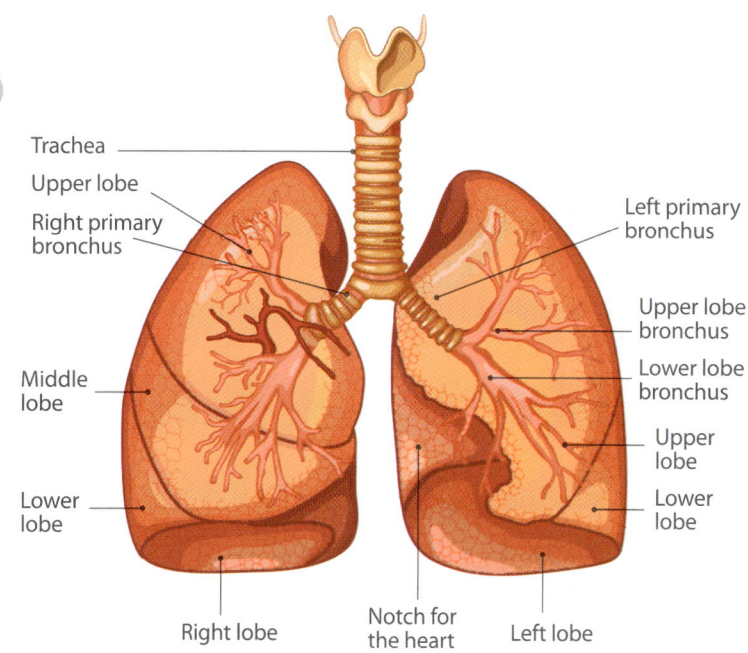

▲ The right lung is broader than the left lung

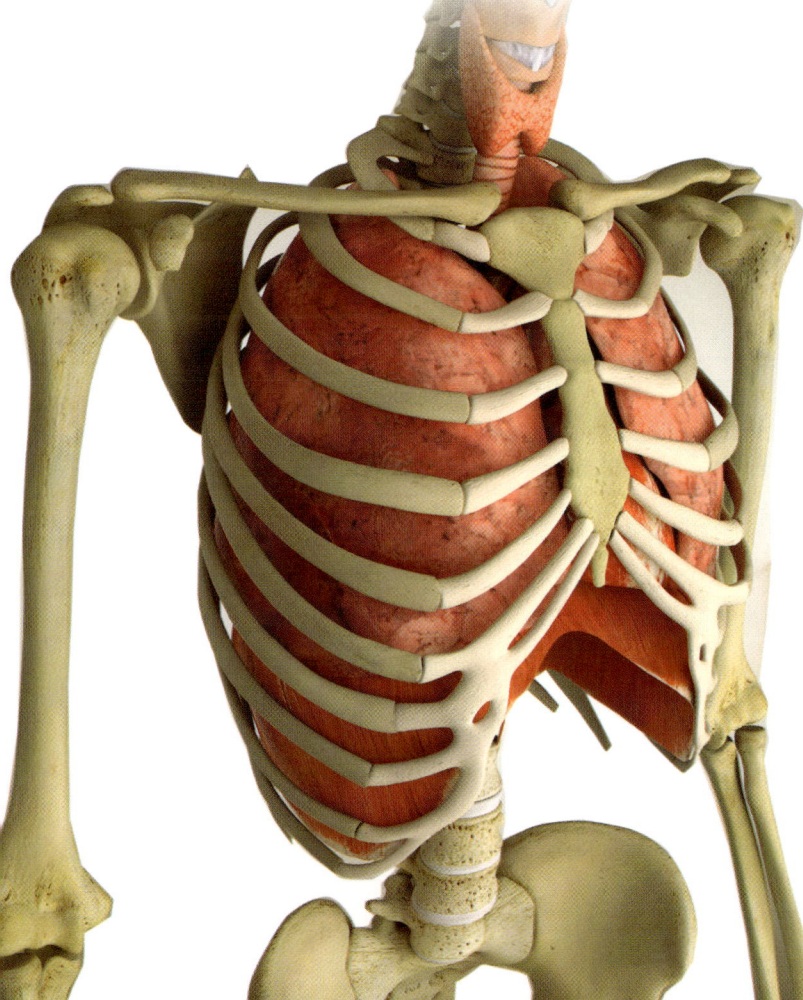

Pleurae

The pleurae that cover your lungs are made of two layers. The inner layer is visceral pleura, that lines your lungs directly, while the outer layer, that is, parietal pleura, links them to the rib cage, sternum, and diaphragm. Between them is the pleural cavity, which is full of air. Pleurae act like a cushion, when the inflated lungs push against the rib cage. They also protect the lungs against infections. The pleurae may swell up due to infection, leading to a condition called pleurisy. It may cause dry cough and difficulty in breathing. Sometimes, the pleural cavity may be filled with too much air, pushing pressure on the lungs. This is called pneumothorax. Injury to the lungs may cause them to be filled with blood, leading to a condition called haemothorax.

◀ The pleural sac cushions the lungs while breathing

Inside Your Lungs

In many invertebrate organisms, the respiratory system does not interact with the circulatory system. It does so in molluscs, where **haemocyanin** molecules in the haemolymph receive oxygen from the gills. Red blood cells are present in fish to carry oxygenated blood from the gills to the heart, then through the body, and bring back deoxygenated blood to the gills again. In reptiles, birds, and mammals, the lungs take over and there are separate arteries for oxygenated blood and veins for deoxygenated blood. This brings a lot more oxygen to the tissues, so they can have a lot more energy.

Blood Vessels and Nerves

Lungs are the only organs that receive deoxygenated blood from the heart's **pulmonary arteries**, which come straight to them from the heart's right ventricle. The artery enters at the hilum and follows the **bronchial tree**, finally branching into arterioles. At the alveoli, the blood flows in tiny capillaries, ready to receive oxygen. They collect into venules, which follow the bronchial tree till the hilum, where the **pulmonary vein** leaves to enter the heart's left atrium.

The lungs are controlled by the parasympathetic and sympathetic nervous systems. The parasympathetic, also known as the 'rest and digest' system constricts the bronchial tree, causing deeper breathing, like when you are sleeping. The sympathetic or 'fight or flight' nervous system dilates the bronchial tree, so you have more oxygenation.

▲ *A cross-section of the alveoli showing walls and capillaries*

In Real Life

The corneas of your eyes get oxygen from the air. This eventually dissolves into tears.

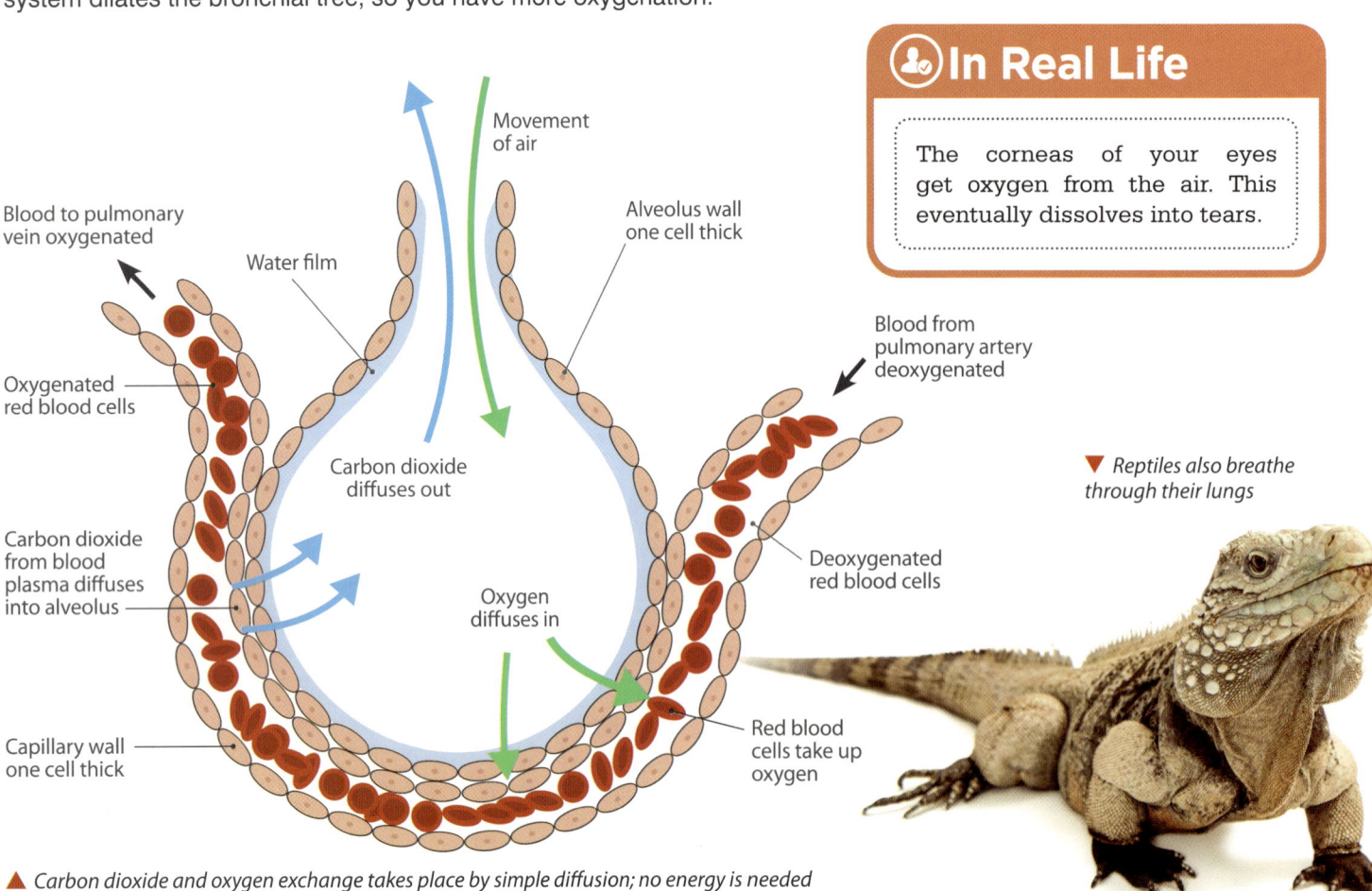

▲ *Carbon dioxide and oxygen exchange takes place by simple diffusion; no energy is needed*

▼ *Reptiles also breathe through their lungs*

Alveoli

These are the working parts of the lungs and look like a bunch of grapes under a microscope. Each alveolus is a sack of respiratory cells, just 0.2 mm thick, with walls that stretch when full of air. They link to each other by alveolar pores, which make the air pressure in them equal. Most of the wall is made of type I alveolar cells, which let oxygen in and carbon dioxide out. A few cells belong to the type II alveolar cell group, which make **pulmonary surfactant**. This is a soap-like substance that makes it easier for oxygen to dissolve in the water of plasma. Immune cells called **alveolar macrophages** roam around the alveoli, snapping up any germs that manage to get there.

Isn't It Amazing!

Mammals have tiny alveoli that are densely packed into their lungs, to increase surface area. For example, while a frog's alveolus is 10 times as wide, it has only 20 cm^2 of gas exchange area per cc, while human beings pack in 300 cm^2 of gas exchange area per cc.

▲ Frogs do most of their breathing through their skin

▼ Dolphins are mammals, not fish. They need to come up for air from time to time

Alveolar Pressure

During quiet breathing, the air pressure in your alveoli, that is, the **alveolar pressure** will match atmospheric pressure, which is 760 mm Hg at sea level. However, it changes with the phase of breathing. When you are inhaling, your lungs expand and air pressure in your alveoli drops. Atmospheric air rushes in to make up the difference. When exhaling, the lungs contract and the alveoli get squeezed. Air pressure rises, and air rushes out to bring down the pressure. This is because of **Boyle's Law**, which says that the pressure and volume of a gas are inversely related.

The Respiratory Cycle

Your doctor's term for breathing is **pulmonary ventilation (PV) or a respiratory cycle**. One PV has two steps—breathing in or inspiration and breathing out or expiration. Inhalation is the term for when you deliberately breathe something in, like the air above food that smells delicious. Exhalation is for breathing out. Most of the time, you breathe quietly. This is called **eupnea**. When you are tired or doing something that needs more breath—like singing—your body switches to forced breathing. This is called **hyperpnea**.

▲ Singers need to control their breathing to carry a tune properly

You use two sets of muscles to breathe—the muscles between your ribs, called **intercostal muscles** and the diaphragm, that is the large muscle stretching between the ribs, sternum, and lumbar vertebra.

▶ Respiration works by using Boyle's Law of gas pressures

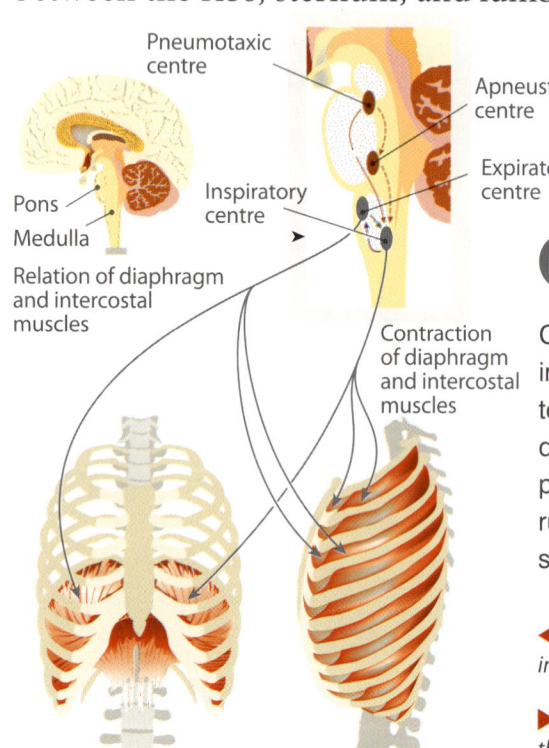

Quiet Breathing

Quiet breathing may be shallow or deep. In the shallow kind, only the intercostals contract and expand. Deep breathing involves the diaphragm too. During inspiration, the diaphragm contracts and pulls the pleurae down. The intercostals contract and pull the ribcage and the attached pleurae, upward and outward. Together, they expand the lung and air rushes in to equalise the alveolar pressure. When the muscles relax, they squeeze the pleurae, which then squeeze the lungs, forcing air out.

◀ In quiet breathing, your skeletal muscles work like involuntary muscles, controlled by the brain stem

▶ Forced breathing helps get rid of the lactic acid built up by the muscles during exercise in the form of CO_2, by oxidation

Forced Breathing

Other muscles join in when you need to breathe in or breathe out more air. This happens when your body needs more oxygen, like when you are tired after exercise or heavy work, or if you experience shortness of breath, or if you have asthma.

When breathing in, the neck muscles pull the ribcage upward, enlarging the chest beyond what the intercostals and diaphragm can do. Put your fingers on the base of your neck and breathe in deeply. Can you feel your shoulders rise?

When breathing out, the muscles of your belly, like the obliques, contract. This squeezes your belly, and the organs in it push the diaphragm up in turn. Pull in your stomach, and you can feel air rushing out of your nose.

Taking Oxygen to the Tissues

Oxygen diffuses from the alveoli into the pulmonary capillaries under atmospheric pressure. But oxygen molecules are not easily soluble in liquids or water, which is what most of your blood is. Only 1.5 per cent of all the oxygen in your blood is actually dissolved. The rest has to be literally carried through the blood to the tissues. That job is done by the molecule haemoglobin, which is present in red blood cells.

Haemoglobin

Haemoglobin is made of three parts—two protein parts called alpha-globins and beta-globins, and an organic molecule called haeme. At the centre of each haeme molecule is an iron atom. The chemistry of haemoglobin makes the iron in the haeme very attractive to oxygen. As blood enters deep inside the lungs, the haemoglobin molecules take up oxygen very quickly to become **oxyhaemoglobin**.

▲ *It is the iron atom that gives haemoglobin, and therefore blood, its bright red colour*

Delivering Oxygen

When oxygenated blood reaches the tissues, it reaches a place low in oxygen and high in acidity. The chemistry of haemoglobin now changes and oxygen is released. It diffuses from the plasma, into the cells, where it is taken up for respiration.

Haemoglobin and oxygen break up faster under higher temperatures. During exercise, muscle tissues release a lot of energy and heat up. Haemoglobin going to the muscles thus gives up oxygen more easily.

Carbon Monoxide

Haemoglobin is attracted even more to carbon monoxide (CO) than to oxygen. CO is a colourless odourless gas that is often found in the exhausts of vehicles and central heating furnaces. It forms carboxyhaemoglobin in the blood, which does not break up in the tissues. When a person inhales CO, they get headaches, feel dizzy, and have pain in the chest. They can die quickly as their tissues, especially those in the brain, die of oxygen starvation. If you suspect a person of having CO poisoning, switch off all flames and electrical devices and take them outdoors.

Incredible Individuals

John Scott Haldane (1860–1936) was a scientist concerned with the safety of coalmine workers, many of whom died of poisoning deep in the mines. He discovered that this was because of carbon monoxide. He published a report on safety in mines which got translated into many languages. He also designed a respirator that would prevent gas poisoning and studied the effect of very high altitude on breathing. (*see pp 18*)

▲ *Carbon monoxide is called the silent killer because it can neither be seen nor smelled*

Oxygen and Carbon Dioxide Cycles

At sea level, the atmosphere has a pressure of 760 mm Hg. All of us are adapted to breathe in this pressure (*see pp 17*). But the atmosphere is made of many gases, each of which has its own pressure, called its partial pressure.

Partial Pressure

Air in the alveolus is a mix of fresh air coming in and carbon dioxide being given out by blood. Therefore, the partial pressures of atmospheric gases are different from those in the open air. Under Boyle's Law, if two chambers filled with a gas are connected, like the atmosphere and your lungs, the gas will move from one chamber to the other to make its partial pressures equal. Even if you did not actively breathe, oxygen would enter the lungs and carbon dioxide would leave. This is how the gills of fish and tracheae of insects work.

Gas	Outside the Body		Inside the Lungs	
	Percent of Air	Partial Pressure in mm Hg	Percent of Air	Partial Pressure in mm Hg
Nitrogen (N_2)	78.6	597.4	74.9	569
Oxygen (O_2)	20.9	158.8	13.7	104
Water vapour (H_2O)	0.04	3.0	6.2	40
Carbon dioxide (CO_2)	0.004	0.3	5.2	47
Others	0.0006	0.5	–	–
Total	100	760.0	100	760.0

Henry's Law

Henry's Law is a law of physics that states that the amount of gas in a liquid depends on its solubility in the liquid, and its partial pressure. Nitrogen has higher partial pressure than oxygen in alveolar air, but it is less soluble in water than oxygen. That is why it does not enter the blood. The partial pressure of O_2 in the alveolus (104 mm Hg) is higher than that in the capillaries bringing deoxygenated blood (40 mm Hg). Oxygen rushes into the capillaries, where it is taken up by haemoglobin. As the capillaries exit, the partial pressure rises up to 100 mm Hg.

In the tissues, the pressures change. For example, the partial pressure of oxygen in muscle tissue is only 20 mm Hg. Oxygen breaks up from haemoglobin and enters the muscle. Fat tissues use up a lot of oxygen, so they have high partial pressure of oxygen; and very little oxygen is released there.

Bohr Effect

Christian Bohr, a scientist, noticed that high acidity makes haemoglobin give up oxygen easily. As tissues make energy from glucose, they release CO_2. This enters the blood plasma, where it becomes carbonic acid:

$$H_2O + CO_2 \rightarrow H_2CO_3 \rightarrow H^+ + HCO_3^-$$

By the **Bohr effect**, the tissues get oxygen in exchange for carbon dioxide. Other acids, like lactic acid, which is produced during **anaerobic respiration** in the muscles, also make it happen.

 ## Carbon Dioxide

Around 70 per cent of the carbon dioxide travels from tissues to lungs as bicarbonate ion. The enzyme carbonic anhydrase in the lungs breaks bicarbonate into water and carbon dioxide, which diffuses out.

$$H^+ + HCO_3^- \rightarrow H_2O + CO_2$$

About 20 per cent binds to haemoglobin as **carbamino-haemoglobin**, while the remaining 10 per cent dissolves in the plasma.

> **Isn't It Amazing!**
>
> While at rest, our bodies consume approximately 250 millilitres (about 15 cubic inches) of oxygen each minute.

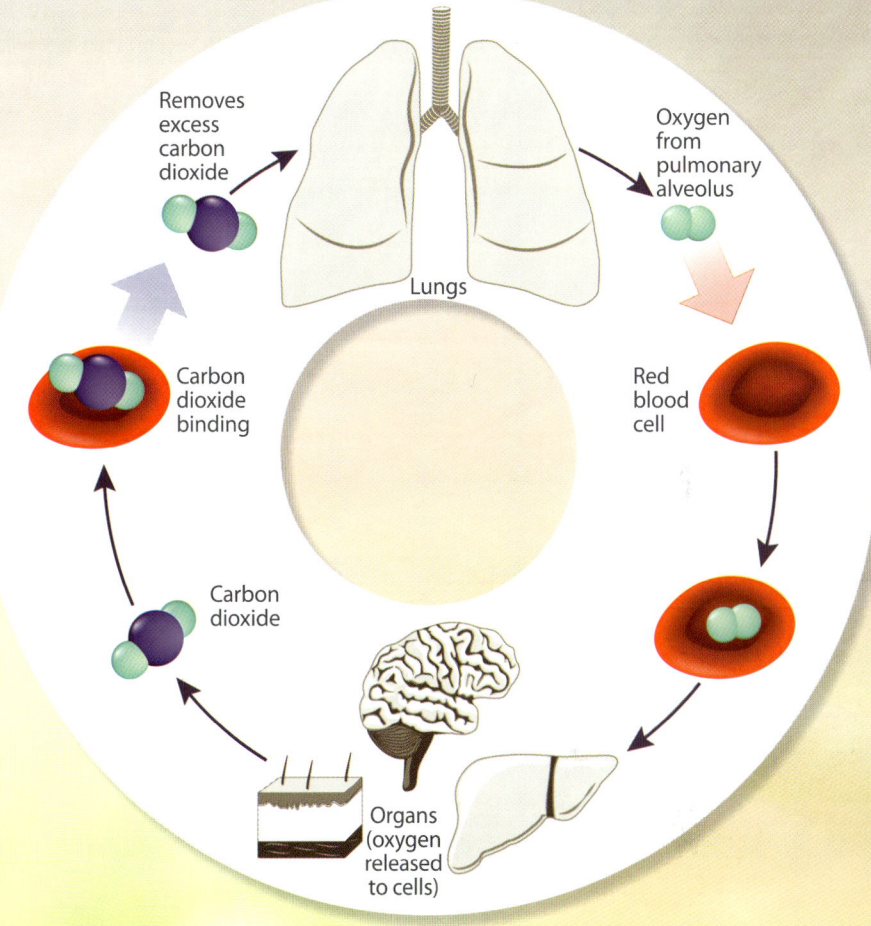

▲ *The diagram shows the respiration process in human beings and the path of the red blood cells*

Cellular Respiration

The word respiration has two meanings. When doctors and scientists talk about the whole body, they mean inspiration and expiration. But when they talk about each cell of your body, respiration has a different meaning. It means how your cells convert their main fuel, that is glucose, into energy. When glucose is chemically broken down by the enzymes in your cells, it releases heat. This heat is used to make another chemical called Adenosine Triphosphate (ATP). Whenever the body needs quick energy, like when you exercise, the ATP is broken up and the heat stored is released again.

▲ Make sure you are breathing well while exercising

Glycolysis

The first step of respiration is glycolysis, in which glucose is broken in a series of reactions into a smaller compound called pyruvate. If the cell has time, it will then carry out aerobic respiration. If the cell is in a hurry, as muscles cells are when you are running, it will carry out anaerobic respiration.

Aerobic Respiration

The oxygen you breathe in is used for aerobic respiration, in which glucose is broken down completely into carbon dioxide and water. Pyruvate goes through the Citric Acid Cycle (also called Krebs Cycle), a series of chemical reactions that turn it into NADH and $FADH_2$. These react with oxygen to form ATP and release CO_2.

Anaerobic Respiration

When the cell needs energy fast, it only carries out glycolysis and the pyruvate is turned into lactic acid instead. Less ATP is made, so you run out of energy and feel tired. The lactic acid is sent to the liver, where it is turned back into glucose, while you are resting.

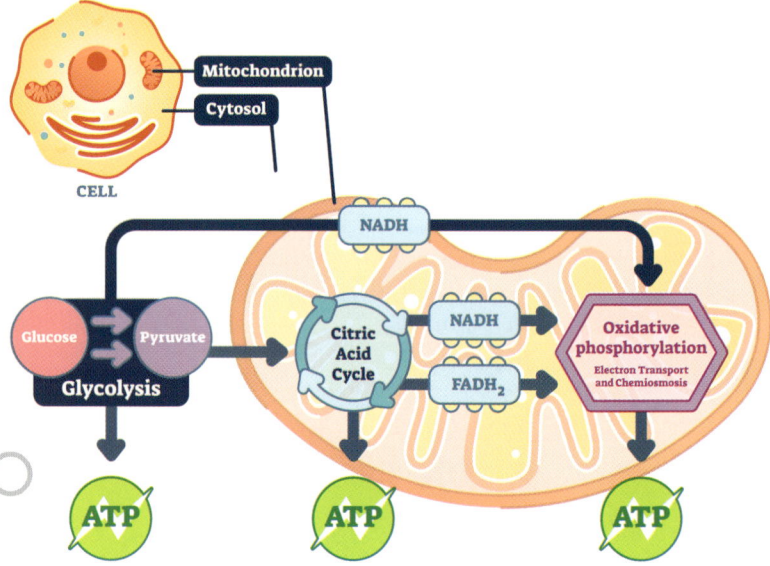

▲ The diagram shows the process of aerobic respiration

▼ Cheeses made around the world are made by the process of fermentation

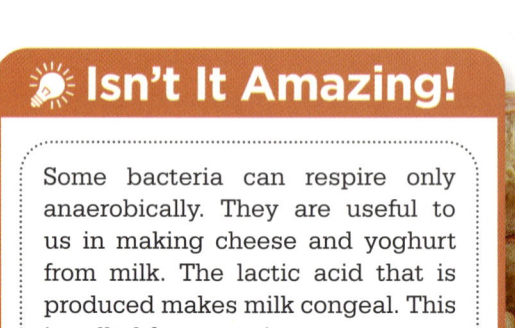 Isn't It Amazing!

Some bacteria can respire only anaerobically. They are useful to us in making cheese and yoghurt from milk. The lactic acid that is produced makes milk congeal. This is called fermentation.

Breathing Underwater

Human beings cannot breathe underwater as water would flood our lungs and drown us. So, what do the creatures that live underwater do? If life is impossible without oxygen, how do they procure it?

Surviving Underwater

If divers go too deep without compressed air, they can get a condition called diver's bends. It is a condition which begins when they go deeper underwater and the pressure on their lungs increases. The nitrogen from their lungs dissolves in their blood and gets into fatty tissues. If they come up too quickly, the nitrogen comes out of the solution, forming bubbles in the body. These bubbles cause joint pains, paralysis, and lack of muscle coordination, called diver's staggers. Doctor's call this decompression sickness. If you go deep water diving, surface slowly by rising about a foot, staying for some time, and rising again, till you are in the safe zone.

▶ *The diagram shows how gills work. Fish also have lungs, but these have evolved into swim bladders that help them float*

Breathing Apparatuses

Animals like sponges and jellyfish filter the oxygen directly from the water, often filling it in a body cavity called **coelom**. Most fish have gills, which are organs that filter oxygen from water. The development of a circulatory system with haemoglobin helped speed up the body's intake of oxygen in fish. They take in water from the mouth and pass it through gill arches. Each arch is made of fine gill filaments, which have blood capillaries that filter oxygen. The gills are covered by a flap of skin called **operculum**. Lungs begin to develop in amphibians when they change from tadpoles to adults. From here onwards, the animal filters oxygen from air rather than water.

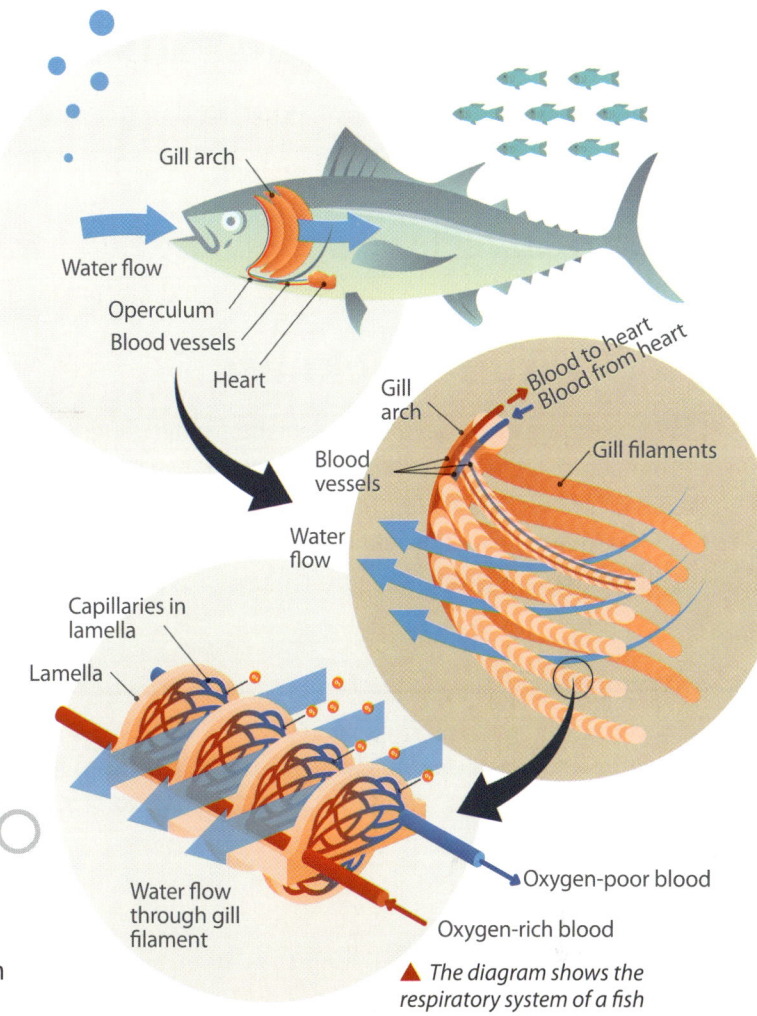

▲ *The diagram shows the respiratory system of a fish*

In Real Life

Octopus gills can extract as much as 80 per cent of the oxygen from seawater.

▶ *If an octopus breathes fast and exhales hard, it can swim backward by jet propulsion*

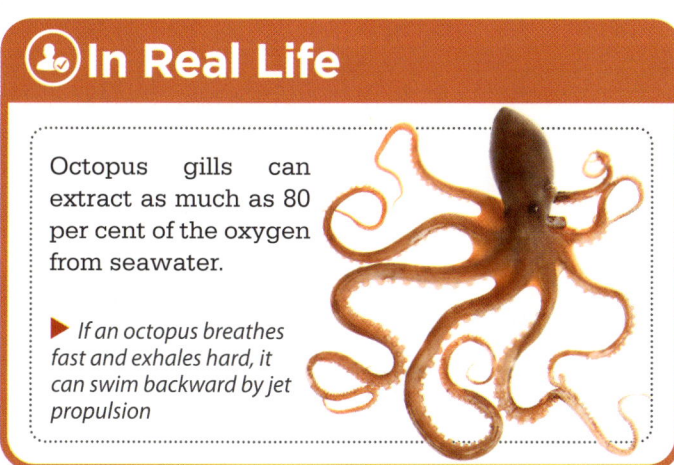

Incredible Individuals

Early divers' equipment required them to be attached to a long tube from which air was pumped from a ship overhead. Jacques Cousteau (1910–1997) invented the portable aqualung in 1943, which made it possible for divers to swim freely underwater. He also made many award-winning movies and TV shows that made underwater diving a popular sport.

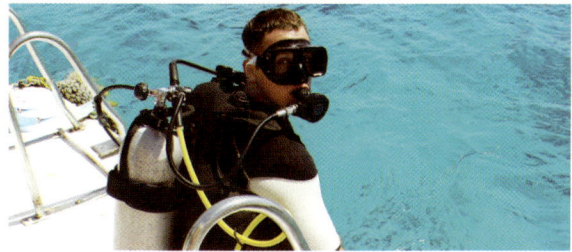

▲ *The longest open saltwater scuba dive till date is 142 hours and 42 minutes*

Breathing at High Altitudes

As we go higher up and above sea level, the amount of oxygen in the air drops. Because of this, the partial pressure of oxygen entering our lungs also falls. At altitudes over 2,400 metres, like in the Himalayas or the Andes mountains, oxygen is 60 per cent less than at sea level. Mountaineers and trekkers experience what they call altitude sickness, if they climb the mountain too quickly.

◀ *Himalayan mountaineers carry pressurised oxygen cylinders to minimise altitude sickness*

Altitude Sickness

Enough oxygen does not enter the blood from the lungs, and so the tissues experience oxygen starvation. Patients experience shortness of breath and panting, headaches, sleeplessness, and reduced vision and hearing. They may get chest pains or vomiting. They must be brought to a lower place immediately. Slowly ascending the mountains helps keep away altitude sickness by letting the body '**acclimatise**'.

Andeans

The people of the Andean plateau in Bolivia also live at very high altitudes. But they have adapted differently. They have more RBCs in their blood, so they can get more oxygen out of the lungs than people at sea level can. They also have more haemoglobin, so they can pull in more oxygen from the lungs. The people of the Ethiopian Plateau in Africa have similar adaptations.

Tibetans

The people of Tibet have lived at altitudes of over 4,000 metres for generations. So how do they not get altitude sickness? They have larger lungs than others and can inhale and exhale more air in a breath than most of us can. This does not tire them out. The lungs thus store more air, and more oxygen goes into their blood.

▲ *Tibetans live with 10 per cent less oxygen in their blood than most people*

▼ *Tibetans have lived at high altitudes for over 25,000 years*

Incredible Individuals

In 1999, Babu Chhiri Sherpa lived on the world's highest point for 21 hours. That is right! He was on Mount Everest without canned oxygen. Doctors call any point above 8,000 metres the death zone.

Asthma

Anyone can get asthma, especially children. When an 'asthmatic attack' happens, the patient gets **bronchospasms**. This is a sudden constriction of the tubes that reduces the volume of air entering the lungs. Allergens like dust, pollen or spores, pet hair, dander, and tobacco smoke, as well as weather changes and heavy exercise, can set off an asthma attack.

Symptoms of Asthma

The main symptoms of asthma are shortness of breath, wheezing, and sometimes coughing. If the face turns a bluish colour, the pulse quickens, and the patient seems confused or anxious, rush them to a doctor. Asthma cannot be cured completely. It can often be managed by avoiding the allergen and using a drug nebuliser.

How Asthma Happens

Asthma is usually an allergic reaction to something that the patient encounters repeatedly. This causes bronchospasms, where the walls and mucosa of the air tubes become inflamed and swollen. Cells of the immune system, such as eosinophils and neutrophils, rush into the lamina propria in high numbers. There is also swelling of the mucus in the bronchi or bronchioles, triggering wheezing.

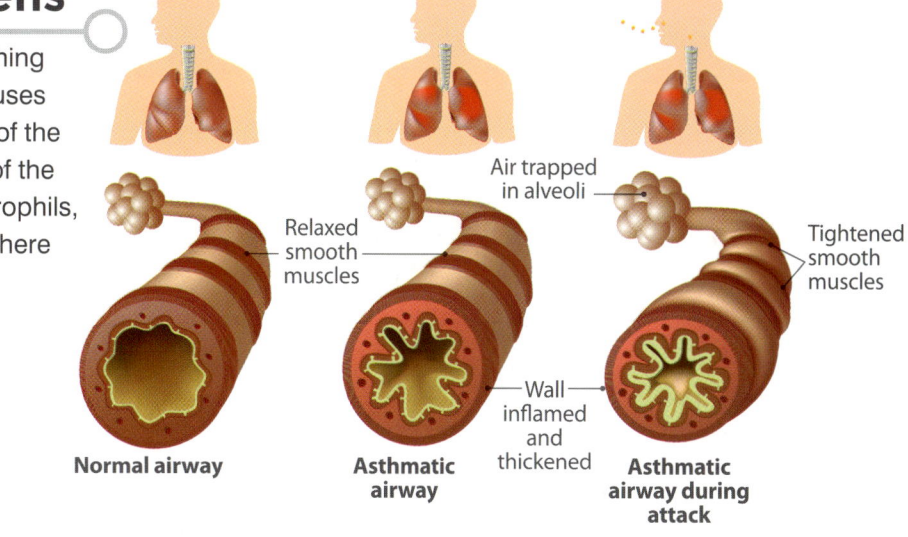

▲ The main symptom of asthma is wheezing, a forced whistling sound from the airpipe

In Real Life

Inhalation as a therapy is as old as 1554 BCE! The ancient Egyptian 'Ebers papyrus', now in the University of Leipzig, records that doctors would heat bricks and then throw cuttings of the herb black henbane on them. The patient would be made to inhale the emitted vapours, which restored their breath.

▲ Henbane is a toxic herb; modern doctors do not recommend it to anyone

Wheezing

Wheezing is the first sign of breathing problems like choking, COPD, or asthma, along with the heaving of the chest. It is also common among smokers and may be triggered by an allergic shock.

If you get wheezing, take your medicines and get to a warm, moist area or breathe in steam from an inhaler. Get to an emergency room if the wheezing goes on for some time, and your skin feels blue, which is a sign that there is not enough oxygen in the blood. You may need blood tests, asthma checks, and lung function tests. *(see pp 22)*

▶ Asthma is one of the most frequent causes of hospitalisation among children

Poison in the Air

Anything in the air that should not naturally be there is a pollutant. Outdoor pollution is the result of burning gasoline and coal. It is also caused by releasing gases like sulphur dioxide, nitrogen oxides, carbon monoxide, ozone, and smoke in high quantities. Indoor pollution happens because of carbon monoxide from central heating, chemical sprays, asbestos, insect droppings, spores, pollen, and most importantly, tobacco smoke. These can cause asthma, chronic obstructive pulmonary disease, heart and blood diseases, and fatal cancers. Pregnant women are often in danger of pre-term birth and giving birth to babies with **congenital diseases**. Climate change is causing an increase in air pollution, with rising dust, mould, and ozone levels.

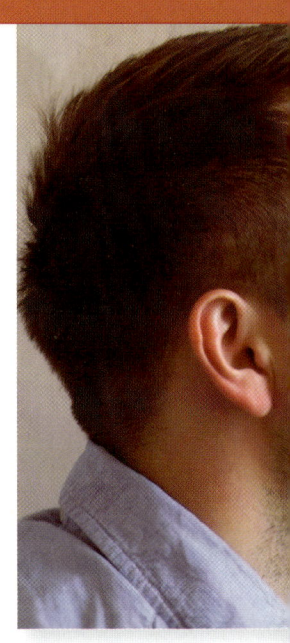

Lung Cancer

Though many people take up smoking, medical research has shown that there is no benefit from it. 3 out of every 10 people who get cancer, get it because of tobacco. The Centers for Disease Control and Prevention says that smoking causes 87 per cent of all deaths because of lung or colorectal cancer. It leaves patients at high risk of heart disease and stroke. Inhaling another person's smoke is no better. 7,300 adults die, each year, of lung cancer because a friend or family member was a heavy smoker.

Lung cancer happens when lung cells go out of the body's control. It develops in four stages and cannot be cured if detected too late. In the fourth stage, it becomes **metastatic**. Here, the cancer cells spread to other organs and damage them, especially the lymph nodes. Small cell lung cancers have cancerous cells that look smaller than others under a microscope. People with this kind of cancer have to be treated with radiation therapy and chemotherapy. Non-small cell lung cancers are more common, where the cells look bigger under the microscope. Doctors can remove them surgically if found early enough.

A person with lung cancer is generally short of breath, wheezes, and coughs all the time, and often coughs up blood. They are always tired, experience pain in the chest, and lose weight very fast.

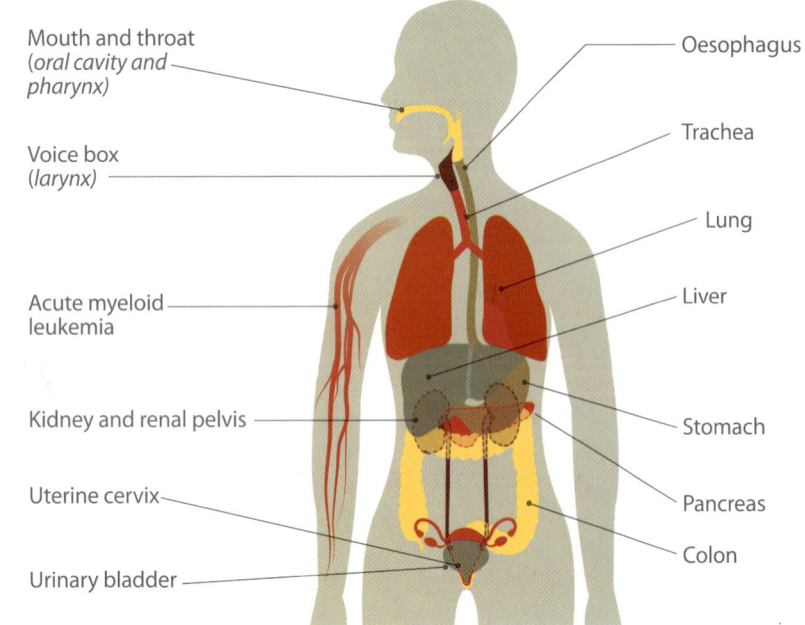

▲ If you know someone who smokes, tell them to stop now

In Real Life

The year 1952 was toxic for the people of London. From 5–9 December, upto 12,000 people died because of fog mixed with smoke from coal fires that descended on the London air.

Chronic Obstructive Pulmonary Disease (COPD)

This is a disease that makes it hard for patients to breathe. It was earlier called bronchitis, but now doctors also include damage to the lung, that is **emphysema** as a condition. Most people with COPD get it from smoking too much. Very few people may also get it genetically.

Air pollution, factory fumes, and cooking on log fires also cause COPD. Patients experience dry cough, tiredness, repeated infections of the lungs or nose, and shortness of breath after even mild work. They may develop cardiac problems if untreated for too long. Doctors use a machine called a spirometer to check for COPD.

COPD cannot be cured. Quitting smoking stops it from becoming worse, and the patient may use an inhaler whenever they get short of breath. Some patients may need to breathe from an oxygen tank. Walking and breathing exercises also help *(see pp 30)*.

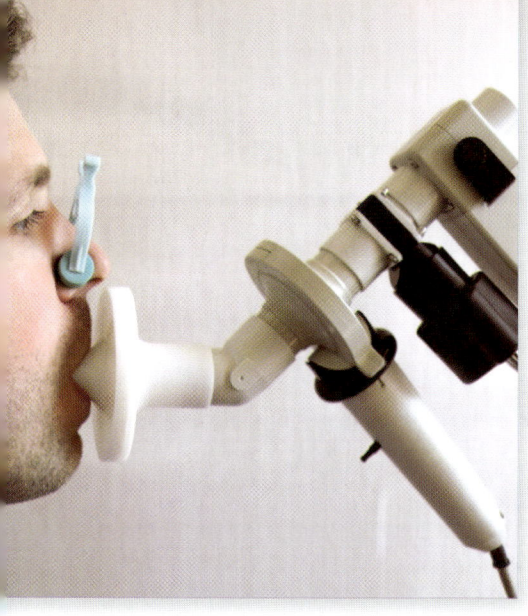

▲ *Spirometers measure how hard it is to breathe. Doctors call it a Pulmonary Function Test*

Respirators and Gas Masks

With a rise in pollution in many cities around the world, doctors recommend using breathing masks. These masks filter out pollutants and dust from the air before they can reach your mouth and nose. N95 respirators are recommended by the US Centers for Disease Control and Prevention for people in dangerous workplaces such as mines, construction sites, and chemical plants.

On the other hand, most gas masks are designed to keep away specific poisonous gases before they can reach you. Many of these masks contain activated carbon, which absorbs certain harmful substances. Gas masks are used by soldiers, firemen, and other emergency workers to go into dangerous areas. High Efficiency Particulate Air (HEPA) masks are suggested for people who may have allergies from pollen and other airborne substances, and for doctors and nurses dealing with airborne viral diseases such as Ebola. Before you put on a gas mask, you must ask your doctor to recommend the one that is right for you.

▲ *Smog causes poor visibility, burning in the eyes, and wheezing*

◀ *Passive smoking is really harmful for children*

Under the Weather

We have all got it, perhaps many times in life. Sneezing, a sore throat, a stuffy or runny nose, and cough—there is nothing uncommon about a cold. While some people believe that a cold will go away on its own, a cold left untreated can become an attack of **influenza**. This can be very serious, requiring you to be admitted to a hospital. Let us understand how and why you get the common cold.

What is a Cold?

Cold is actually your immune system's common response to different types of viral infections. You catch them by breathing in infected air, touching unclean surfaces that have been touched by an infected person, or by being in contact with a person with a cold.

When you have a cold, your respiratory system makes a lot of mucus to trap the germs in. When you cough, it comes out as **phlegm**, a thickened form of the mucus. Make sure that you wash away the phlegm or incinerate the tissues you have coughed into. You may also get a lot of tears as they remove germs from your eyes. You get a sore throat because the lamina propria swells up with immune cells and lymph as they fight the germs.

Dealing with Cold

Regularly wash your hands and keep your home, school, or office clean to keep all types of cold away. If you have a cold, stay at home, get lots of rest, and do not meet people. Drink a lot of soup and juices, but not coffee or colas. Gargle out the soreness in your throat with warm, salted water. Antivirals, medicines that kill viruses, may help you control the infection. Cold often makes way for bacterial infections like bronchitis or sinusitis, so you may need antibiotics too. Other medicines, which your doctor might suggest, are:

- Expectorants, which loosen the mucus and make it easier to breathe.
- Cough suppressants that stop you from coughing.
- Nasal decongestants, that unblock the nose.
- Antihistamines, which reduce swelling of the throat and stop sneezing.
- Pain relievers, which soothe headaches and throat pain.

▶ *Mercury thermometer was invented by German physicist Daniel Gabriel Fahrenheit*

Influenza

A cold can easily lead to influenza or flu. Patients of influenza are often tired and get fevers, chills, and body aches. They feel grumpy all the time and may feel nauseated. Often they also have a runny nose. Most people can get over flu without medicine, but children under the age of 5 or people over 65, pregnant women, and people with asthma, diabetes, or lung diseases need treatment.

▶ *Small children with colds should not be given aspirin or cough medicine*

HUMAN BODY | LUNGS & RESPIRATORY SYSTEM

Viruses and Vaccines

There are four types of influenza viruses, namely A, B, C, and D. Influenza viruses also exist as hundreds of strains. Influenza A is the most infectious and has caused many **pandemics** or outbreaks of disease all over the world. The Spanish flu of 1918 killed nearly 100 million people! Luckily, there are vaccines against them now, and you should get them every year.

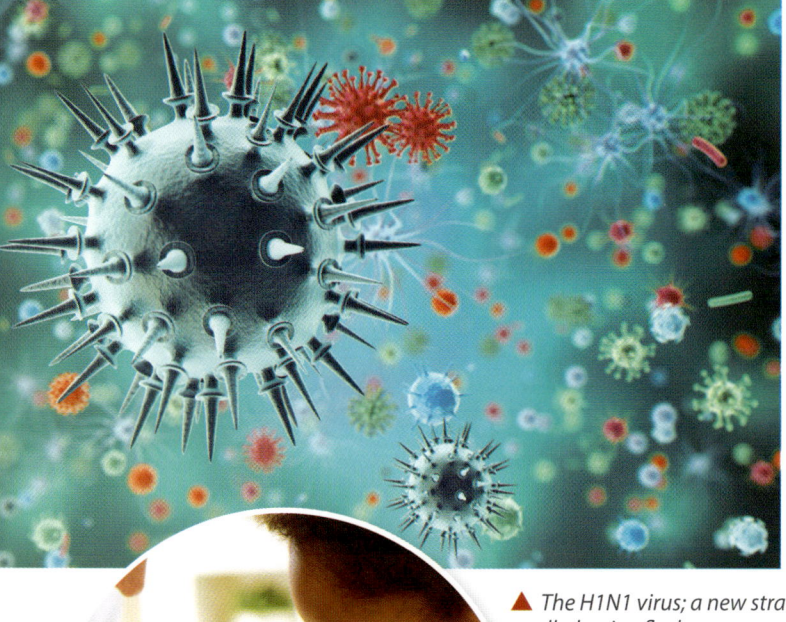

▲ The H1N1 virus; a new strain called swine flu, became a pandemic in 2009

Incredible Individuals

During the 2009 swine flu pandemic, the World Health Organisation had predicted that 2 billion people would get infected worldwide. Former US President Barack Obama got himself vaccinated to encourage people to get the vaccine. It was also a way to discourage a growing 'anti-vaccine' movement, where people believed that vaccinating themselves or their children would lead to autism.

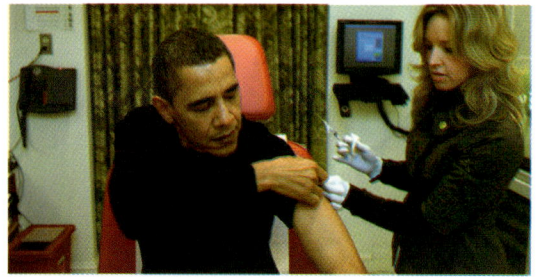

▲ Former US President Barrack Obama getting a vaccine injection

Measles

Fever, red and watery eyes, a dripping nose, and severe congestion of the throat and nose—these are the signs that somebody might have measles. It is easily confused with common cold and most people know it only when a rash develops on the skin, and bluish-white spots appear on the tongue. There is no effective cure for it, but babies who have been vaccinated will never get it. In the USA, Europe, and Australia, the disease has mostly been eliminated, though it exists in Africa and Asia.

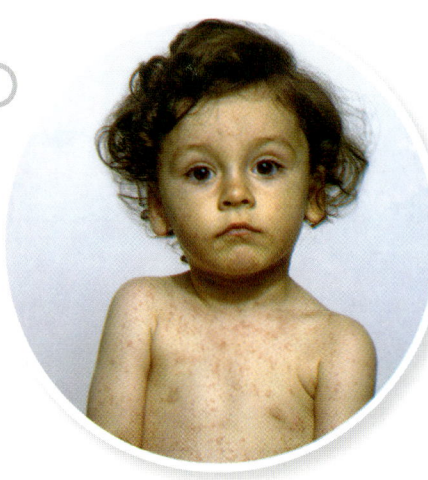

▲ Measles is the most infectious yet vaccine-preventable disease

◄ Flu masks are important for preventing the spread of infection

In Real Life

Psittacosis is a disease that you can get by inhaling germs from the dried-up droppings of pigeons and other birds. It can cause a cough, headache, and vomiting.

Reye Syndrome

Reye Syndrome is a condition in some children who have suffered from influenza or chicken pox. There is a higher risk if the child was treated with aspirin. It causes swelling in the brain, vomiting, and a feeling of tiredness all the time. Children with this syndrome feel confused and dizzy and may be left with permanent brain damage. In very serious cases, they may get seizures, inability to breathe, and may even go into a coma. Sadly, there is no cure for it.

Invaders of Your Lungs

The respiratory system has many defences against germs—the mucus in the lining of the inner nose, the pharynx, trachea, and bronchial tree trap germs. The cilia brush them out of the respiratory system. Special tissues called NALT and BALT, and immune cells in the bronchi act as spies against germs. They bring in more immune cells to destroy them. Yet, some bacteria and viruses, such as the influenza virus, can make it past them. Let us see what some of them do, and how the body fights back.

▲ Immune cells called macrophages in the alveoli destroy any germs that still make it to the lungs

▼ You shouldn't ignore your coughs. If you keep coughing for 2 days or more, visit the doctor

Tuberculosis (TB)

Historically called consumption, this disease is caused due to an infection of the lungs by *mycobacterium tuberculosis*. Once a major scourge of the world, it is on the wane, though still deadly. You can get it if you have a weak immune system, and come in contact with someone who has the disease already when they cough, sneeze, or talk to you without covering their face.

You may have TB if you have been:
- coughing for three weeks or more,
- coughing up blood,
- having fever and sweating at night,
- losing weight and not feeling hungry,
- feeling weak and tired.

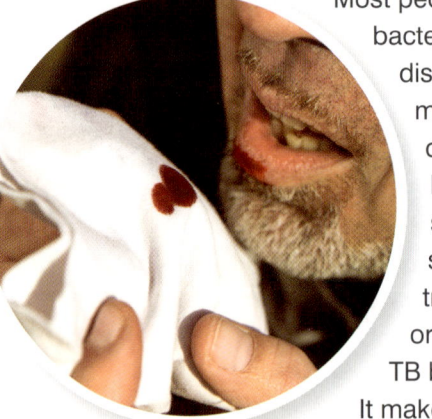

▲ Coughing up blood is a symptom of tuberculosis

Most people who get the bacterium will not get the disease again. But they may not be able to get rid of it either. This is called latent TB. If their immune system becomes weak, such as after an organ transplant, a major illness, or HIV infection, then the TB bacterium starts growing. It makes holes called lesions, in the lungs and other organs.

💡 Isn't It Amazing!

Cattle can also contract tuberculosis. Some scientists are of the view that human beings got tuberculosis from cows, especially in the Neolithic Period when they domesticated cattle. However, some other scientists disagree with this observation and think that they may have got it separately.

The Poets' Disease

In the 19th century, without a cure available, people died a slow death from TB. Poets and writers wrote of their impending death and suffering, from John Keats and P.B. Shelley—who died of it, to Leo Tolstoy and Anton Chekhov, whose characters usually had it. After Robert Koch discovered the bacterial cause of TB, and the BCG vaccine was developed later, literature became more hopeful again. TB is now curable through a combination of drugs that you need to take for six months to a year.

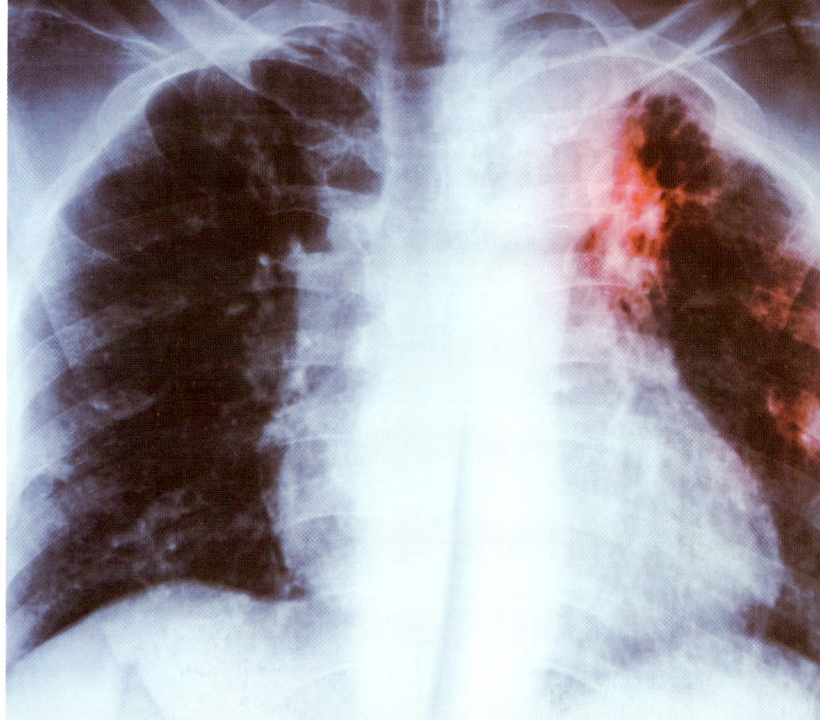

▲ An x-ray of the chest, showing TB lesions in the lungs (coloured in)

▲ John Keats caught tuberculosis from his brother, Tom

▲ P. B. Shelley had contracted TB, but was in remission

Pneumonia

Pneumonia is caused by many bacteria invading your lungs, especially *Streptococcus pneumoniae* or pneumococcus. It is very serious in smokers, old people, and children under the age of five, though others get it too. Pneumonia feels like a cold at the start; with coughs, fever, and wheezing. If the fever continues with chills, chest pain, and cough with phlegm, you may have gotten pneumonia. The disease gets severe if the bacteria enter the blood, from where they can get to the brain, causing meningitis.

What keeps colds away *(see pp 16–17)* also keeps pneumonia away. Doctors treat pneumonia with antibiotics and recommend proper rest.

▶ Wearing facemasks can help you avoid breathing in germs

Incredible Individuals

President William Henry Harrison (1773–1841) of the USA caught a cold on 26 March 1841. Unable to rest, his cold turned to pneumonia, and he died nine days later. He had only become president on 4 March.

▲ William Henry Harrison is the shortest-serving US president in history

Pertussis

Pertussis is also called whooping cough. It is a major cause of child deaths in some countries of Africa and Asia, though it has been nearly eliminated in the USA and Europe. It is caused by the bacterium *Bordetella pertussis*. Its symptoms are similar to that of a cold—a dry cough and fever. Then they get 'whoops', when they cough terribly followed by a whistling sound when they breathe. In small children, it may lead to periods when they cannot breathe at all. However, babies who are given the DPT vaccines never get it.

Exercises to Breathe Better

Stress, poor sleep, and lack of exercise affect the way you breathe. This is also true for diseases like COPD. Over time, stale air collects in your lungs and leaves you gasping for breath. By doing breathing exercises, you can clear out your lungs, while inhaling fresh air. This eases your muscles and brings more oxygen to your tissues.

Breathing exercises do not replace medical treatments, but can help ease discomfort and help healthy people avoid illnesses. They are a good way to take time off work or study, and relax your muscles. This practice of relaxation is called meditation. It is widely practised in Asian countries. Good ventilation, regular sleep, and a balanced diet also help your respiratory health.

▲ Breathing exercise are effective in dealing with depression, anxiety, and stress

Pursed Lip Breathing

Get to a calm, quiet place. Sit down comfortably and relax your arms and legs. Inhale through your nose. Exhale through your mouth, with the lips held close, for twice as long.

Pranayama

This is an ancient breathing exercise from India. Here is the simplest way to do it—put your finger on one nostril and breathe in deeply and slowly till your chest has expanded fully. Hold onto the air you have taken in for a minute, then exhale it slowly till your chest is compressed fully. Breathe in again. Pranayama helps remove CO_2 effectively, but you must do it only when guided by an instructor till you are trained enough.

◀ Pranayama is part of an Indian system of mind and body training called yoga

Belly Breathing

Lie down comfortably on your back and loosen your muscles. Keep your hands on your belly and press them down gently. Breathe in deeply and slowly, through your nose. Breathe out through the mouth for twice as long.

▼ Belly breathing helps exercise your diaphragm

In Real Life

Some people begin to breathe very fast when they are worried or excited. This is called **hyperventilation**. The effect of hyperventilation can be reduced in a person by practising the following exercise with them: ask them to breathe while you say slowly '1, 2, breathe in', and '3, 4, breathe out'.

Your Body's Sound Box

Just below the epiglottis, where the trachea begins, is your larynx. It is also known as the voice box. Made of cartilage and muscles, it has an interesting structure that allows you to make sounds. Human beings have among the most complex larynges, which allow them to make dozens of different sounds. They can string the sounds together to make words and sentences. This includes all the grunts and hisses you make when you are too busy to reply properly to someone. But did you know that your larynx develops from the same tissue that, in fish, becomes gills? Such organs are called evolutionary homologues.

Structure

If you feel your throat, you will realise it is hard, though not as hard as a bone. This is where the voice box is, made of three pieces of cartilage—the epiglottis *(see pp 6–7)*, thyroid cartilage, and cricoid cartilage. The thyroid cartilage is bigger in men and can be felt easily. It is called Adam's apple. The cricoid cartilage is thicker and ring-shaped. Three small pairs of cartilage are attached to the epiglottis—the arytenoids, corniculates, and cuneiforms—that help move the vocal cords.

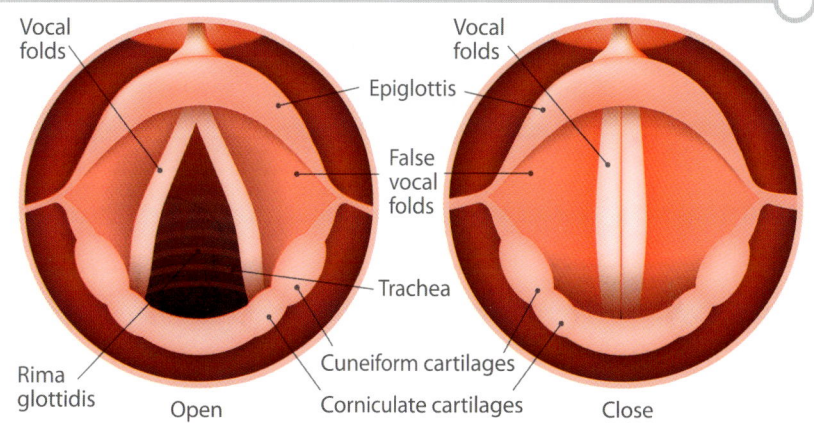

Below the epiglottis is the glottis, made of folds of tissue called the true vocal cords and the false vocal cords. True vocal cords are thin and membranous, held in place by the cartilages and vocal muscles. As air passes, they vibrate like guitar strings to make sound. Men have broader cords that give them a deeper voice, while women have narrower ones, so they sound shriller than men.

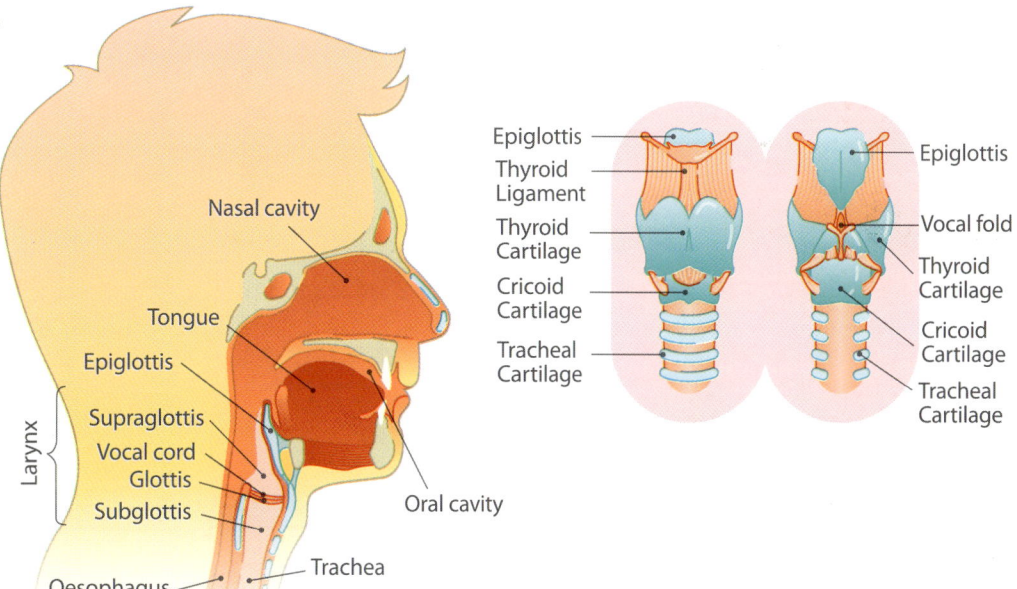

▲ *The vocal folds close when we speak, so that the air going out makes them vibrate*

In Real Life

As teenage boys grow older, their larynx cartilages become ossified. Bone minerals deposited in them make them harder, making the voice box more resonant. Their nasal sinuses also grow bigger, giving more internal echo and that is also why your voice sounds deeper to you than it does to others.

▶ *Adolescent male voice changes happen under the influence of the hormone testosterone*

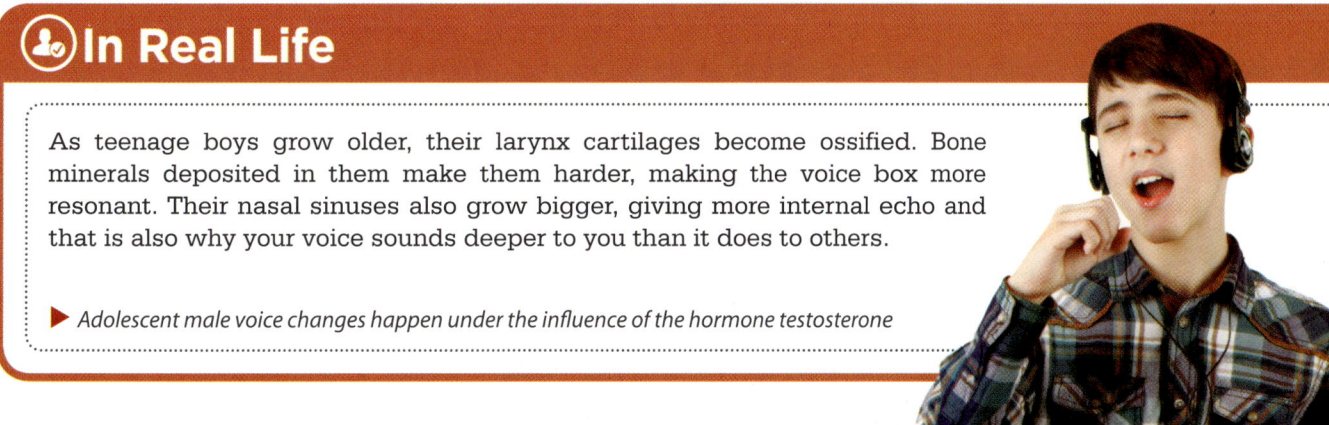

Making Words, Making Sense

Our larynx helps us make sounds, but how do we turn that into words and sentences? How does each language we speak have different sounds and different ways to combine those sounds into words and paragraphs?

👤 Larynx

It turns out the larynx is not just complex, but very finely tuned, like a piano, to make dozens of sounds. It interacts with our tongue, pharynx, mouth, nasal cavity, and even sinuses. This is how it makes sounds, pitches, and tones. It can change them ever so slightly to make the same words tinged with different emotions. So, the same sentence can be a question, a statement, sarcasm, or mockery.

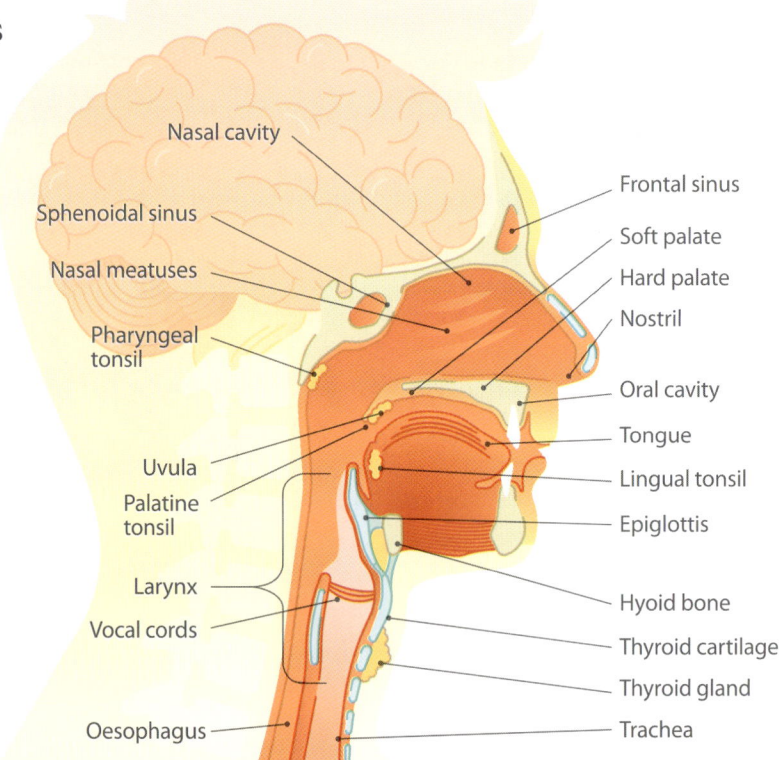

Different parts of the brain and respiratory system work together to make speech.

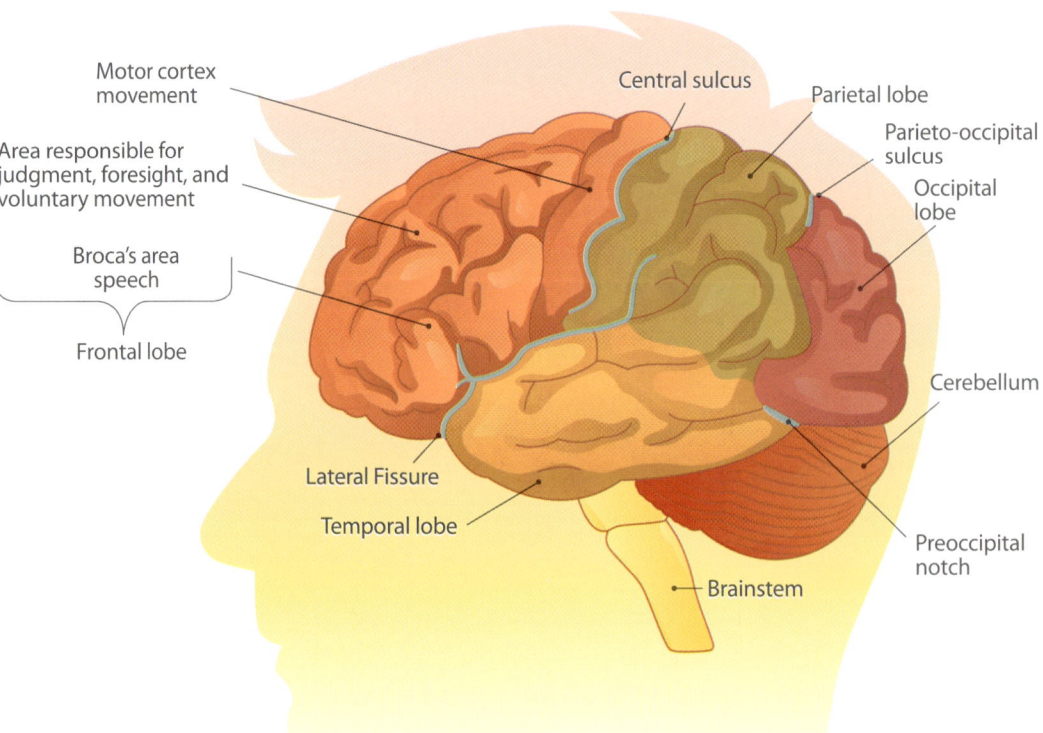

HUMAN BODY — LUNGS & RESPIRATORY SYSTEM

 Laryngeal Muscles
The laryngeal muscles work with the brain to make sounds. The cricothyroid muscle pulls the cricoid and thyroid cartilages to the front, stretching and thinning the vocal cords. This raises the pitch of the sound as we talk. Other muscles contract the vocal cords to make deeper sounds, and a whole lot of other adjustments.

 Vocal Folds
Your vocal folds are open when you breathe in. This allows air to go into your trachea and lungs. When you breathe out, the folds open and let the air out of your lungs. This lets you control your breathing when you are speaking.

 Pharynx
As the vocal cords vibrate, the air passing through them vibrates with the same frequency. This air then resonates in the laryngopharynx above, before passing into the mouth. This helps you make unvoiced sounds, like 'cut', 'cheap', 'tub', 'thin', etc. It also helps make rough sounds like 'hat' and 'heat', or smooth ones like 'at' and 'eat', by restricting or easing the flow of air.

Organ	Name	Sound	Example
Lips	Labial	p, b	**p**at, **b**at
Tongue and Teeth	Dental	θ, ð	**th**ick, **th**e
Tongue and palate (press hard)	Palatal	ch, j	**ch**oke, **j**oke
Tongue and palate (press gently)	Liquid	l, r	**l**oom, **r**oom
Back of throat	Velar	k, g	**c**rab, **g**rab

Tongue and Lips
Your tongue is very important in shaping the sounds that come out by the way it touches the pharynx, the roof of the mouth or the palate, and the teeth. Labial sounds arise when the air is forced through the lips and comes out as a burst.

 Diaphragm
It might be surprising, but your diaphragm also helps to make sounds. It turns your chest into a resonating cavity like the box of a guitar. It creates sounds like 'gut', 'jeep', 'dub', 'this', or 'bun'.

 Nose and Sinuses
The nose and its associated sinuses help you make the humming sounds 'm' and 'n'. To make the first, you keep the lips pressed and breathe out, you get a sound like 'mum'. Breathe out with your lips open and your tongue pressed to the teeth and you get 'nun'.

 Cheeks and Jaws
Your cheeks and jaws shrink or expand your mouth cavity, and help you make vowels.

	Jaws shut	Jaws open
Cheeks shut	b**oo**t	b**oa**t
Cheeks open	b**ee**t	b**ai**t

 Brain
The Broca's Area and Wernicke's Area in the cerebral cortex control speech and interact with the hippocampus and frontal cortex to regulate memory, vocabulary, grammar, and sentence formation. They send out motor signals through the midbrain and cerebellum to control how sounds are finally made.

Fluid in Your Lungs

Did you know that you could drown and die at the top of very high mountains? This is because of a condition called high altitude pulmonary oedema, in which the vessels in the lungs constrict, causing fluid to leak from the blood vessels to the tissues of the lungs. This condition occurs in lowlanders who ascend to higher altitude regions, usually above 2,500 to 3,000 feet, at a rapid pace. Let us find out what happens to a person when fluid enters their lungs, and how their life can be saved.

Pulmonary Oedema

Doctors refer to any tissue becoming full of plasma or fluid as oedema. Pulmonary oedema or water in the lungs, happens because of too much blood pressure, too little alveolar pressure—like at high altitudes, or damage to the capillaries of the lungs due to injury or infection. Plasma from the blood fills the alveoli. The pulmonary surfactant is disturbed, and oxygen does not dissolve anymore. The patient will make a gurgling sound as he breathes and feels heavy in the chest. Without urgent medical help, the patient may die within 10 minutes.

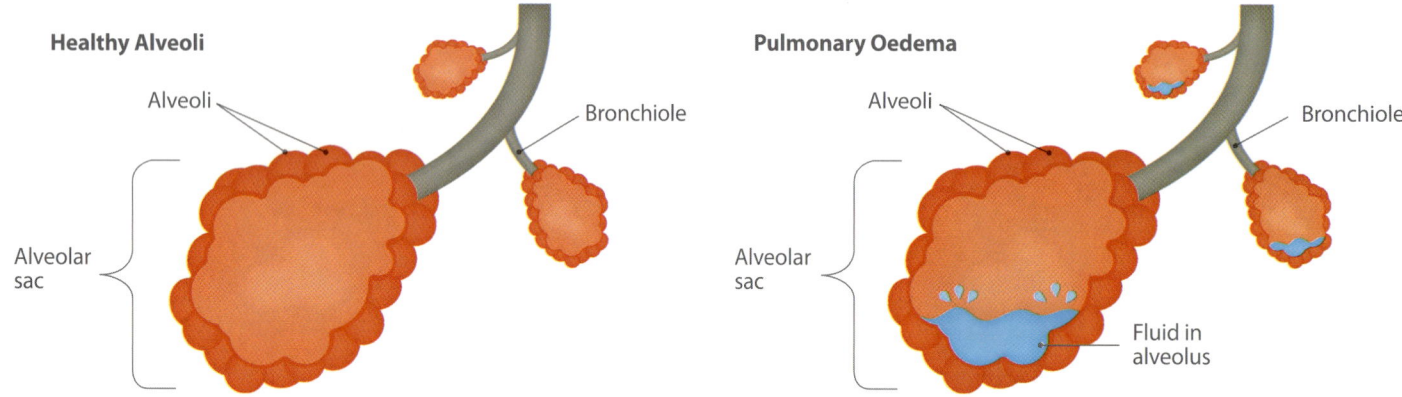

▲ Oedema is sometimes called internal drowning, but doctors do not use that term

Congestion

Congestion happens when blood enters the lungs due to injury or infection. It may happen because of high blood pressure or failure of the left side of the heart.

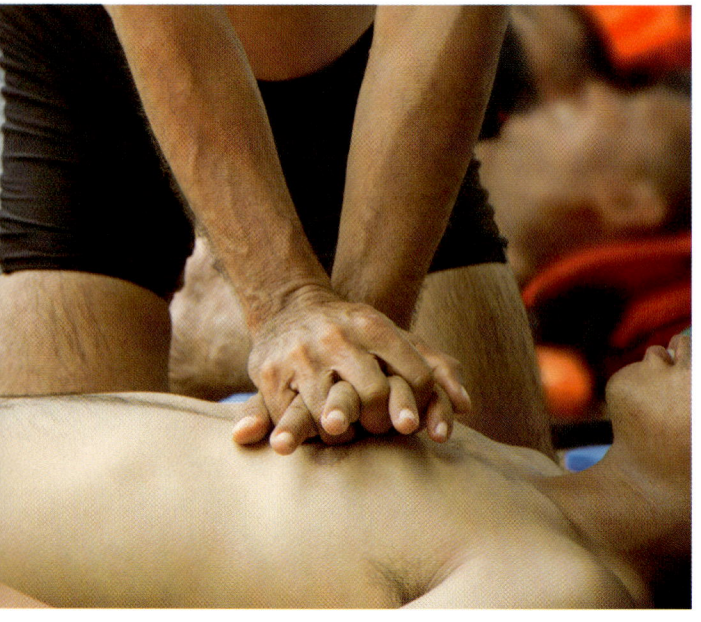

◀ Cardiopulmonary Resuscitation (CPR) is the first response for a drowning victim. Do not try to pump water from their lungs

Isn't It Amazing!

Diving into cold water triggers the mammalian diving reflex. Blood flows to the limbs and belly. More blood goes to the heart and brain. The heart rate also slows down.

Drowning

Doctors define drowning as the process of experiencing respiratory impairment from submersion or immersion in liquid. Unlike oedema, water does not enter the lungs when a person is drowning, because the body triggers the **laryngospasm**. In this, the vocal cords shut off the airway, if they sense water coming in. Instead, the person dies from lack of air if they remain underwater for too long. If they become unconscious, the laryngospasm may relax, and water may enter their lungs, causing oedema.

Bringing Someone Back to Life

Many people can be brought back from near death by a few techniques, known as resuscitation. While all medical professionals are trained to do them, a few are simple enough that you too can learn to do them. Be ready to help anyone who may be in need. But you have to be careful and do it right, so it needs some practice. Remember that some may not be suitable for children, who may be hurt rather than helped. Always remember to call the emergency number immediately after you have done the procedures, so that the patient gets the right medical help.

Heimlich Manoeuvre

This was invented by Dr Henry Heimlich for helping people who have swallowed something and are choking. If the object is not removed from their throat, they can die. Do this only if the patient cannot help themselves. Grab them from behind and make a fist of one hand, thumb inward. Put it on their belly, put the other hand over it and squeeze hard. It pushes the diaphragm up, forcing air out of the lungs. Whatever was choking them gets spat out.

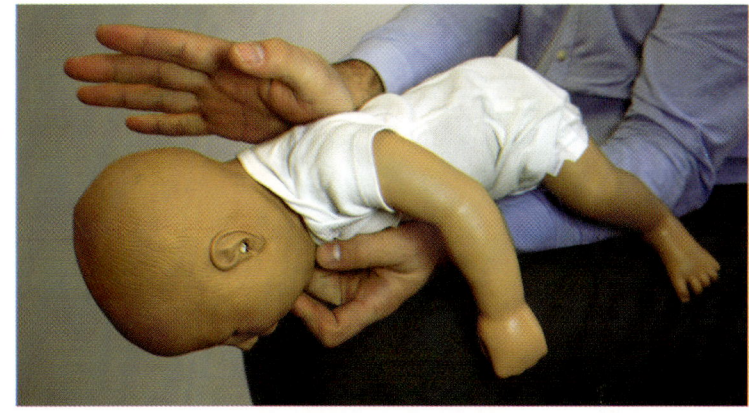
▲ Do not try the Heimlich manoeuvre on a healthy person. It can cause them cramps

Cardiopulmonary Resuscitation

Do this for someone who has fainted because of a heart attack. If their brain does not get oxygen soon, they can die. Lay them on the floor and press their chest down with both your palms, forcing them to gasp. Start compressions till they are able to breathe again, or until medical help arrives. If the victim still does not breathe, hold their mouth open, pinch their nose, and blow air into it. This is mouth-to-mouth resuscitation, sometimes called the kiss of life.

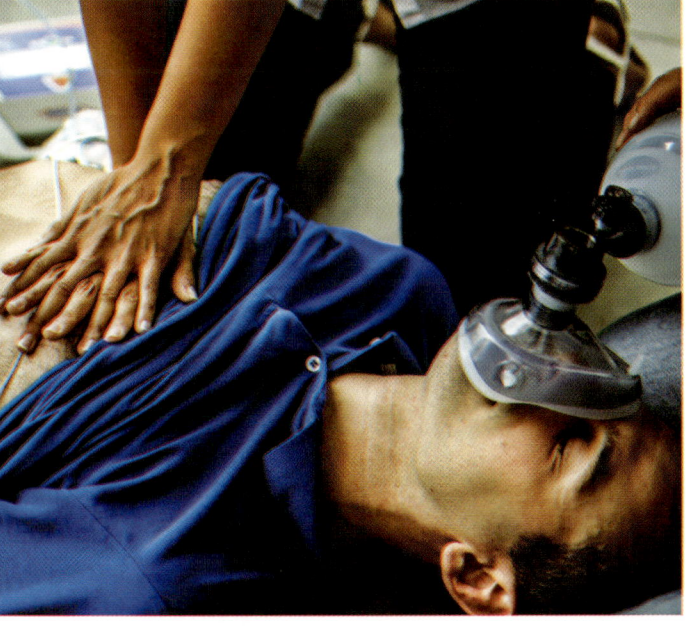
◀ Follow A-B-C during CPR: Open the airways, trigger breathing, and let circulation happen

In Real Life

In extreme cases, doctors may recommend oxygen therapy for people with COPD or viral diseases. During this therapy, the patient is made to breathe pure oxygen, as opposed to air, which only has 21 per cent of the gas. Oxygen is stored in a pressurised cylinder and delivered through a special breathing tube. It does not reduce the damage to the lungs and other tissues caused by pollution, smoking, or viral infections. But, it helps overcome shortage of breath and oxygen starvation in the tissues.

Incredible Individuals

When famous artist Pablo Picasso (1881–1973) was born, he was not breathing. His uncle and doctor, Salvador Ruiz Blasco, had taken a smoke break. He rushed in and blew the smoke into Pablo's face, which brought him to life!

▶ Picasso's first word was 'lapiz', Spanish for 'pencil'

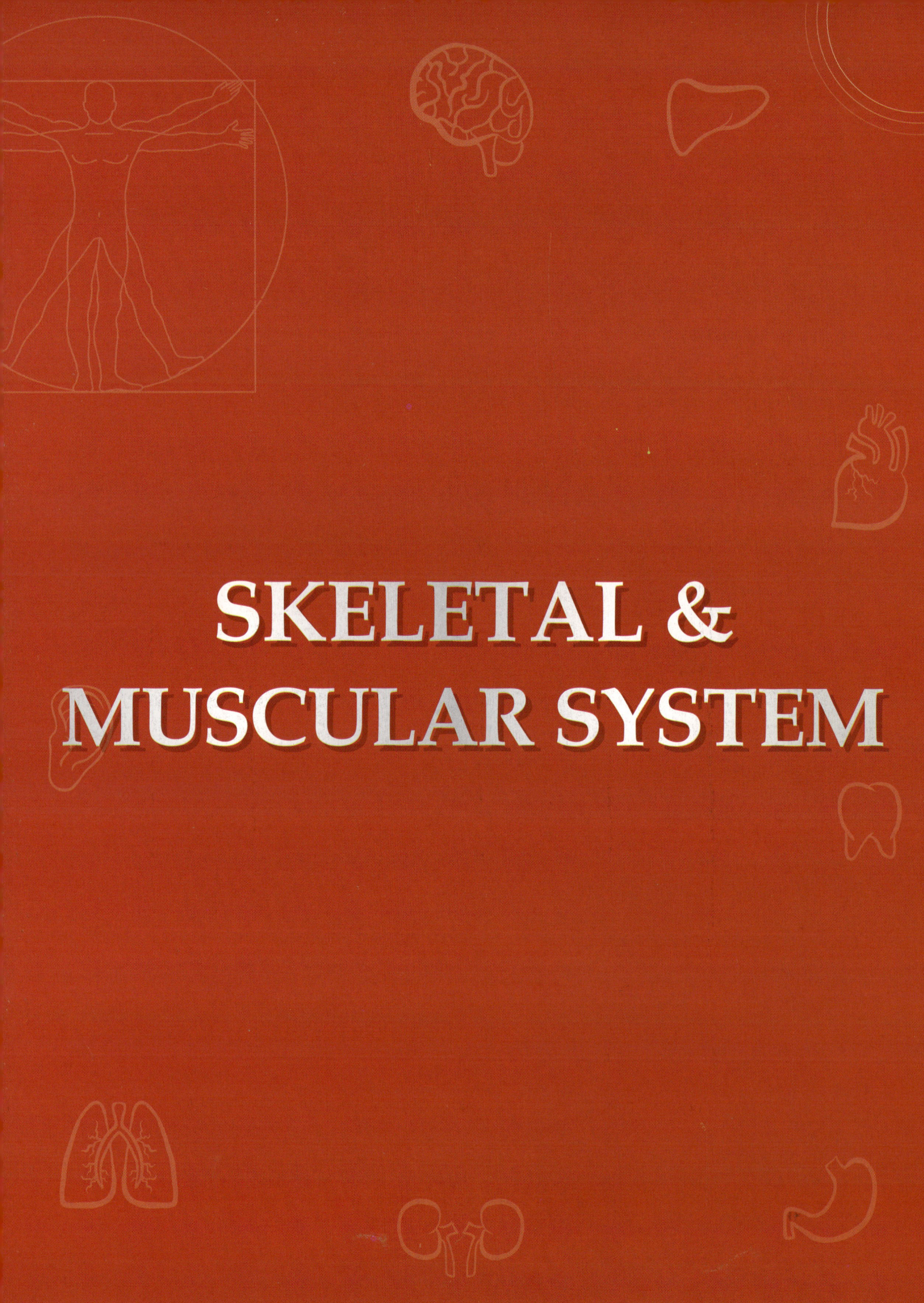

THE BONES THAT MAKE US

When you look into a mirror, you see the person that your bones and muscles made. You might notice that you are tall or short, stocky or gangly. In the long course of evolution, bodies needed to adapt to protect themselves better and become stronger, faster, and more agile. Muscles came first. You see muscles in animals as simple as nematodes. They help them move, find food, and mate with partners. They also help digest food.

Bones helped create a stiff **endoskeleton** to which the muscles and other organs could attach. Other creatures, like insects and clams have **exoskeletons**, which are hard coverings outside their bodies. But exoskeletons do not give them the freedom of movement of limbs that our bones give us. The earliest 'bones' were actually made of cartilage, and then animals evolved to add mineral deposits of calcium phosphate to make them harder.

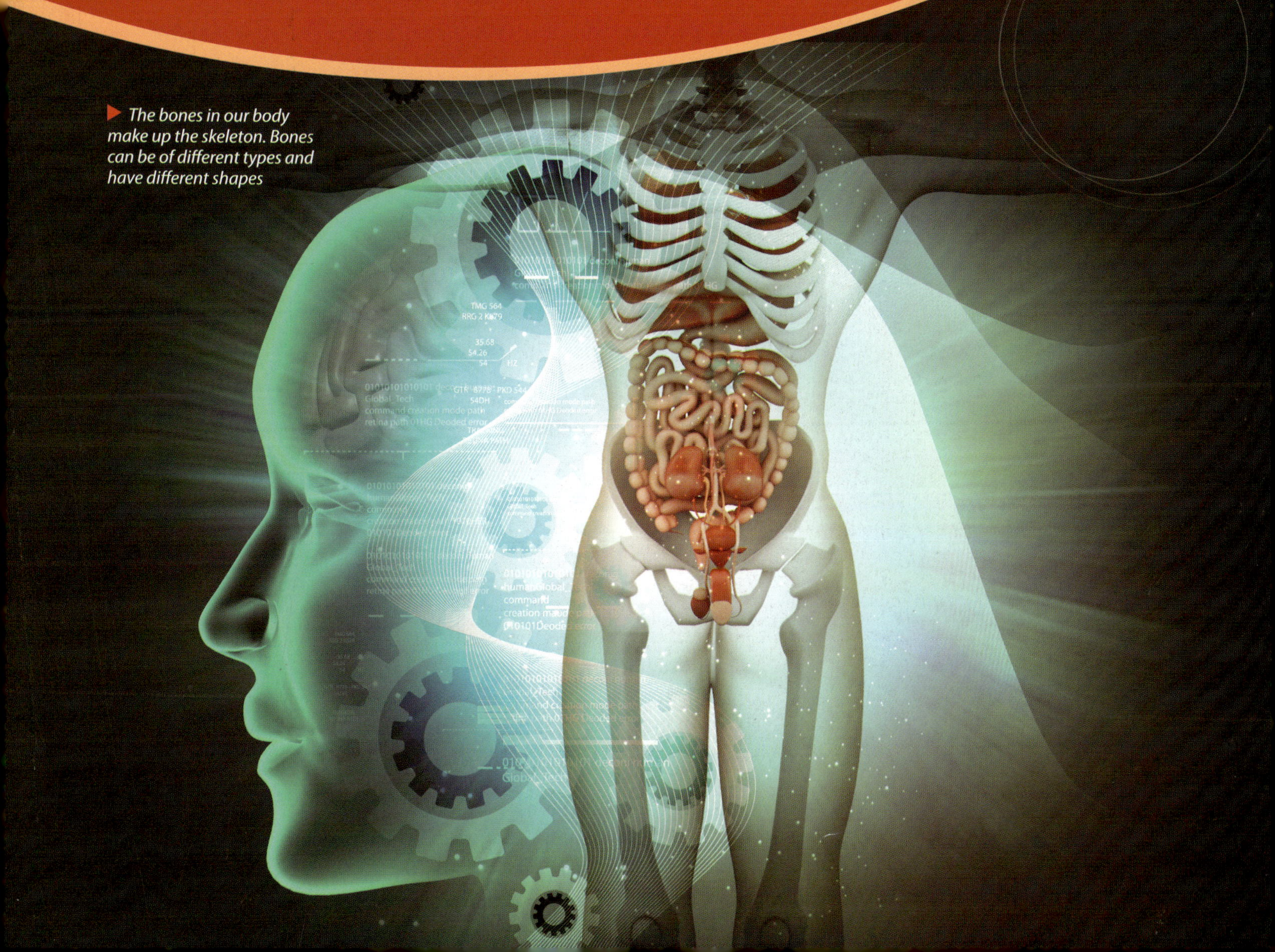

▶ The bones in our body make up the skeleton. Bones can be of different types and have different shapes

Protecting the Brain: The Skull

The skull has fascinated people from the beginning of time. It is a symbol of death and danger in many cultures. This could be attributed to the fact that the brain and the head are considered to be representative of a person. Brain damage can seriously impair one's life, and the functioning of any and all other parts of the body.

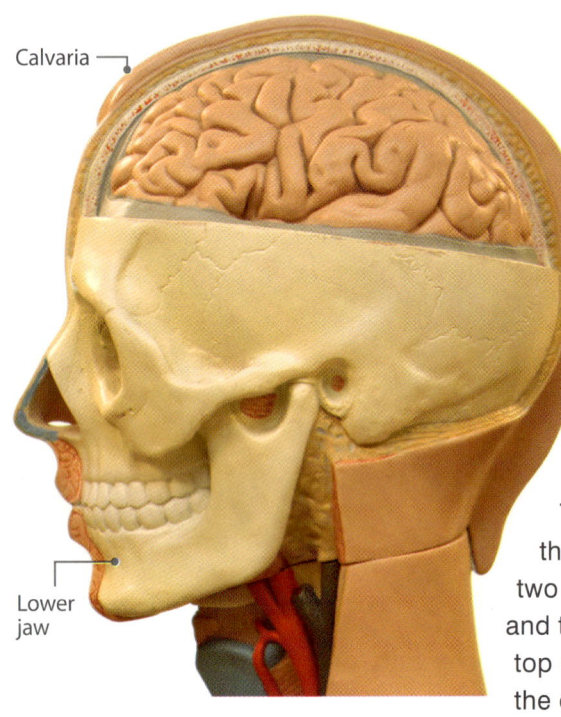

Calvaria

Lower jaw

The Brainbox

Human life wouldn't be the same without the skull. It protects the brain and the organs of the face, such as the eyes, nose, inner ear, and mouth. It is harder than most bones, so if you have a bad fall, your skull does not fracture easily.

The skull, also known as the cranium, is made of two parts—the brainbox and the facial bones. The top of the skull is called the calvaria, its sides are called temples, while the bottom is called the base. Large plate-like bones enclose the brain. They are the frontal bone—the forehead, a pair of parietal bones—at the back of your head, a pair of temporal bones—on the sides, and the occipital bone—at the back of your neck. They correspond to the lobes of the cerebrum.

The temporal bone has a hole to let in the ear canal and the mandibular fossa, into which the lower jaw fits. The occipital bone has a large hole in it, through which the spinal cord passes from the brain to the body. The sphenoid makes the floor of the brain box. It is full of small holes to let in cranial nerves and arteries and let veins out. The last is the ethmoid bone, which also makes the roof and septum of the nasal cavity.

In Real Life

The human brainbox—in proportion to the rest of the body—is among the largest in the animal kingdom. It has to be, for it evolved to have a brain that is also the largest in proportion to the body. The brain takes up over two-thirds of the skull.

◀ The illustration shows the size of the human brain compared to other animals

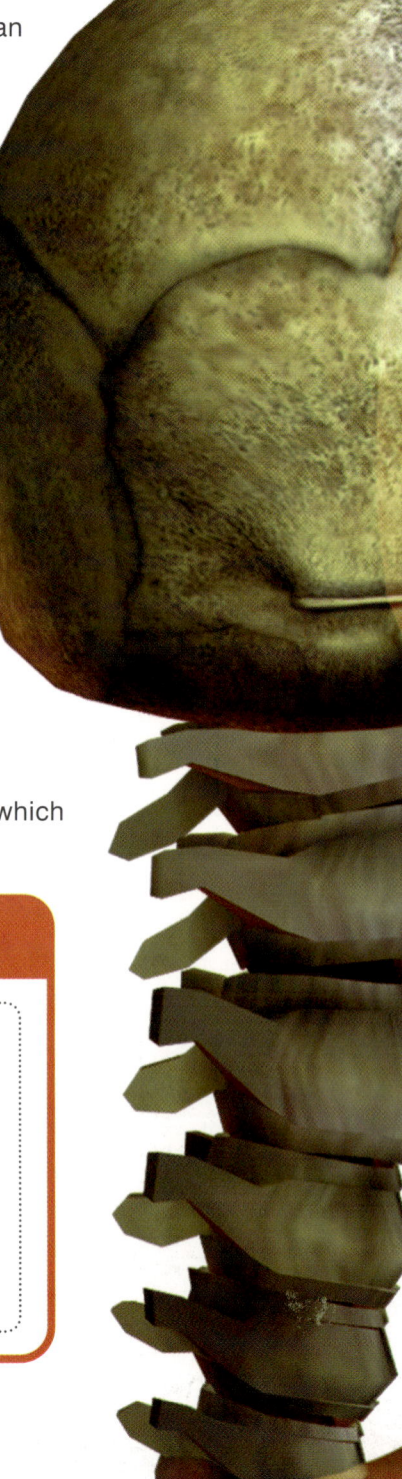

Incredible Individuals

Have you seen someone in a plaster cast? They might have needed it because of an injury that broke their bones. This technique of using plaster to grip broken bones in a plaster cast was invented by a Dutch army doctor named Antonius Mathijsen (1805–1878) in 1851. He realised that gypsum plaster, when mixed with water, would set very quickly, keeping bones in place.

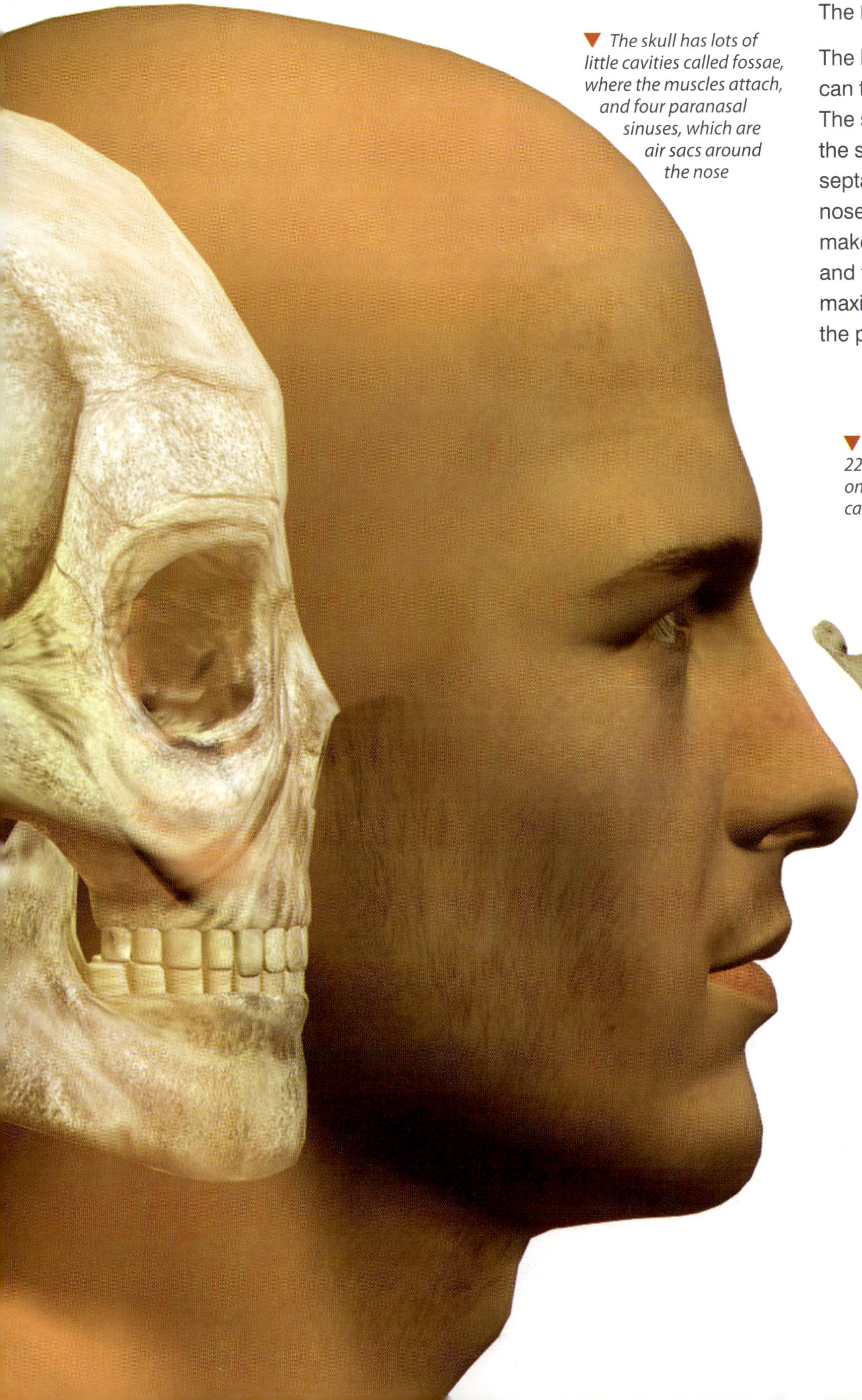

▼ The skull has lots of little cavities called fossae, where the muscles attach, and four paranasal sinuses, which are air sacs around the nose

Eyes and Nose

Your eyes sit in two cavities in the skull called the **orbit**. The frontal bone makes up the upper half of both the orbits. Your cheekbones, that is, zygomatic bones, make up its outer sides. The optic nerve passes through the sphenoid in their back. The tiny lacrimal bones, and parts of the ethmoid, make up the inner sides. The **maxilla** makes up its floor.

The bony part of the nose that you can touch is made of the nasal bones. The single vomer and the ethmoid make the septum, which divides the nostrils. The septal cartilage, which makes your visible nose, attaches to the ethmoid. The palate makes both the floor of the nasal cavity and the roof of the mouth. It is made by the maxilla, and a small pair of bones called the palatines.

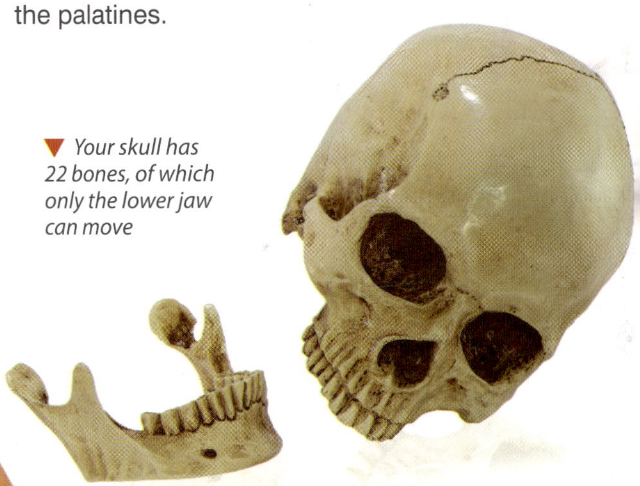

▼ Your skull has 22 bones, of which only the lower jaw can move

The Mouth

Your mouth is made of the jaws and the palate. The jaws contain your teeth. There are 24 milk teeth in children and 32 teeth in adults. Your upper jaw is called the maxilla, which is actually a pair of bones joined together. The lower jaw is called the **mandible**. It is attached to the temporal bones on both sides of the mouth. The hyoid is a special bone, which is not attached to any other bones. It is behind the mandible and the tongue attaches to it.

Stiffening the Back
The Vertebral Column

▼ Let's take a look at the vertebra

The vertebral column, also called the spine, divides the animal kingdom into two—those without it—**invertebrates** and those with it—**vertebrates**. Fish, amphibians, reptiles, birds, and mammals are vertebrates, with a bony spine. In some fish, like sharks and rays, the spine, like the rest of the skeletal system, is made of cartilage. It develops from the notochord, a stiff rod-like organ in the embryo.

 Spine

The spine is made of 33 bones called vertebrae (singular: **vertebra**). It protects the spinal cord, and provides an anchor for the ribs, the shoulders, and the hips. It also helps you flex your body in many ways, so you can run around and play with ease.

The spine has five regions. As you grow, the vertebrae of the sacrum merge into each other to form a single bone which becomes part of the hip. The four tail vertebrae also merge to form the tiny coccyx, which curves inwards. Animals with tails have many more vertebrae in them.

 **In Real Life**

Sometimes the gel-like part of the intervertebral disc gets crushed and comes out of the spine. This is called a slipped disc. It causes a lot of pain.

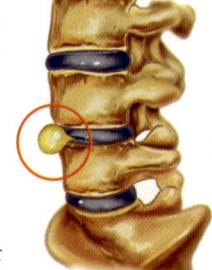

▶ An illustration of slipped disc

Region	Type of Vertebrae	Number	Names of Vertebrae	Attaches to
Neck	Cervical	7	C1–C7	The skull
Upper Back	Thoracic	12	T1–T12	The ribs
Lower Back	Lumbar	5	L1–L5	The sacrum
Hip	Sacral	5	S1–S5	The hip bones
Tail (Coccyx)	Coccygeal	4	Tailbone	Bottom of the sacrum

HUMAN BODY SKELETAL & MUSCULAR SYSTEM

Vertebra

Each vertebra has two parts—an anterior segment or the vertebral body, and a posterior part or the vertebral arch. Each vertebra has 'processes', which is an outgrowth of tissue or cell. Between the bodies of two vertebrae are **intervertebral discs,** made of cartilage. The outer part of each is fibrous, while the inner part is jelly-like. These discs act as cushions when the vertebrae bend backwards or forwards, so that their bodies do not rub against each other. The vertebral arch extends in two arms that meet behind the vertebral foramen. The foramina of all the vertebra form a tube-like canal, through which the spinal cord passes.

Finally, the processes act as places where skeletal muscles attach. Vertebrae in different regions of the spine have different processes. The spinous process sticks out of each vertebra and points downwards. This stops you from twisting yourself too much and dislocating your spinal cord. The gaps between the vertebra are also spaces from which the spinal nerves pass from the spinal cord to various parts of the body.

◄ The S-shape of the spinal cord allows us to walk and run

▲ The segmentation of the spine gives your body incredible flexibility

Regional Vertebrae

Cervical vertebra has extra foramina, through which the arteries to the brain pass. The first of them (C1), is called atlas. It bears the weight of the skull. Its body is hollow. The second (C2), is called axis. Its body has a pivot called the dens, which fits into the hollow of atlas. This is what lets you turn your neck around.

The thoracic vertebra has processes to the side called transverse processes. Both they and the vertebral arch have dimples on both sides called facets. Here is where the ribs attach. Lumbar vertebrae have additional processes called articular processes, where muscles can attach. The fused arches of the sacral vertebrae make up an ear-like surface, to which the hip bones attach.

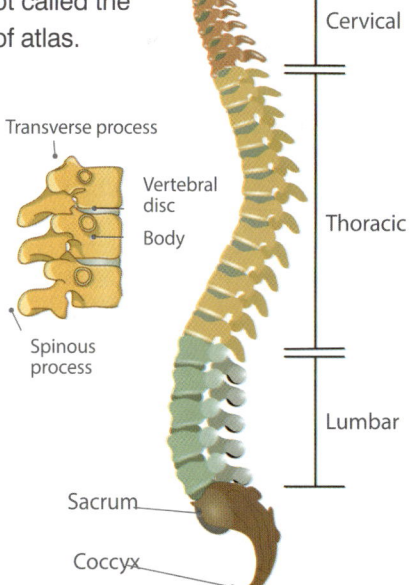

▼ The five regions of the spine

Isn't It Amazing!

The bony plates of a stegosaurus were not part of its spine but grew from the skin covering the spine.

◄ Skeleton of a stegosaurus

At the Heart of It
Shoulders and Ribs

Your chest has two of the most important organs of your body, your lungs and your heart. As we evolved, our skeleton developed a tough protective cover for them—the ribcage or **thoracic basket**. It has your ribs, as well as the breastbone or **sternum**, which protects your heart and lungs from the front. The ribs end in the 12 thoracic vertebrae of the spine.

The top of the ribcage is covered by your collarbones, or clavicles, while the back is covered by the shoulder blades, or scapulae. These two make up the **pectoral girdle**, which allows you to move your arms and shrug your shoulders.

Pectoral Girdle

The clavicle is the easiest bone to figure out. You can feel it starting from the base of the neck to the beginning of the arms. The two clavicles make movable **joints** with the breastbone, so you can move your arms forward or back. When you move your forearms back, you can feel your shoulder blades.

Three powerful muscles attach to the pectoral girdle. The levator scapulae attaches it to the base of the skull and the cervical vertebrae. The rhomboid major and rhomboid minor muscles attach it to the thoracic vertebrae. Together, they allow the shoulder to bear a great amount of weight, whether on the back or the front.

▼ Close-up view of the pectoral girdle

Acromioclavicular joint
Clavicle
Glenohumeral joint
Scapula
Humerus
Scapulothoracic joint

▲ Clapping hands and flapping wings—both actions require the pectoralis major and the pectoral girdle

Breastbone

This is a very tough bone in the front of your chest, and it looks like a bony tie. It is also called the sternum. The upper part is the manubrium, to which the clavicles and the first pair of ribs attach. The lower part is the sternum proper, to which the remaining true ribs (2–7) attach. The pectoralis major muscle connects the breastbone, ribs clavicle, and scapula to the forearm. Heart surgeons have to break the breastbone to reach the heart during surgery.

Isn't It Amazing!

In birds & bats the breastbone is huge, compared to the body. This in turn helps the pectoralis major become powerful enough to flap the wings to make them fly.

◀ A bird's skeleton is notable for both its strength and lightness

The Ribs

You have three kinds of ribs. Ribs 1–7 are true ribs as they attach to thoracic vertebrae T1–T7 and to the sternum. Ribs 8–10 are false ribs, named so because they do not attach directly to the sternum, but to rib 7. Ribs 11–12 do not attach to the sternum at all and are called floating ribs.

The head of the rib is the part that attaches to the arch of a thoracic vertebra. A little bump near the head, called the tubercle, attaches to the transverse process. After this, the rib turns sharply toward the chest, forming a C-shape. The ribs do not attach to the sternum directly, but taper into cartilage, which attaches them to the sternum. This allows the rib cage to expand with the lungs when you breathe in air, and contract with the lungs when you breathe out.

◀ The 7 true ribs attach to the first 7 of the 12 thoracic vertebrae

▲ Ribs sticking out of the chest shows that a person is severely underweight

Inside Your Bones

Our skeleton comprises our bones. But what exactly exists inside each bone? Even though they often feel like nothing but hard, white rods, bones, in fact, have a complex internal structure. Not all bones are the same. Some are much harder than the others, like the skull plates. Other bones are spongy and lighter than they feel. Many bones are hollow inside. However, they are filled with a fatty substance called **marrow**. Did you know that this marrow is the place where your blood cells are born?

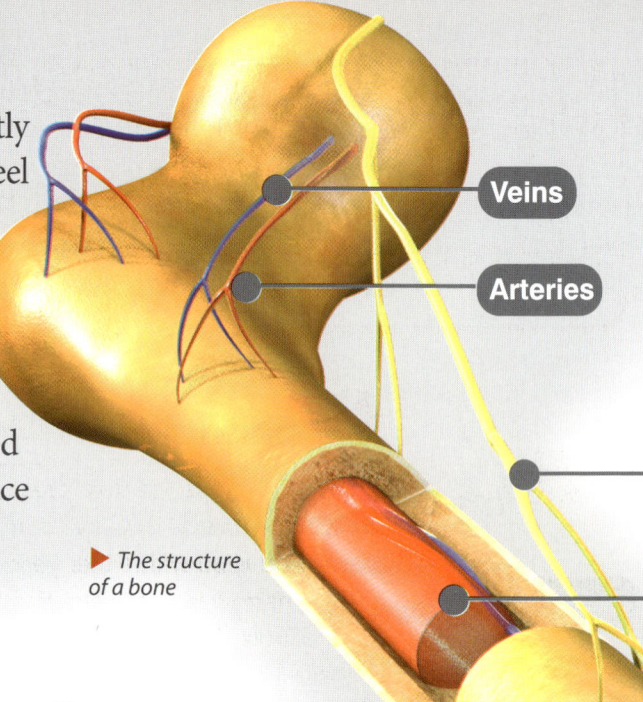

▶ The structure of a bone

The Making of a Bone

We start out as babies with a cartilage skeleton. Each 'bone' has a tissue that covers it, called the periosteum. Special cells, called osteogenic cells, migrate to the bone being formed. Some stay as they are until the day that the bones need to be repaired. Most will become **osteoblasts**, which are cells that make **collagen**. Collagen is a protein that forms threads easily, and is seen in **tendons** and **ligaments**, as well as in hair and nails. In the bones, it forms a three-dimensional 'matrix' of woven fibres.

Osteoblasts fill this matrix with calcium hydroxyapatite, a mineral compound made of calcium and phosphorus ($Ca_{10}(PO_4)_6(OH)_2$). The bone matrix may be filled completely with this compound to make compact bone, or leave air gaps to make spongy bone. As the osteoblasts lay down minerals, they get trapped in it and become osteocytes. Osteocytes keep the bone nourished and lay down more mineral if needed.

As we grow, so do our bones. Special cells called **osteoclasts** help in this process. They 'resorb' the minerals so that the bone can become wider and longer, before osteoblasts lay down fresh matrix.

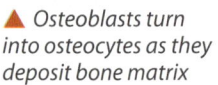

▲ Osteoblasts turn into osteocytes as they deposit bone matrix

◀ Bones do not stay the same throughout life but are constantly remodelled to grow in size and weight as you grow

💡 Isn't It Amazing!

Did you know that shark bones are actually made of cartilage? Cartilage is a light, flexible material made of chondroitin sulphate, hyaluronic acid, and collagen fibres. It is also found where bones meet to form a joint.

▶ Sharks are called cartilaginous fish

HUMAN BODY | SKELETAL & MUSCULAR SYSTEM

Incredible Individuals

Scientists believe that the ancestors of our species (*Homo habilis*) first started out as scavengers. They ate meat left behind by other predators. They started making tools out of stone to get to the marrow of left-behind bones, and that was one of the first tools that humankind made.

Bone Marrow

Many bones, like the vertebrae, ribs, sternum, hip bones, and bones of the arm and leg are hollow inside and filled with either yellow or red marrow. Yellow marrow is a jelly-like tissue made of fat-storing cells. Red marrow is more complex and is made of stem cells that make the cells of the blood—Red Blood Cells (RBCs), White Blood Cells (WBCs), and platelets.

Until you are seven years old, almost all your marrow is red. After that, most marrow becomes yellow and stops making blood. But if a bad injury or fever with lots of blood loss occurs, yellow marrow can become red again.

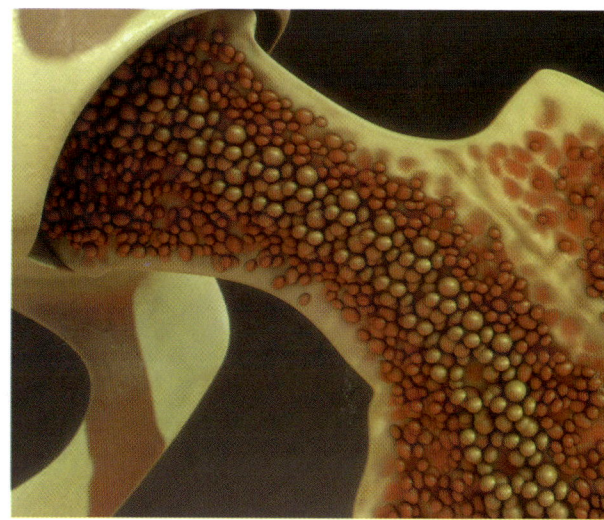

▲ A 3D illustration of bone marrow

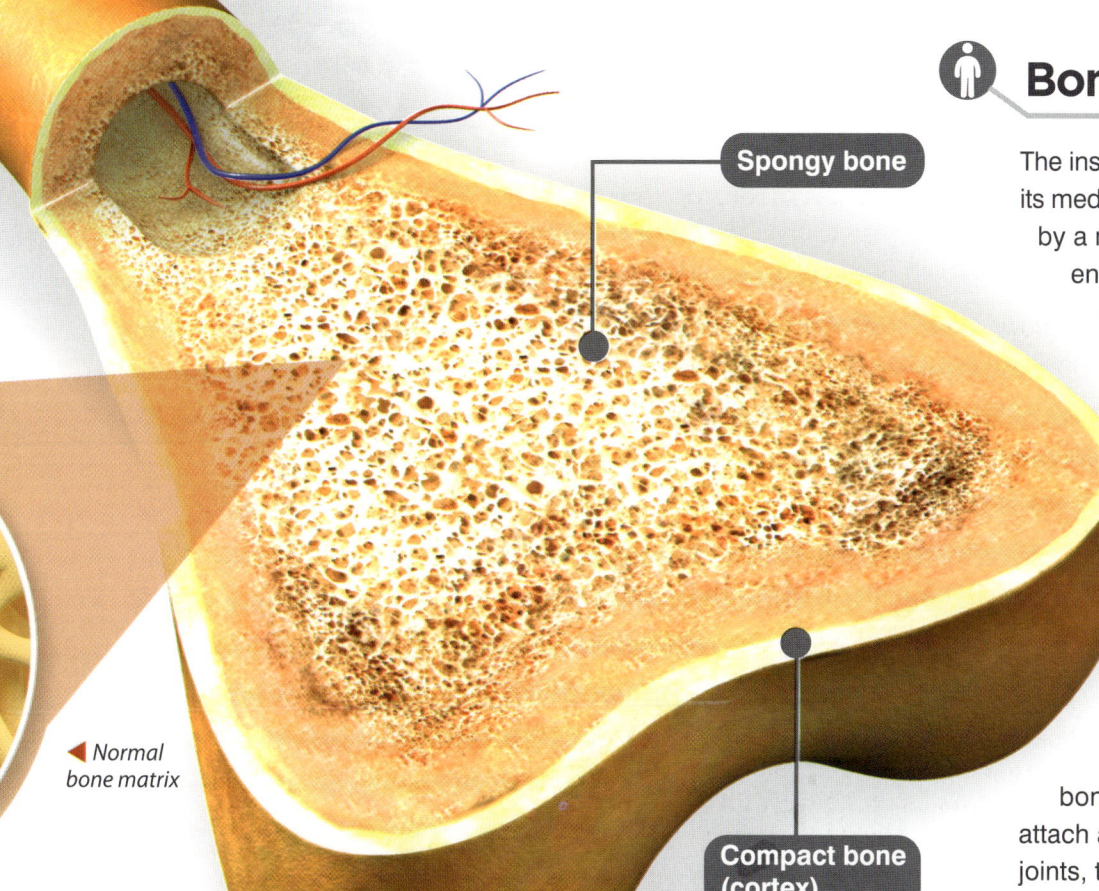

◀ Normal bone matrix

Bone Membranes

The inside of a bone is called its medullary cavity. It is lined by a membrane called the endosteum. It contains the bone cells that help the bone grow and also repair it. The outside of a bone is also covered by a membrane called the periosteum. Blood vessels, nerves, and lymphatic vessels come to the periosteum, from where nutrients diffuse into the bone. Tendons and ligaments attach at the periosteum. At the joints, the periosteum is replaced by articular cartilage.

The Bones that Keep Us Going

Every animal needs its arms and legs to move from one place to another. The bones of the arms—**humerus**, **radius**, and **ulna**; and those of the legs—**femur**, **tibia**, and **fibula** have knobbed endings and long, cylindrical shafts that help in our movement. Let us see how they fit into the limbs, and how they work.

Upper Arm and Forearm

Your arm is made of three parts—the upper arm, the forearm, and the hand. The upper arm extends from your shoulder to your elbow. It has only one bone—the humerus. It is attached to some very powerful muscles like the latissimus dorsi in the back and the pectoralis major in the chest, which help raise or lower your arms, and move them forward or back. It also has two major muscles that run alongside it, the **biceps brachii** and the **triceps**. The biceps pull the forearm towards you, while the triceps extend it again.

The head of the humerus bone meets the shoulder blade to form the glenohumeral joint. At the other end, it has the trochlea, a pulley-shaped area that joins with the ulna, and the capitulum (little head) that joints with the radius. Together, these make the elbow.

The forearm has two bones, the shorter radius and the longer ulna. They join at the elbow to make the proximal radioulnar joint, and at the wrist to make the distal radioulnar joint. These let you rotate the forearm. The tricep attaches to the ulna, while the bicep attaches to the radius.

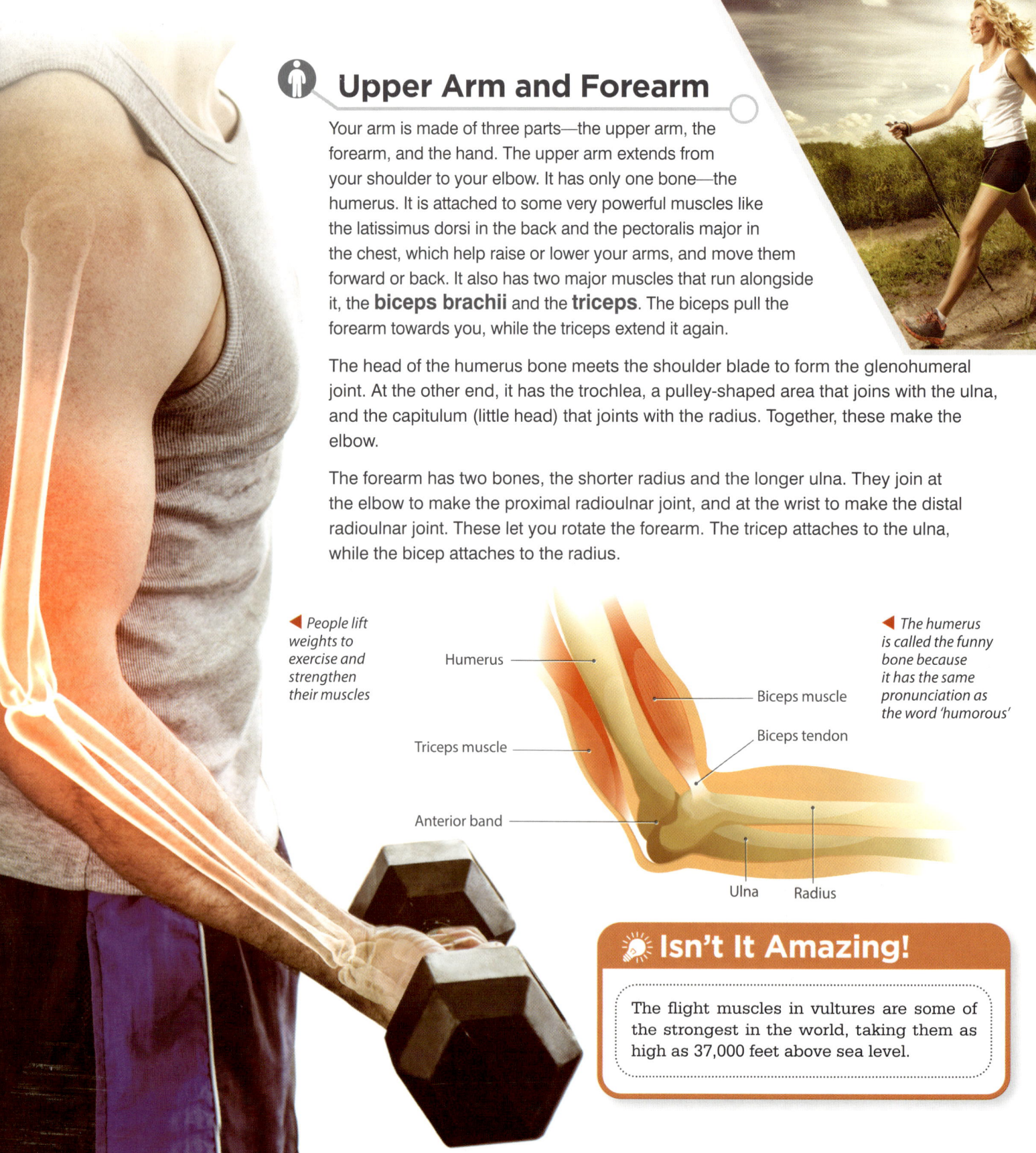

◀ People lift weights to exercise and strengthen their muscles

◀ The humerus is called the funny bone because it has the same pronunciation as the word 'humorous'

Humerus
Biceps muscle
Triceps muscle
Biceps tendon
Anterior band
Ulna   Radius

Isn't It Amazing!

The flight muscles in vultures are some of the strongest in the world, taking them as high as 37,000 feet above sea level.

HUMAN BODY | SKELETAL & MUSCULAR SYSTEM | 137

◀ Our nervous system helps the bones and muscles of the arms and legs coordinate with each other, and also with the spine, for smooth body movements

In Real Life

The femur bears the weight of the body when we stand. It can bear a load of up to 1,100 kg.

▶ The position of the femur and tibia while the body is in action

Thigh and Foreleg

Like your arm, your leg too is made of three parts—the thigh, the foreleg, and the foot. The thigh extends from your hip to your knee and has only one bone in it—the femur. It is attached to some of the most powerful muscles in your body. These are the **gluteus** muscles in the back and the adductor muscles in the front, which raise or lower your thighs, and move them forward or backward for walking. It also has two major muscles that run alongside it, the biceps femoris and the quadriceps. The biceps pull the foreleg towards you, while the quadriceps extend it again.

The head of the femur meets the hip to form the hip joint. At the other end, it has the epicondyles, a pulley-shaped area that joins with the tibia to make the knee. It does not meet the fibula. A small bone called the patella caps the knee.

The foreleg has two bones, the shorter, fibula, and the longer, tibia. They join below the knee to make the proximal tibiofibular joint, and at the ankle to make the distal tibiofibular joint. These let you rotate the leg. The calf muscles run alongside them, from the femur to the anklebone.

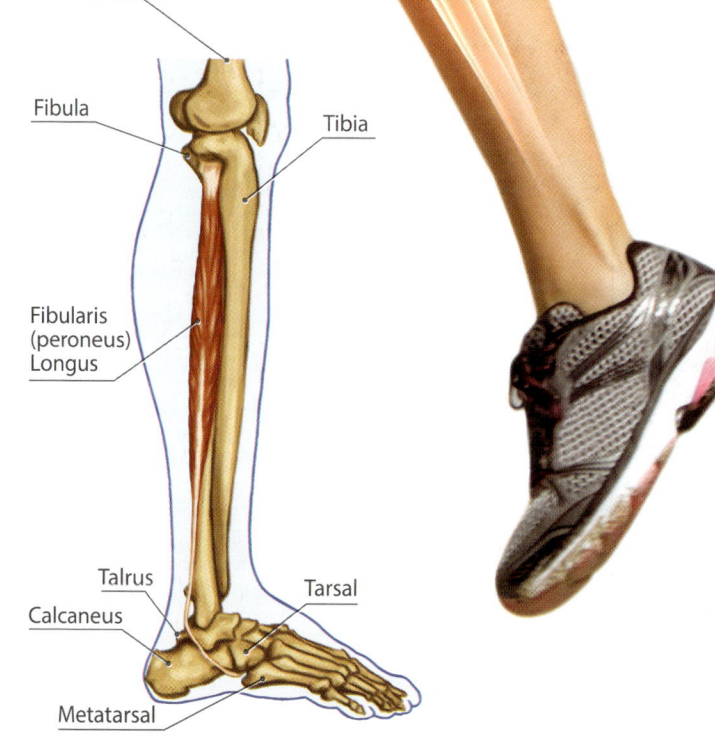

▲ The femur and tibia are the longest and second-longest bones of your body, respectively

In Your Hands and Feet

The bones of the hands and feet are similar in structure, because we evolved from animals that used them both for walking. But as they evolved to do utterly different things, small changes happened over millions of years to make them look quite unlike each other. The biggest difference was the change in the way the thumb works. It can 'oppose' the other fingers, which helps you pick up things, hold chopsticks, write letters, or catch a basketball. The big toe on the foot cannot do all this, but it helps you balance while you stand on tiptoe, as your hands reach out for a jar from the top shelf.

Isn't It Amazing!

Chimps have thumbs too, but these are not as flexible as ours. We owe it to the flexor pollicis longus, the muscle in our thumbs that makes it possible to pinch and grip the tiniest of things.

▶ Chimps also have opposable big toes that help them climb trees

The Hand

While the arm has only 3 bones, the hand has 27. All these bones give the hand the flexibility it needs to do the hundreds of things that you do in a day. You can handle a mobile phone, type on a computer keyboard, use a spoon and fork, climb into the school bus, hold your books, or even bite your nails.

Each finger has three **phalanges**, while the thumb only has two. They are jointed so you can make a grip, but you cannot bend them outwards. If you feel the back of your hand, you can make out the five **metacarpals**, which have the little muscles that help you move your fingers. The wrist is made of eight **carpals**. These little bones give the wrist an amazing level of flexibility as they make many joints and attach many little muscles, so you can rotate your hand, move it sideways, or up and down. One of the carpal bones, which is located beneath the thumb joint is known as the trapezium bone.

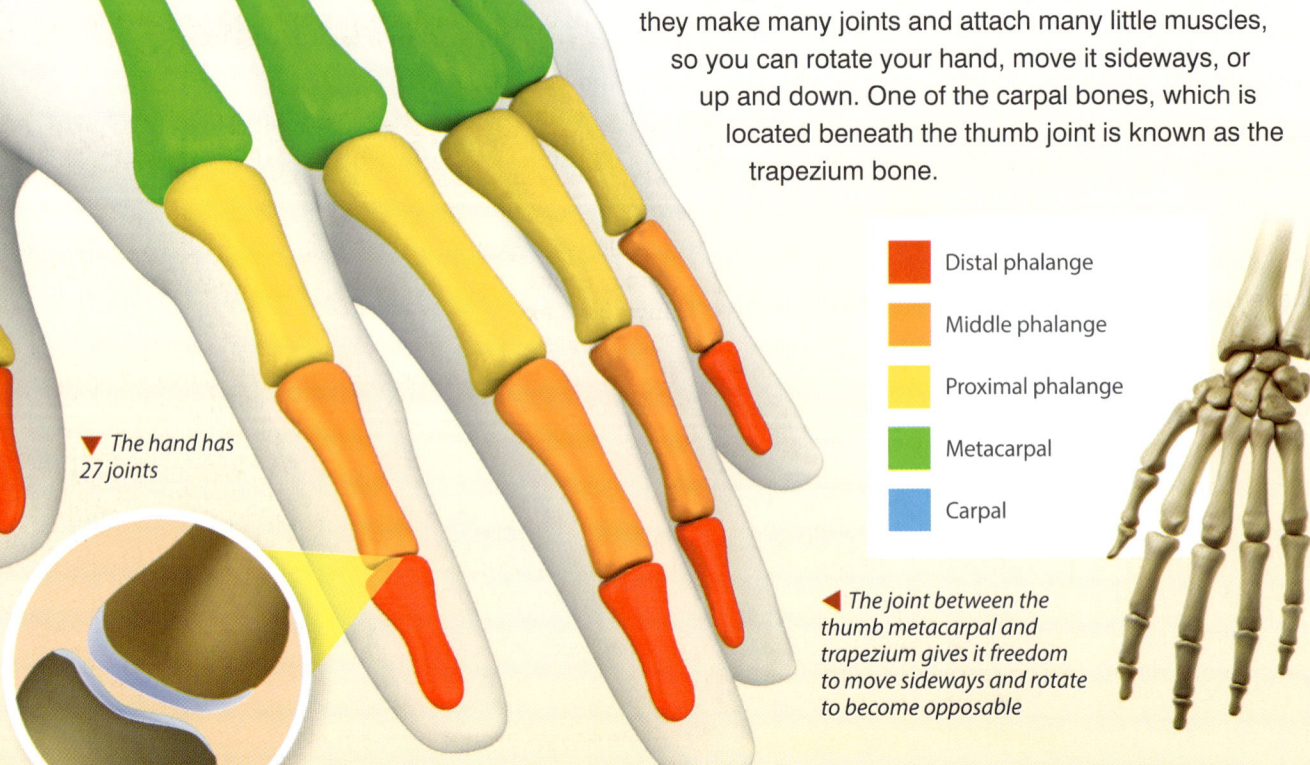

▼ The hand has 27 joints

- Distal phalange
- Middle phalange
- Proximal phalange
- Metacarpal
- Carpal

◀ The joint between the thumb metacarpal and trapezium gives it freedom to move sideways and rotate to become opposable

HUMAN BODY | SKELETAL & MUSCULAR SYSTEM

The Foot

The foot has 26 bones. Unlike the hands, these bones do not make the feet flexible. But they help them bear the enormous load of your body, whether you are standing, walking, or running.

Each toe has three phalanges, but only the first and part of the second are free. The rest are locked inside the foot. Like the metacarpals of the hand, the **metatarsals** make up the sole of the foot. They are long and make up the 'flat feet' that help you balance while standing. The short, angular bones, that make up the ankle are called **tarsals**. The seven tarsals help transfer the weight of the body to the ground, and also help with rotating your feet, and moving them up, down, or sideways. The largest tarsals are the talus, which makes your ankle bone, and the calcaneus, which makes your heel. The calcaneus, tarsals, and metatarsals make up the longitudinal arch of the foot. The arch absorbs energy when the foot is pressed down when walking and releases it when the foot is raised—giving you the 'spring' in your step.

- Calcaneus
- Talus
- Navicular
- Cuboid
- Cuneiforms

▲ The illustration shows the bones of the flat foot

▶ The bones of the feet have powerful ligaments attached to them, which act as shock absorbers

Becoming Bipedal

Bipedal refers to animals that walk on two feet. Among the apes, we are the only ones to have become exclusively bipedal. For this, the bones of the legs had to be modified in three ways. First, the femur, fibula, and tibia modified to form a pillar upon which the body's weight is balanced while standing. Second, the bones of the heel modified to act as a counter-balance to those of the feet. As we walk, the body transfers its weight from one leg to the other through the thighs and heels. Third, the joints modified to allow the legs to stride without buckling while walking.

Incredible Individuals

Achilles, a legendary Greek hero, could not be hurt anywhere except in his heel. The phrase "Achilles' heel" now means an unsuspected weakness.

▶ According to the myth, Achilles' mum dipped him in the Styx—a mythological Greek river—when he was a baby, to make him immortal, but since she held him by the heel, that part remained mortal

Some Very Hip Bones

Today, most people lead a sedentary lifestyle. You might find yourself sitting a lot while writing tests at school, playing mobile games in the bus, having dinner, watching TV, or doing homework. Our hip takes all the stress of this sitting. Originally meant to support the spine and the hindlegs, your hip bones evolved to attach powerful muscles. In women, they also need to bear the weight of a whole baby as it grows within them. Then the baby comes out through the birth canal, parting the hip bones.

▼ The gluteus muscles attached to the hip provide both strength and flexibility

Structure

When you are growing up, your hip is made of three pairs of bones—the large ilium, the ischium, and the pubis. By the time you are 18 years old, they will have fused to form just two hip bones that together make the **pelvic girdle**. The ilium is the largest, uppermost part. You can feel its top, that is the iliac crest, just below your waist. These fleshy muscles also make up your bottom and cushion you while you sit. The insides of the ilium are cup-shaped and hold the inner organs in them.

The ischium and pubis make arches that merge at the bottom of your torso. Between them, they contain the obturator foramen, which lets the blood vessels and nerves of the legs pass through, and also the muscles that connect to the thigh. The outside of the ilium, ischium and pubis make a cup-like shape called the acetabulum, where the femur makes a joint.

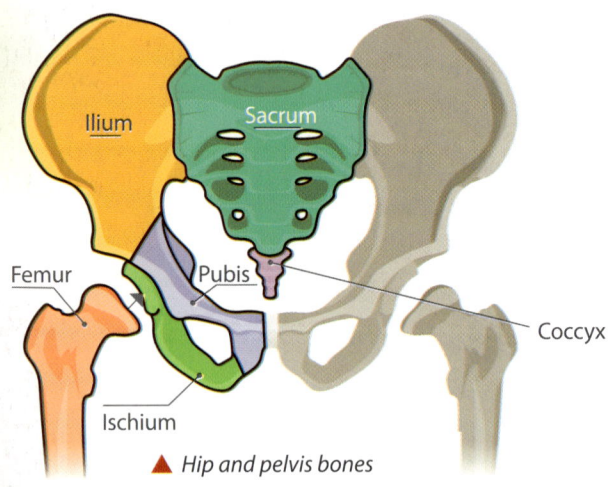

▲ Hip and pelvis bones

Pelvis

The hip bones and the lower spine—sacrum and coccyx—together make up the pelvis. They are linked by a number of ligaments that allow you a lot more freedom of movement, compared to the legs. Without the pelvis, you wouldn't be able to dance. It also allows your bottom to spread out when you are sitting, so that you do not feel cramped up.

Helping You Move
Skeletal Joints

Your skeleton would make you as stiff as a statue if there weren't any joints in it. The joints give the bones the space they need to move. They work on the principle of levers, and act as the fulcrum in which the load, that is the bone, is moved by the effort—the muscle. Joints can be classified according to the amount of mobility they provide—functional classification, or the kind of material that links the bones—structural classification. Some joints are temporary. They vanish as you grow up and the bones fuse together.

Functional Classification

Joints can be functionally classified into three types:

- Synarthroses, which do not allow the bones to move at all, like the sutures of the skull.
- Amphiarthroses, which allow some movement, like the intervertebral discs and the symphysis pubis between the hip bones.
- Diarthroses, which allow a lot of movement between the bones. Diarthroses are further classified into:
  - Uniaxial joints, which allow up and down movements only, such as fingers, elbows, knees.
  - Biaxial joints, which allow up and down, and sideways movements, such as wrists and ankles.
  - Multiaxial joints, which allow up and down, sideways, and rotational movements, such as hips and shoulders.

In Real Life

Joints and muscles work as type II levers, the joint as fulcrum and the muscle as effort, with the bone serving as load.

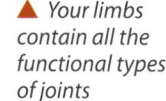
▲ *Your limbs contain all the functional types of joints*

Structural Classification

Joints are classified as fibrous, cartilaginous, and synovial, depending on how they are made. Fibrous joints are connected by collagen fibres, such as those between teeth and jaws. In cartilaginous joints, cartilage fills the spaces between bones like the **cranial sutures**. Synovial joints are complex, with **synovial fluid** lubricating the joints.

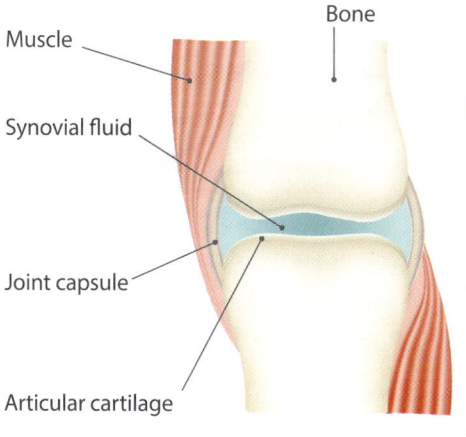
▲ The structure of a joint

How Joints Work

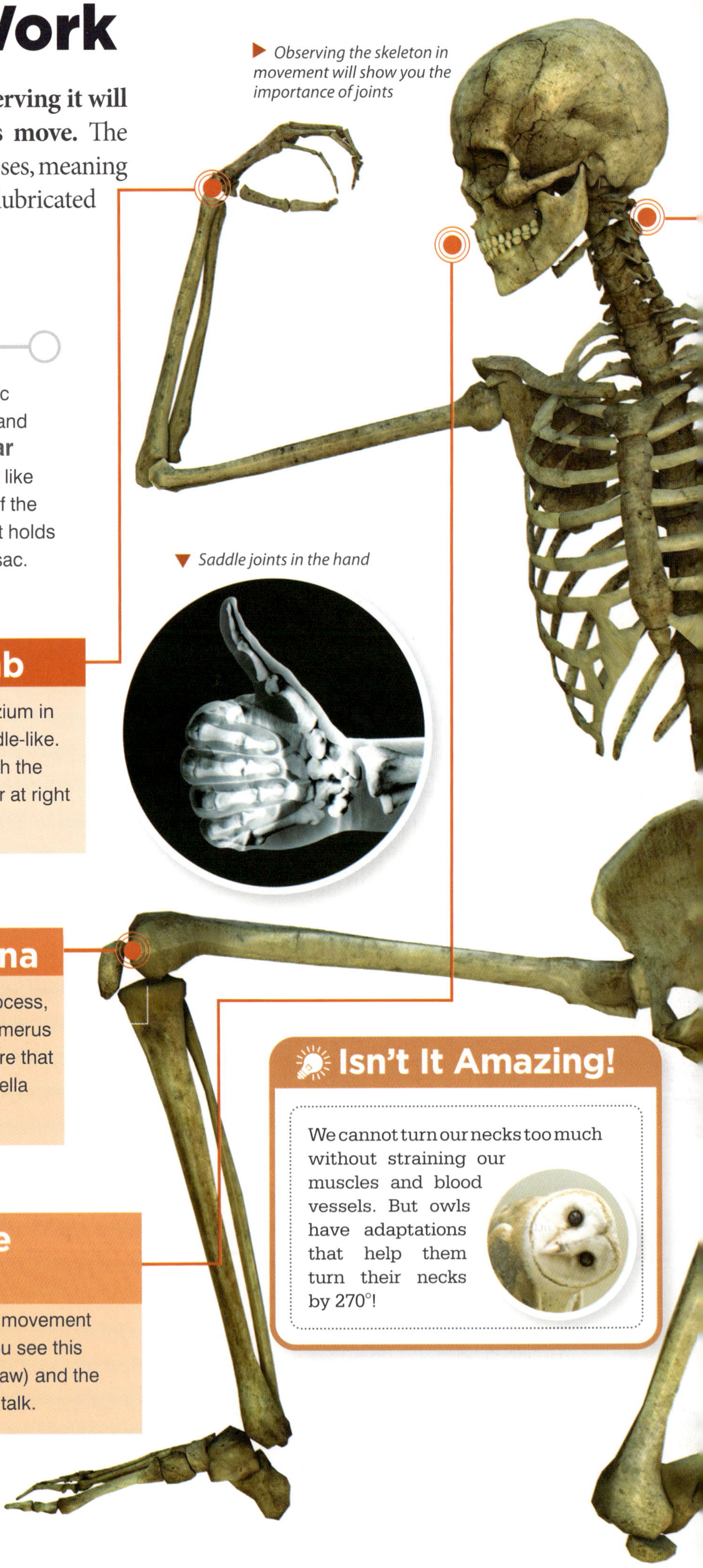

▶ Observing the skeleton in movement will show you the importance of joints

A skeleton may appear creepy, but observing it will help you understand how our bodies move. The joints you see here are all synovial diarthroses, meaning they allow plenty of movement and are lubricated by synovial fluid.

More about Joints

Synovial fluid is made of plasma from the blood that is rich in nutrients. It also contains hyaluronic acid, which acts a bit like engine oil. It is elastic and viscous, preventing friction in the joint. **Articular cartilage** covers the ends of the bones, acting like a shock absorber. The rest of the joint is made of the periosteum of both the bones, making a sac that holds the synovial fluid in. So, it is called the synovial sac.

Saddle: Wrist & thumb

The metacarpal of the thumb meets the trapezium in this joint. Both the faces of the bones are saddle-like. Hold your thumb and forefinger apart to stretch the web between them. Bring both hands together at right angles and you can see how this joint works.

▼ Saddle joints in the hand

Hinge: Humerus & Ulna

The upper end of the ulna is the olecranon process, which locks into the olecranon fossa of the humerus when the arm is stretched fully. This makes sure that the forearm does not bend backward. The patella does the same job in the knee.

Isn't It Amazing!

We cannot turn our necks too much without straining our muscles and blood vessels. But owls have adaptations that help them turn their necks by 270°!

Bicondylar: Mandible & Temporal Bones

This joint allows both hinge-like and sideways movement to the extent that the skin can be stretched. You see this best in the joint between the mandible (lower jaw) and the temporal bone, which lets you chew food and talk.

HUMAN BODY — SKELETAL & MUSCULAR SYSTEM

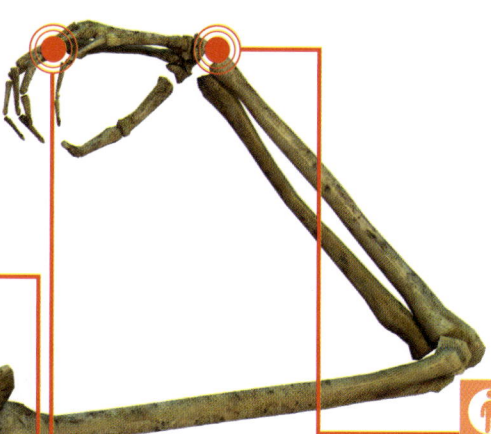

⭐ Incredible Individuals

Tommy John's baseball career was finished in 1974 when his elbow ligament was torn. A new surgical procedure got him back in play. He went on to win 164 matches. It is now called the Tommy John Surgery in the USA.

Gliding: Wrist bones

The eight carpals of the wrist have flat faces facing each other. This allows them to glide over each other a bit. Nevertheless, the wrist helps your hand move in all three planes—sideways, up-down and rotational. The tarsals also work this way.

Condyloid & Ellipsoid: Carpal and Forefinger

This joint allows a little movement up and down and a little sideways. You see this in the joint between your palm and your forefinger. It allows all movement except axial rotation. An example of the condyloid joint is the ovoid head of one bone moving into the elliptical cavity of another. This joint is seen in the wrist and the base of the index finger.

Pivot: Axis & Atlas

The dens of the axis—second cervical vertebra—is the pivot onto which the hollow body of the atlas—first vertebra—fits, making a perfect pivot that lets you turn your neck about 100° on each side.

Ball & Socket: Femur and Hip

The head of the femur is shaped like a ball, which fits into the acetabulum of the hip. Several ligaments and muscles help the joint move in three ways—forward and backward (walking), raising and lowering the leg sideways, and rotating the leg horizontally. The humerus and shoulder joint works the same way.

Ligaments

Ligaments connect two bones to each other at the joints. They are made of tough sheets of collagen, so that they don't tear when the muscles stretch out those bones. Proximal ligaments are close to the synovial sac, while remote ligaments attach to their bones further away.

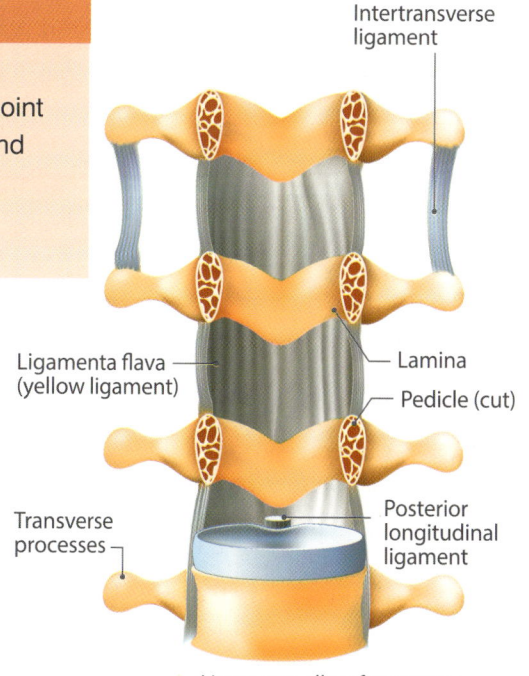

▲ Ligaments allow for twisting movements of the spine.

Breaking Point
Bone Fractures

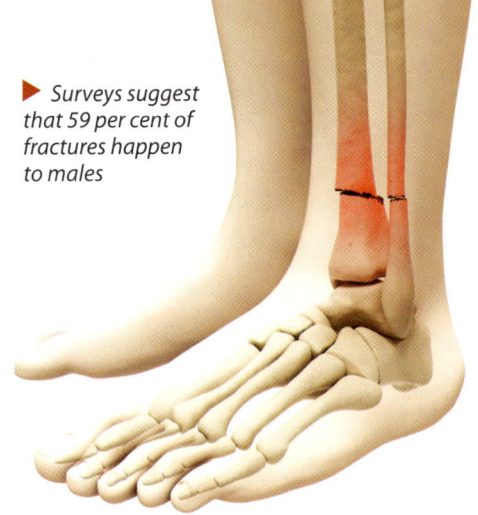

▶ Surveys suggest that 59 per cent of fractures happen to males

Although our bones are strong enough to take the stresses and strains of daily life, they are also brittle. This means that though they are hard, under great force, they can break or shatter. Depending on how you get hurt, fractures can happen in different ways. Fractures often occur in childhood, when the bones have not become hard enough. **Osteoporosis** and bone cancer leave them likely to break in an accident. Excessive force and trauma suffered during accidents can also lead to fractures.

Types of Fractures

Doctors must consider how a fracture happens before they can set it right and put it in plaster. A closed or simple fracture stays within the skin; but if you get hurt so badly that the bone breaks and tears out of the skin, it is an open or compound fracture. These are harder to set right as they may pierce organs and may require complicated surgery. Doctors also classify fractures in different ways.

Isn't It Amazing!

Boa constrictors do not have enough force to crush bones, as was thought earlier. But they can definitely suffocate their prey.

▶ Boas kill their prey by squeezing them hard enough to stop their hearts

▼ Transverse fracture
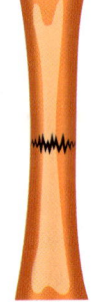
It happens across the width of the bone

▼ Longitudinal fracture

It happens along the length of a bone

▼ Oblique bone fracture

It happens at an angle

▼ Spiral fracture
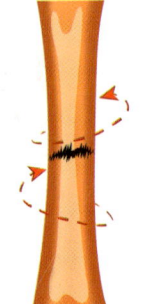
It happens when the broken parts of the bone get twisted

▼ Greenstick fracture

It happens when only one side is broken

▼ Comminuted fracture

It refers to many breaks in the bone leading to splinters

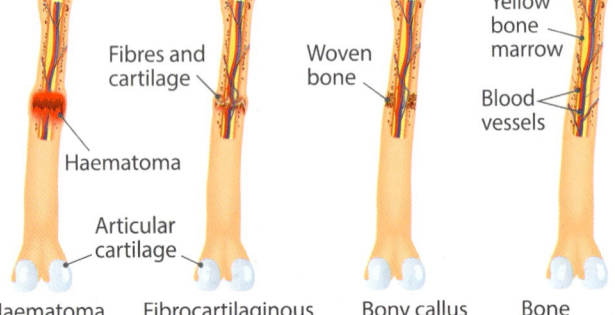

Haematoma formation — Fibres and cartilage, Haematoma, Articular cartilage
Fibrocartilaginous callus formation — Woven bone
Bony callus formation — Yellow bone marrow, Blood vessels
Bone remodelling

▲ Bone repair is carried out by osteogenic cells, which multiply to make new osteoblast cells

Bone Healing

Bones heal very quickly after fracturing. The blood vessels that get broken in the fracture make a thick clot between the bones, called a haematoma. Osteogenic cells from the periosteum secrete fibres and cartilage to 'stitch' the bones together, forming a callus. In the next few weeks, new bone cells grow inside the callus, the blood vessels reconnect, and the bone marrow also regrows.

Moving Our Bodies
Muscles and Tendons

Say muscle and you immediately think of the bulging biceps and triceps of weightlifters, or the powerful muscles in the legs of athletes. But did you know that the organ with the most muscles is your face? It has hundreds of tiny muscles that allow you to make lots of expressions, and convey emotions, from smiling to frowning, and from emoting anger to fright and even laughter.

In Real Life

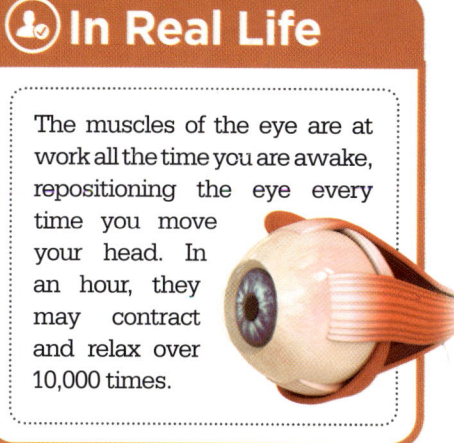

The muscles of the eye are at work all the time you are awake, repositioning the eye every time you move your head. In an hour, they may contract and relax over 10,000 times.

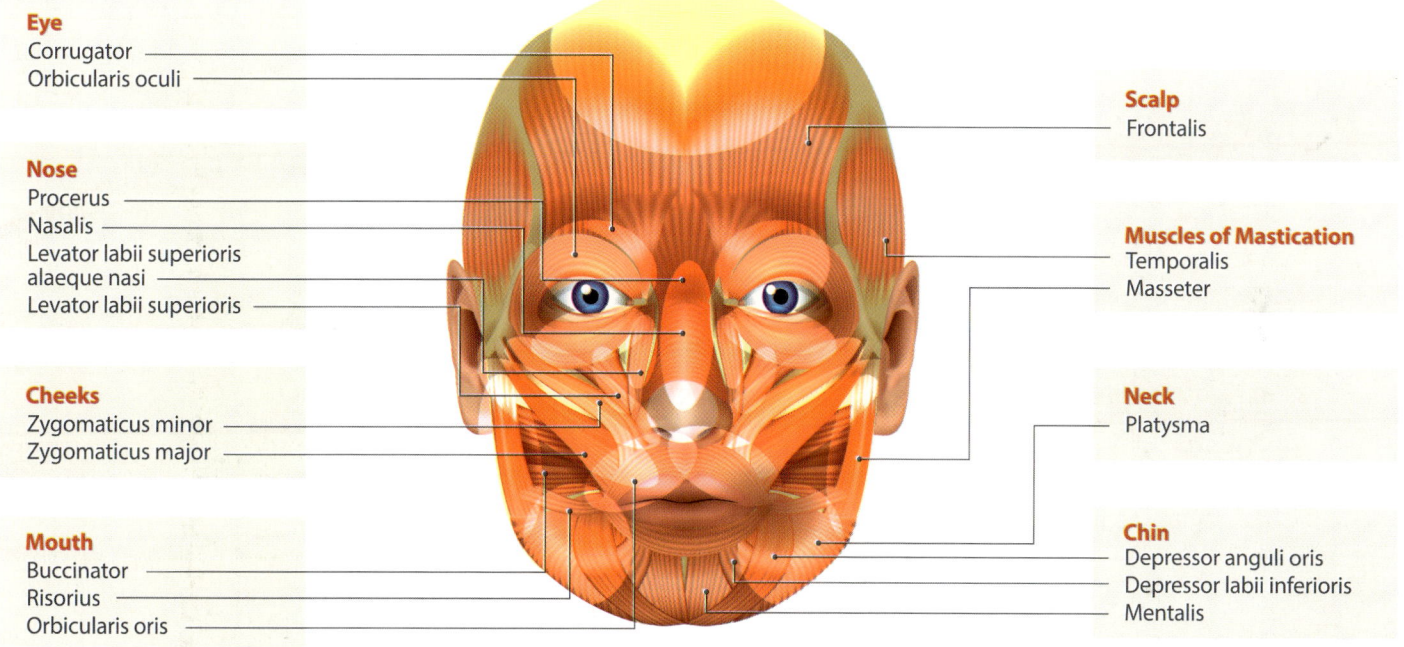

Eye
Corrugator
Orbicularis oculi

Nose
Procerus
Nasalis
Levator labii superioris alaeque nasi
Levator labii superioris

Cheeks
Zygomaticus minor
Zygomaticus major

Mouth
Buccinator
Risorius
Orbicularis oris

Scalp
Frontalis

Muscles of Mastication
Temporalis
Masseter

Neck
Platysma

Chin
Depressor anguli oris
Depressor labii inferioris
Mentalis

▲ The diagram shows some of the muscles of your face

Muscles in the Head

There are other muscles too, which do their work without you ever finding out. They help you with breathing and digesting food, with keeping blood flowing through your veins and arteries, and urine flowing out of your body. A very special kind of muscle keeps your heart pumping throughout life. Let us find out what muscles are, and how they work.

Types of Muscles

After many years of study, doctors and scientists say that muscles come in three types:

- **Striated muscles** are fibre-like in shape and are attached to your bones. You move them as you wish.
- **Smooth muscles** are circular in shape. They are attached to your internal organs, and move on their own.
- **Cardiac muscles** are halfway between striated and smooth muscles. They are present in the heart and also move on their own.

Tendons

Striated muscles hold onto their bones with tissues called tendons. A tendon is like a piece of elastic, able to stretch and hold a lot of strain as the muscles contract and expand. They are made of collagen. On one side, the tendon merges with the muscle, and on the other, it merges with the coating of the bones—periosteum.

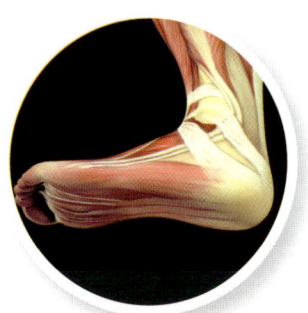

▲ Tendons (white) reduce strain on your muscles (red)

▶ Regular exercise keeps your muscles healthy

The Muscles We Can Control

Much of your body's weight is made of muscle. Most of it is made of skeletal muscles. Whether you reach out an arm to catch a ball, pull your stomach in to exhale, smile at someone, or run away from a cockroach, your skeletal muscles are doing the majority of the work. That is why they are called voluntary muscles. They act when you want them to.

Muscle Tissue

Under a microscope, muscle tissue looks like long streaks of threads. The scientists' name for this is striation, these are therefore called striated muscles too. Most of the meat you eat is made of the same kind of muscle.

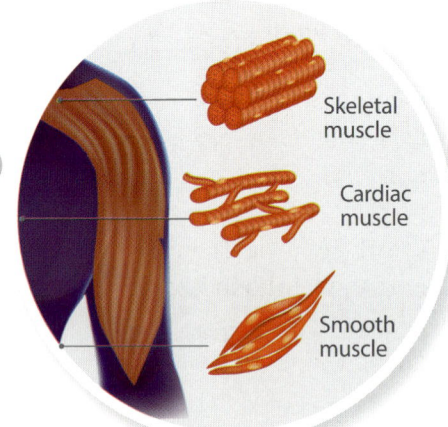

▲ Types of muscles

Under the Microscope

The part of the muscle that does all the work is the **myofibril**. It is made up of proteins called actin and myosin. These myofibrils are inside muscle cells called sarcomeres and are attached to the covering membrane of the cell, called the sarcolemma. Each muscle cell has hundreds of mitochondria to give it instant energy. It is also connected to the nervous system by **neuromuscular junctions**, which allow you to control the muscle.

The sarcomeres are bound together in bundles called fascicles. Lots of fascicles together make a muscle, along with blood vessels and connective tissue. Connective tissue is made of cells that help feed cells to the muscles and also make the muscle flexible. It is made of three layers—the endomysium surrounding the fascicles, the perimysium connecting them, and the epimysium wrapping around the muscle.

Isn't It Amazing!

Cheetahs do not have stronger muscles than any other big cats, but the muscles of their forelimbs pack in more sarcomeres, end-to-end, in each fascicle. This helps them attain speeds of up to 29 mps when chasing prey, and that is what makes them the fastest animal on Earth.

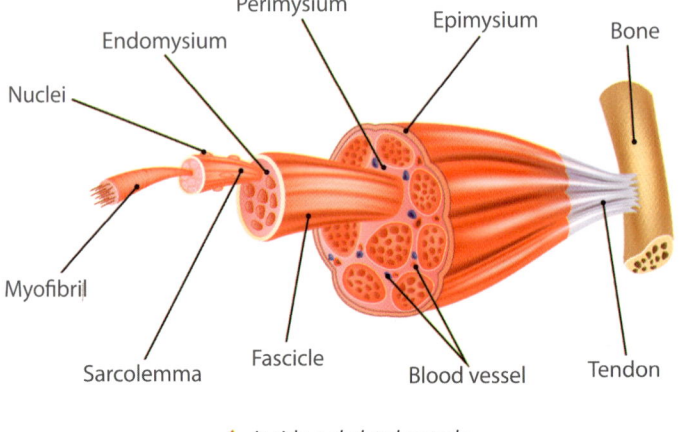

▲ Inside a skeletal muscle

The Muscles We Cannot Control

The big muscles that attach to our bones and make up most of our body are important. However, alongside them are thousands of tinier muscles that work throughout our lives without resting. They are the involuntary muscles. They are connected to our sympathetic and parasympathetic muscles, so we generally do not have control over them, although some, like the muscles of our eyelids, are both voluntary and involuntary. They do not have stripes when you see them under a microscope, so they are also called smooth muscles.

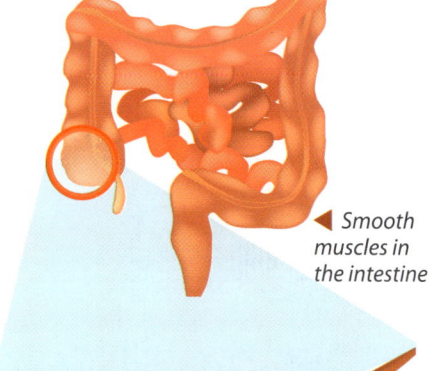

◀ Smooth muscles in the intestine

Finding Smooth Muscles

Many of these muscles are present in the walls of our intestines, making food go along; our arteries and veins, making blood go along; and in the urinary bladder, making sure urine does not leak until we are ready. Some control valves in the blood vessels. Other muscles, like sphincters, act like valves themselves in the digestive system. As these are ring-shaped, they are called circular muscles. Other smooth muscles help the lungs contract and expand. Muscles in the uterus allow women to give birth.

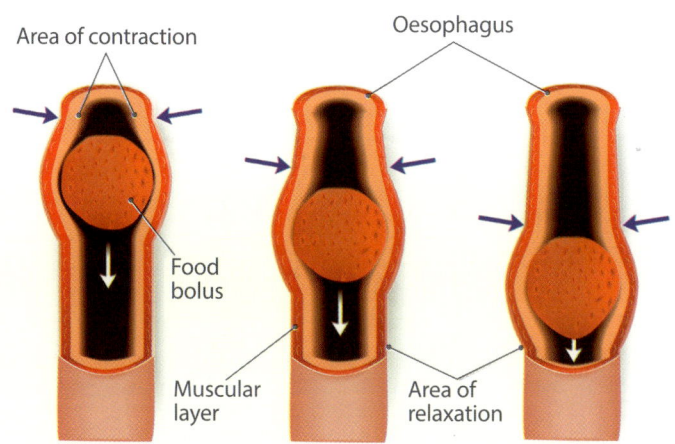

▲ Rings of smooth muscle in the food pipe contract and expand to push food along. This is called peristalsis

In Real Life

Most muscles are relaxed when not at work. Sphincters are different. They are usually closed, so that things do not leak from one organ to another, like from the stomach to the food pipe, or from the small intestine to the stomach.

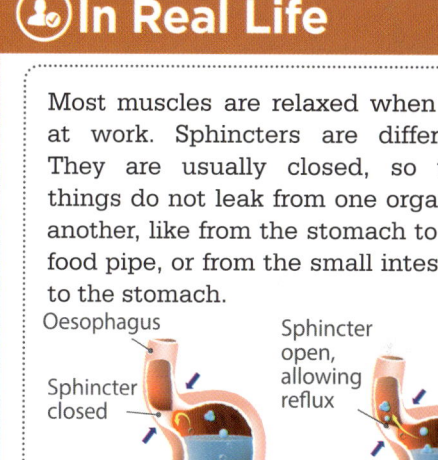

▲ When sphincters cannot close, they cause upsets like acid refluxes

A Closer Look

Smooth muscles look a bit like stretched rugby balls or insect cocoons. They are not organised into fascicles like skeletal muscles, but are surrounded by endomysium. Dense bodies help tether the myofibrils to the sarcolemma and intermediate filaments connect all the dense bodies. This helps pull all the thick filaments (myosin) and thin filaments (actin) together when the muscle contracts. Motor neurones of the autonomic nervous system make junctions with all the muscle cells to help them coordinate their contraction and relaxation.

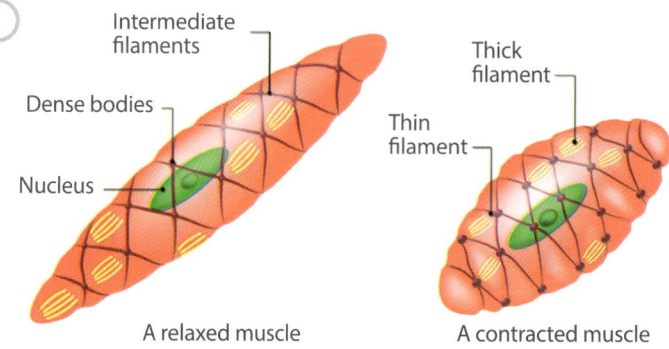

▲ Smooth muscles do not need as much energy as skeletal muscles, but they do not need to produce as much force either

How Muscles Work

Bodybuilders and weightlifters love showing off the bulge in their biceps as they contract. They have exercised a lot for those big muscles, but what are they made of? Why do they appear so strong? How come the muscle bulge is not there when their hands are stretched out or relaxed?

◀ *Building muscles also builds and strengthens the bones and tendons*

 ## Why Muscle Bulge

Muscles cells contain proteins that do all the hard work. The two main ones are actin and myosin, which form thread-like filaments. Myosin has 'heads', which can stick to actin. When the muscle is at rest, the actin and myosin filaments are apart, and the muscle is flat. When the muscle contracts, millions of actin-myosin pairs are pulled together and the muscle bulges out from the body.

 ## Neuromuscular Junctions

Your muscles are connected to motor neurones through tiny synapses called Neuromuscular Junctions (NMJ). When your muscles have to contract, they get a signal from the brain or spinal cord in the form of a tiny electric current. This current reaches the NMJ, and a chemical called a neurotransmitter is released. It creates a new current across the muscle fibres.

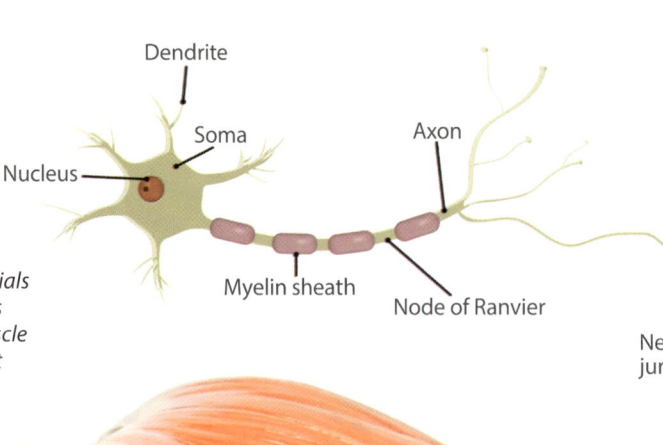

▶ *Action potentials from the nervous system tell a muscle when to contract*

Action Potential

This is a tiny electric current that passes along the length of a muscle fibre. When there is no message from the brain, there is sodium (Na^+) outside the neurone and potassium (K^+) and chloride (Cl^-) inside the muscle cells. This is the resting potential. When a signal arrives, Na^+ rushes in and K^+ rushes out, creating the action potential. As the muscle relaxes, the two slowly switch places again.

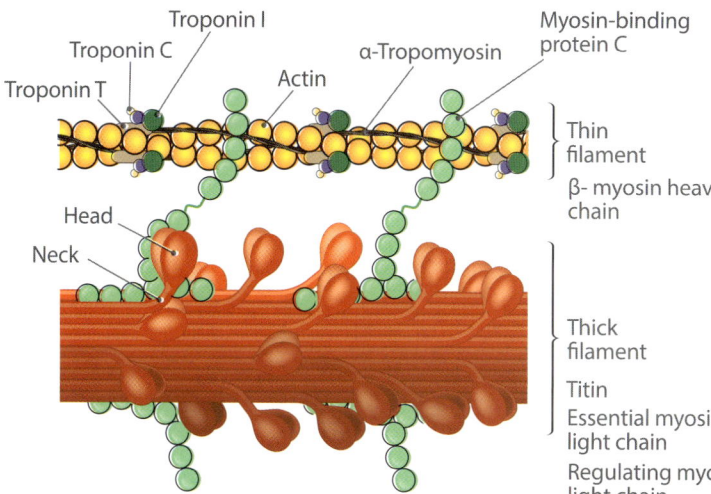

▶ Lots of different proteins work together to make a muscle contract and relax

A Muscle in Action

When the brain sends a signal to the muscle to work, it contracts. Like two magnets that stick to each other, the myosin heads and actin filaments stick to each other. The filaments are pulled closer. The sarcomeres become shorter but thicker. As this happens all over the muscle, it bulges out of the body.

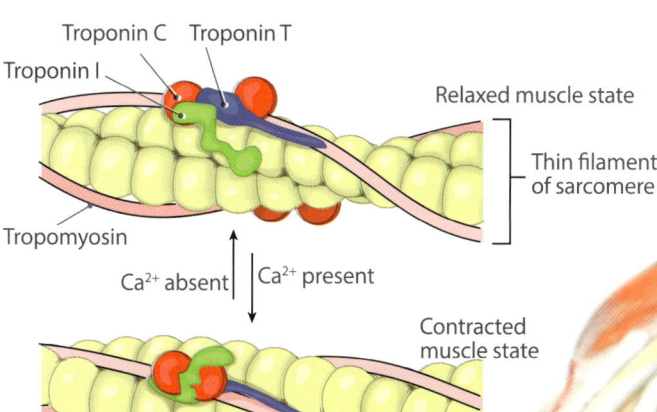

▲ Muscles contract and relax due to the action of calcium ions on the troponin complex

Isn't It Amazing!

Muscles work faster at high temperatures and slower at low temperatures. Warm-blooded animals such as birds and mammals maintain a constant body temperature, whatever the weather is, so that their muscles can react quickly to signals from the brain. This helps them run or fly away quickly if faced with danger. Cold-blooded animals like insects and reptiles depend upon the weather being warm to be active.

Relaxing the Muscle

To relax the muscle again, the body needs to carry out respiration. Enzymes in the muscle cells convert glucose into carbon dioxide and water. This gives out a lot of energy. Some of it is trapped in a molecule called ATP, while the rest goes out as body heat. ATP attaches to the heads of the myosin filaments, causing them to detach from the actin filaments. The filaments come apart, the sarcomeres flatten, and the muscle thus relaxes.

▲ Yoga can relieve muscular tension

Keeping the Heart Going

After the brain, the heart is our body's most important organ, pumping blood throughout our lives without a rest. Most of the heart tissue is made of a thick, muscular layer that doctors call the myocardium. The muscle cells in it (cardiac muscles) are like voluntary muscles in some ways, and like involuntary muscles in other ways. This makes it possible for them to expand and contract without ever getting tired. Cardiac muscles are controlled by special cells called pacemaker cells, that are connected to the parasympathetic nervous system, which make sure that you get a regular heartbeat. They get blood from the coronary artery.

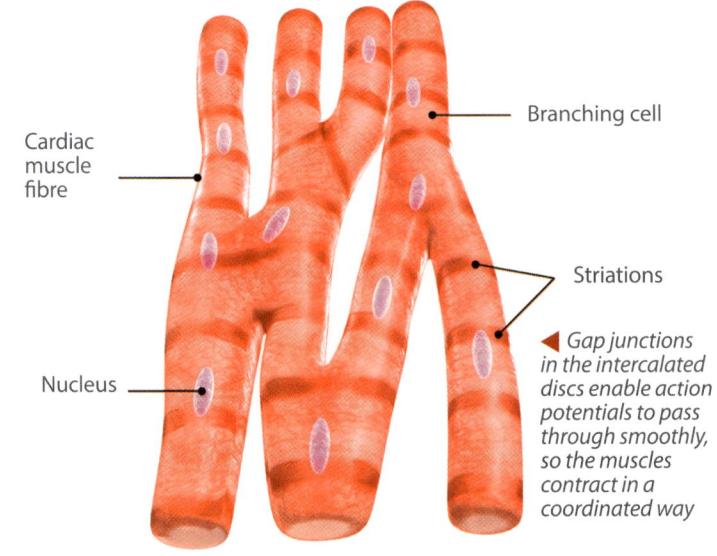

 ## Cardiac Muscle

Cardiac muscles are made of sarcomeres like skeletal muscles, but these are shorter and more branched. The branches connect them to each other, through intercalated discs. These allow the cells to pass action potentials between them and also contain desmosomes, which act like clips that hold the muscles together. This makes it possible for the muscles to contract and expand together. The muscles are organised in layers around the heart, so that it pumps smoothly every time it gets a signal from the pacemaker.

Cardiac muscle fibre

Branching cell

Striations

Nucleus

◀ Gap junctions in the intercalated discs enable action potentials to pass through smoothly, so the muscles contract in a coordinated way

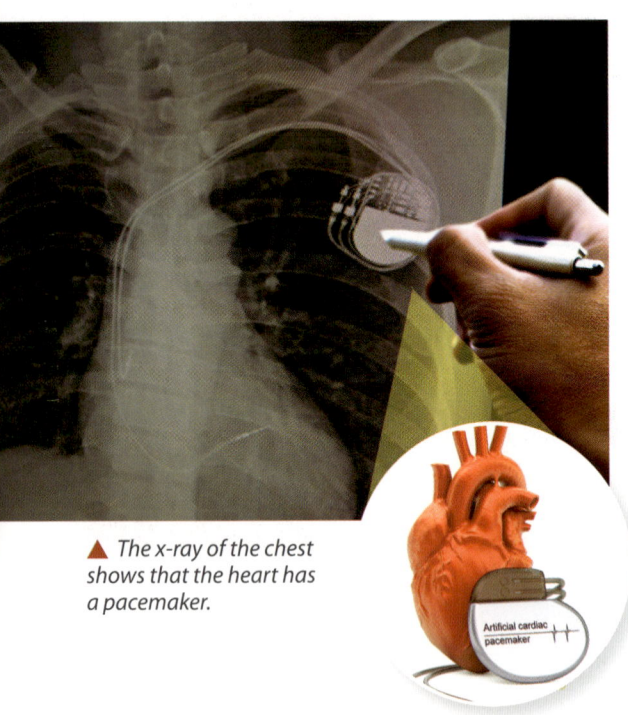

 ## Pacemakers

The vagus nerve connects the heart to the parasympathetic nervous system. It forms neuromuscular junctions with the pacemaker cells. The nervous system gives regular electric pulses to these cells in the form of action potentials, which travel throughout the heart through the gap junctions.

Artificial pacemakers work in a similar way to the vagus nerve. They are placed in the chest of people with heart arrhythmia, so they can get a tiny electrical current that makes the heart beat with normal rhythm.

▲ The x-ray of the chest shows that the heart has a pacemaker.

★ Incredible Individuals

Willem Einthoven got the Nobel Prize in 1924 for inventing the electrocardiogram, which can measure the electric currents in the myocardium.

Exercise in Moderation

Regular exercise or play keeps your bones and muscles healthy. Your muscles keep their strength and flexibility, your circulation improves and oxygen reaches all tissues. Exercise helps you put on more muscle mass, as the protein you eat is turned into new muscle cells and muscle fibres.

▲ *Team exercises not only improve physical health, but also allow you to make friends and improve coordination*

Exercise Rules

It is important to warm up before you begin to exercise. Your muscles spend little energy while at rest, but when you start exercising, playing or working, they need to quickly speed up, so they can burn a lot more sugar and get energy. A brisk walk makes them spend thrice as much energy and heavy exercise makes them spend up to 12 times of energy. After exercise, you must do some cool down exercises to bring your body back to its resting state.

▼ *Playing or exercising without warming up could give you a cramp.*

 Cramps

Too much strain on your muscles, or sudden stretching can give you muscle cramps. This happens when the muscle contracts but cannot expand. It is painful and stops you from using the muscle. Swimming in cold water can give you swimmer's cramp, which can be dangerous as it may cause drowning.

Why We Get Tired

When the muscles are at work, they need a lot of energy quickly. Instead of waiting for oxygen, they switch to anaerobic respiration. Glucose is broken down into lactic acid and the energy released is used for contracting and expanding the muscles. The increase in lactic acid causes biochemical changes in the muscle tissue, and it begins to slow down after some time. This is tiredness. It makes the body rest, and in that time, the lactic acid is broken down to carbon dioxide by aerobic respiration. The muscles come back to normal.

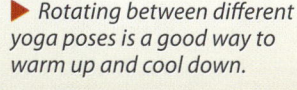

▶ *Rotating between different yoga poses is a good way to warm up and cool down.*

Getting Older

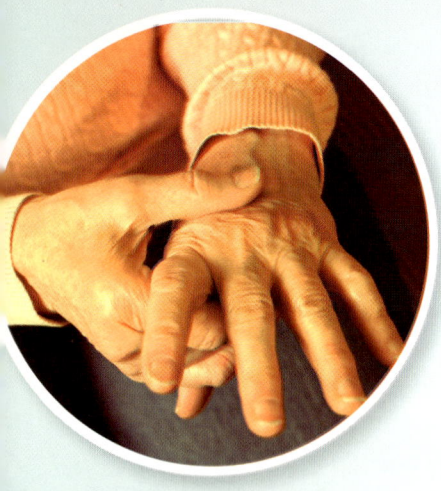

Arthritis means inflammation of the joints, while rheumatism is a condition that affects tendons and ligaments. Spondylitis is an inflammation of the vertebral column. The inflammation affects the cartilage that surrounds the joints and sometimes also the nerves of the muscles that move these joints. These diseases cause severe disabilities because they make it painful to move the joints. It gets difficult to lift weights and walk. These diseases usually affect older people, as their joints wear away due to tissue degeneration. But younger people may get them due to infection, obesity and nutritional deficiencies.

Osteoarthritis

Osteoarthritis or degenerative joint disease is very common. Patients suffer from a lot of pain in their joints of the knees, hip and fingers, because the cartilage that cushions the joints has worn away. The pain sets in after they have been active for a few hours and lasts for a long time. Sometimes, there is a crackling sound (crepitus) as the patient moves.

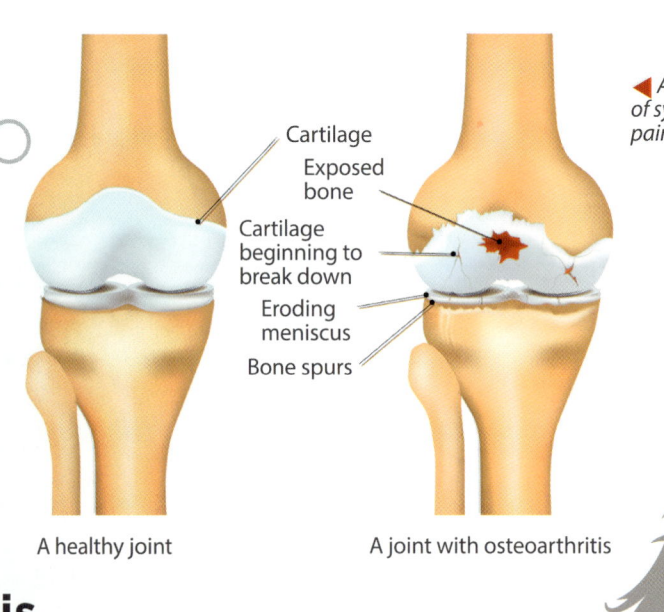

◀ Arthritis causes loss of synovial fluid, causing pain in moving joints

A healthy joint   A joint with osteoarthritis

Infectious Arthritis

Some bacteria, like staphylococcus and pneumococcus, infect the joints and cause the formation of pus. This kills cartilage cells, leading to arthritis. People who have tuberculosis and leprosy may also get arthritis, as the diseases affect their bones when they become severe. Rubella also affects the joints if the patient has not been vaccinated for it.

Rheumatoid Arthritis

This is an autoimmune disease, that is, it occurs when the immune system attacks the cells of your joints. It particularly affects the joints of the knees and elbows.

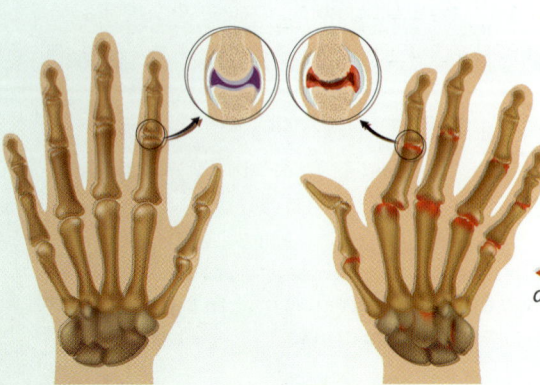

◀ Arthritis causes deformity of the hands

A normal hand   A hand affected by rheumatoid arthritis

Gout

This bursitis-like disease affects the knees, elbows, fingers and toes. It causes them to pain at night. This happens because uric acid crystals build up in the synovial sac. Uric acid is a waste product of protein metabolism in the body and is removed from the blood by the kidneys. It builds up if the patient eats too much protein-rich food, drinks too little water or has kidneys that do not work properly.

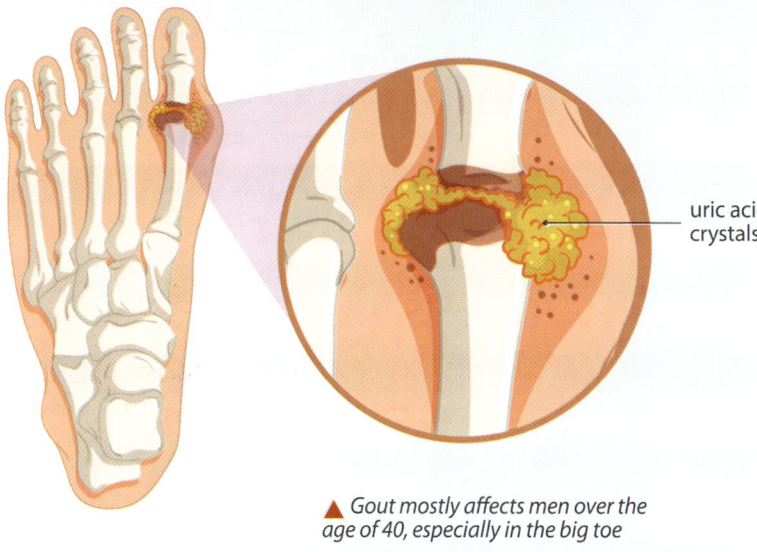

uric acid crystals

▲ *Gout mostly affects men over the age of 40, especially in the big toe*

Bursitis

Bursitis is a disease that affects the synovial bursa, tissue that connects the bones of the joint and contains the synovial fluid. It leads to a gradual deposition of insoluble calcium salts in the synovial fluid, leading to the joints becoming stiff.

Ankylosing Spondylitis

This is a disease of the vertebral column (spine). Men get it more often than women, often between the ages of 15-40. This disease causes arthritis in the joints between the vertebrae and also the joints between the vertebrae and the hip bones. Inflammation starts in the lower back and slowly moves up.

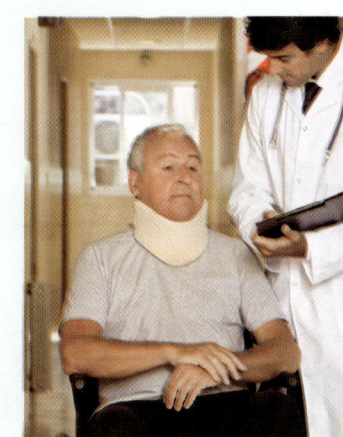

▶ *People with spondylitis often need to wear a brace to keep their neck from turning, as it can cause a lot of pain*

Rheumatic Fever

This disease is caused by bacteria called haemolytic streptococci. It affects the heart, nervous system and joints, causing inflammation and pain. It usually happens if a throat infection is left untreated and the bacteria are able to enter the bloodstream.

In Real Life

In the past, gout was called the 'disease of kings', since it was caused by a rich lifestyle of consuming protein-rich foods and alcoholic drinks.

Osteoporosis

As we get older, our bodies are not able to absorb enough calcium and phosphorus from food. Instead, our body makes up for what it needs by taking them from the bones. As the bones lose calcium, their structure changes and they become more sponge-like. The pores within the bones expand, making them weaker. This is called osteoporosis and makes the bones more likely to fracture in case of an accident or a fall. It also makes the spine bend over, causing a condition called hunchback. Women are more prone to developing osteoporosis.

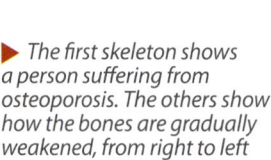

▶ *The first skeleton shows a person suffering from osteoporosis. The others show how the bones are gradually weakened, from right to left*

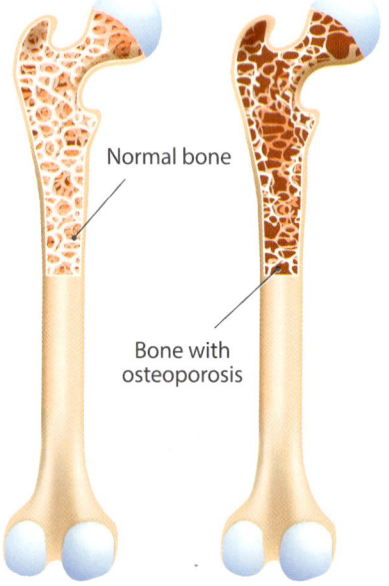

▲ *Osteoporosis weakens bones, making them unable to bear the body's weight*

ⓘ Bone Mineral Density

This is how doctors tell whether you will get osteoporosis or not. They will take a special X-ray called a DXA scan, which tells them the density of minerals (calcium and phosphorus) in 1 cm2 of bone. They compare this with the density of other healthy people of your age and gender to get a T-score. If the T-score is -2.5, that means your bones are dangerously below the density they need to be.

ⓘ What to Do about Osteoporosis

First, you need to know how you can get it. If a person does not get enough calcium from food, does not exercise enough, is underweight or smokes or drinks too much alcohol, then that person has a very high risk of getting osteoporosis. If you have dense bones to begin with, they might not become very weak. That is why men have lower chances of getting osteoporosis than women. Some kinds of cancers and intestinal disorders may also lead to osteoporosis. This disease is hard to cure but easy to prevent. Eating food rich in calcium and Vitamin D (which helps your body absorb calcium) and taking long morning walks or jogs throughout life is good enough.

✓ In Real Life

Not getting enough Vitamin D in their diet causes rickets in children. Such children grow up with bent bones, a soft skull, low height and their teeth do not erupt in time.

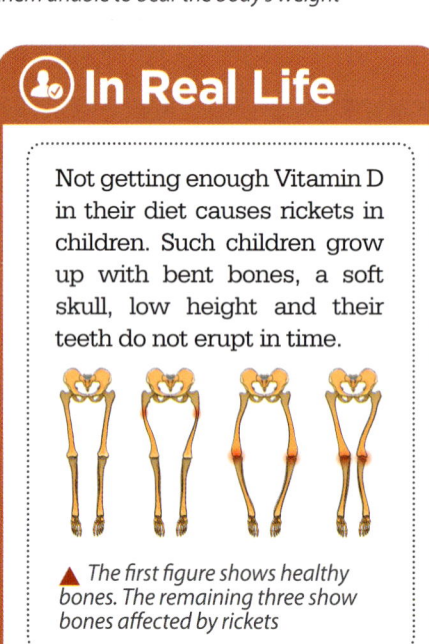

▲ *The first figure shows healthy bones. The remaining three show bones affected by rickets*

▶ *Foods rich in calcium and phosphorus fight osteoporosis*

Muscular Diseases

These are diseases that cause muscles to weaken. In muscular dystrophy (MD), patients show weakening of muscles, as muscular tissue is slowly replaced with fatty tissue. Patients slowly become unable to balance themselves, walk or lift their arms and also suffer from breathing and heart problems. This disease is hereditary, that is, it is because of defects in the genes you get from your parents. There are many types of MD, depending on how the disease progresses. Another disease that affects muscles is myasthenia gravis, in which your immune system attacks your muscles.

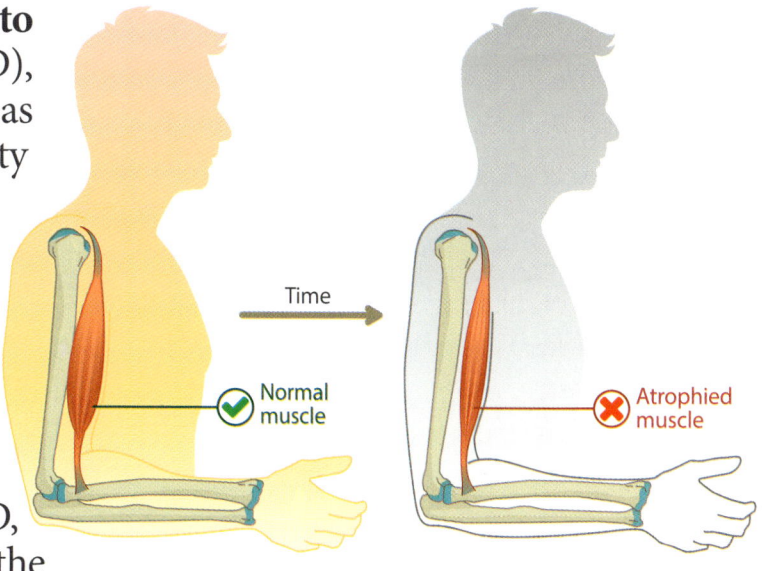
▲ Muscular dystrophy causes progressive weakness of muscles as muscle cells die off one by one

Duchenne Muscular Dystrophy (DMD)

In DMD, a gene on the X-chromosome is mutated, so that the muscles do not make dystrophin. Dystrophin is a protein that the muscles need to keep the muscle fibres bound together. Without it, muscle cells get damaged and die easily.

Because men have one X-chromosome while women have two, men who inherit the gene from their mothers will get DMD. Women do not get the disease, but there is a 50 per cent chance that they will give it to their sons. Few men who have this disease make it past the age of 20.

▶ Duchenne Muscular Dystrophy (DMD) occurs in one of every 3,300 males

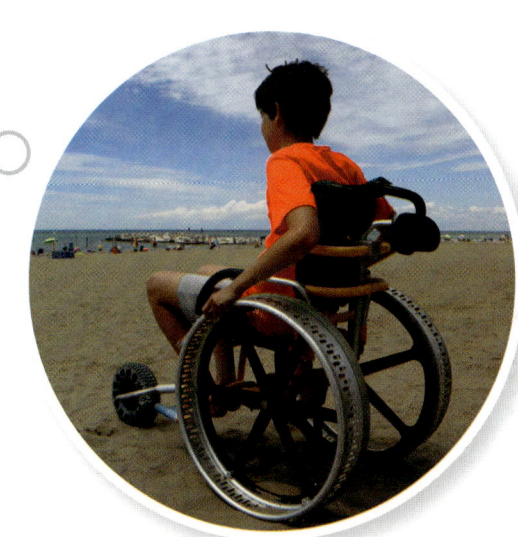

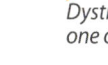 Myasthenia Gravis

You can get this disease if your immune system develops antibodies against acetylcholine receptors in the junction of the motor neurones and muscles. Doctors call myasthenia gravis an autoimmune disease. It causes muscles in the face, neck, throat and limbs to become weak, so patients are often unable to swallow food, or even close the mouth. It is even difficult for them to smile. Over time, muscles may weaken and die. There is no cure, but patients are given steroid drugs to suppress the immune system.

◀ Droopy eyelids and an inability to show facial expressions are symptoms of myasthenia gravis

Incredible Individuals

'Darius Goes West' is an award-winning documentary about Darius Weems (1989–2016), a teenager with DMD, going on an 11,000 km road trip across America to raise awareness about the disease.
The movie raised over 2 million US dollars, which were given to Charley's Fund set up for DMD research. Darius gave it the $52,000 he was given in prizes.

STOMACH & DIGESTIVE SYSTEM

WE ARE WHAT WE EAT

We are the atoms and molecules of the food that we eat. We need to spend a lot of energy and time getting food from the environment. We have a dedicated organ system in our body to turn this food into something our body can use. Our **digestive system** evolved to convert the food we eat into energy. The other organ systems like the circulatory system, nervous system and respiratory system, exist to support the digestive system.

The digestive system is a complex system, beginning with teeth, which break down food in smaller bits; the stomach, which dissolves these pieces into a semi-solid mass; the intestine, which breaks them down into molecules and absorbs them into our body; and the anus, which removes the undigested bits.

Scientists who study the evolution of animals suggest that our body plan is really a tube-within-a-tube. They even suggest that this body plan is the same in all organisms, from the tiniest worms to the biggest whales. The inner tube is our digestive system, from mouth to anus, while the outer tube is our body enclosed by the skin, with our arms sticking out. Between these is the space in which all our organs function.

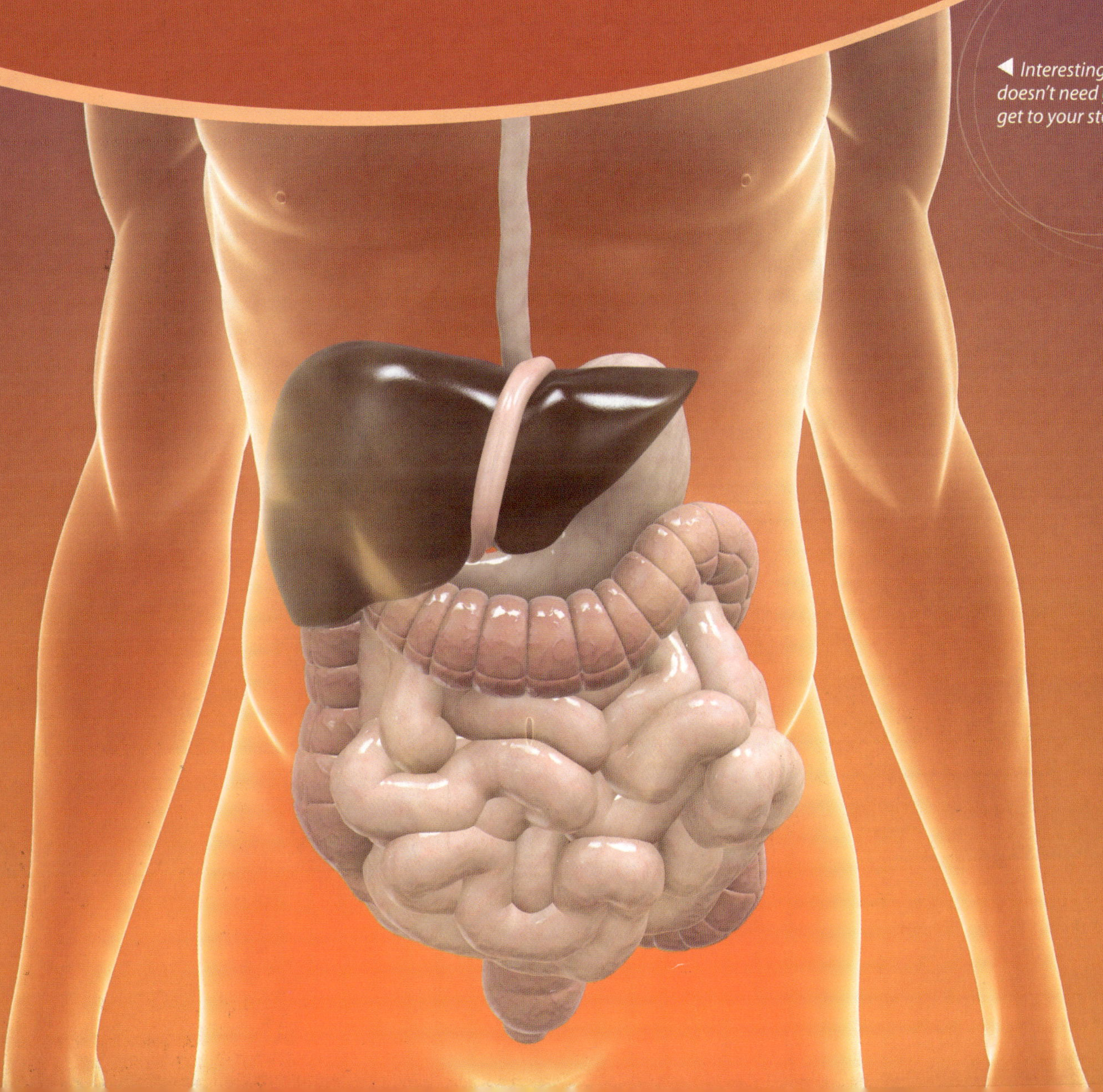

◀ *Interestingly, food doesn't need gravity to get to your stomach*

The First Bite

As you take the first bite of your meal, several parts of your body get to work. The first of these are your jaws, teeth, and tongue. You use them to chew the food you eat into smaller bits. This makes it easier to digest the food. If you do not chew your food, some of it might remain undigested, so your stomach and intestines would have to work harder. So, do not eat too quickly or just swallow your food, but take the time to chew.

▲ Like your fingers, your lips also have prints that are unique to you

The Lips

The lips close the mouth and prevent food from falling out. Our lips allow us to talk properly. If you eat spicy food, you might feel a burning sensation on your lips. Similarly, while drinking a hot beverage, you might feel the warmth on your lips. This happens because your lips have nerve endings so that you can feel the food you are eating. The skin on your lips is very thin, so it can become dry and cracked in winter. During this season, it is a good idea to use a balm to provide moisture to the lips.

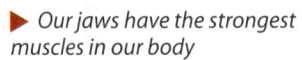
▶ Our jaws have the strongest muscles in our body

The Jaws

Though we bite and chew with our teeth, they cannot move by themselves. We move our jaws to make the teeth move. By opening and closing the mouth, we help our teeth bite, crush, and grind the food. The upper jaw is called the **maxilla**, while the lower is called the **mandible**. The jaw muscles are very powerful. The mandible closes with a force of 112 kg.

Isn't It Amazing!

The Australian Saltwater Crocodile has the world's strongest jaws, with a force of about 16,458 Newtons or 1,678 kg. But, the jaw muscles of this crocodile are so weak, that you can hold them shut with a rubber band!

▲ Crocodiles can quickly snap their jaws shut with the powerful closing muscles of their jaws. However, their jaws are too weak to open them as quickly

The Tongue

Our tongue presses food against the roof of the mouth, crushing it. It senses how hot or cold food is. We taste the food that we eat because it has thousands of tiny cells called taste buds. We have different kinds of taste buds to check the taste of sweet, salty, sour, and bitter food.

The tongue is the most flexible organ of our body, made up of several tiny muscles. When the vocal cords in the throat that is the larynx make sounds, the tongue's movement along the floor of the mouth, the roof, and the teeth, turn them into the sounds we recognise, such as A, B, C, D, and so on, arranged into words. The tongue muscles are controlled by special areas of the brain that enable them to move correctly so that we say the words right.

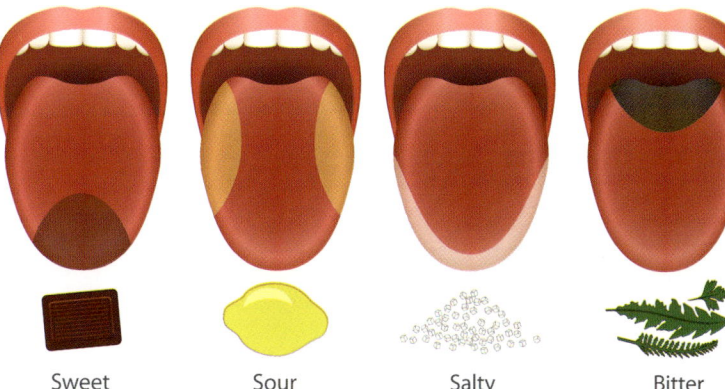

▶ Different taste buds on different parts of the tongue help taste distinct flavours

Sweet   Sour   Salty   Bitter

Drool

Drool, or saliva, which is the scientifically correct term, helps us chew our food. It is made in salivary glands, which are found below the tongue and behind the throat. If your mouth did not have saliva, you would get a very dry mouth, making it difficult to chew and swallow food.

▶ Babies cannot control their drooling. As adults, we may drool while we are sleeping

The Throat

The throat is at the front of the neck. It has the thyroid gland, the food pipe, the wind pipe, and the voice box. The back of the neck has the spinal cord. The **thyroid gland** makes the **thyroid hormone**, which tells your body to make energy from the food you eat. The food pipe carries food from the mouth to the stomach, while the wind pipe carries air from your nose to the lungs. They cross each other at the **pharynx**. The pharynx has a little lid-like organ called the epiglottis, which closes the windpipe when the food you just swallowed is being sent to the stomach.

Incredible Individuals

Some conjurers perform a special trick of swallowing a whole sword! They can do this because they can overcome their **gag reflex**. This is your body's way to stop choking when something touches the back of your mouth.

Teeth & Gums

We need our teeth not only to bite and chew food, but also to be able to talk to people properly. From the time we are 1-year-old, till about 12 years, we have 24 milk teeth. Once those fall out, we get a new set of 32 teeth which we keep throughout most of our lives. They begin to fall out when we become old, but not if we are healthy. The teeth join the tongue for another job—making many different kinds of sound, so that we can talk to our friends!

Teeth

Our teeth are arranged in two rows, in the upper and lower jaws. They do many things for us.

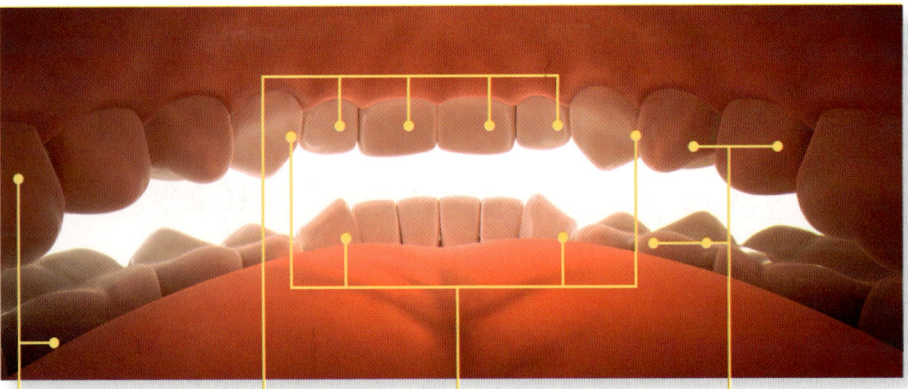

The last three teeth on each side of both jaws—making twelve in all—are the molars, that grind food into a paste that can be swallowed.

The front eight teeth—four above, four below—are called incisors. They help bite off the food.

In each jaw, there is one tooth to the left of the incisors and one to the right. Together, they make up four teeth called canines. These tear off hard bits of food.

The next two on each side of both jaws—making eight in all—are premolars.

▲ Enamel, the top surface of your tooth, is the hardest part of your entire body

▶ Healthy gums make for healthy teeth

Gums

The gums are the parts of the jaw which hold the teeth in place. Keeping your gums healthy by brushing everyday will make sure that your teeth do not fall out!

A 1.8 million-year-old skull found in Georgia (the country) by scientists in 2005, had only one tooth left. This showed that the person had lived to a very old age till he had lost his teeth and was eating with his gums.

Gingivitis

Gingivitis is what doctors call a painful inflammation of the gums, which may cause bleeding and bad breath (halitosis). It is caused by bacteria which thrive in your gums if you do not brush them properly.

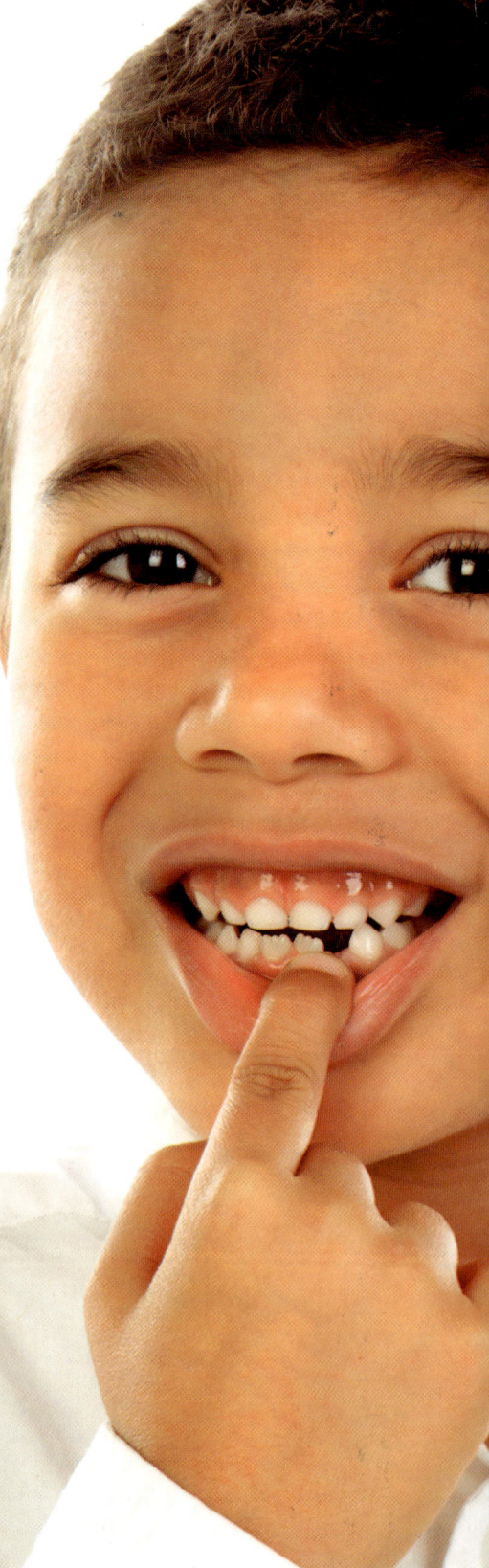

HUMAN BODY | STOMACH & DIGESTIVE SYSTEM

Parts of a Tooth

Each tooth is made of two parts—a root and a crown. The root is deep inside the gums and has nerves and blood vessels in it, which link to the jaw through the root canal. The crown of the tooth, made of calcium carbonate, does all the work. It has an inner part called dentin and an outer, harder part called enamel. It is important to take care of your crowns, by brushing them every day. Otherwise bacteria grow in the teeth and make holes called **cavities**, which can become very painful. If the cavity goes very deep, you need a **root canal operation** to fill it in.

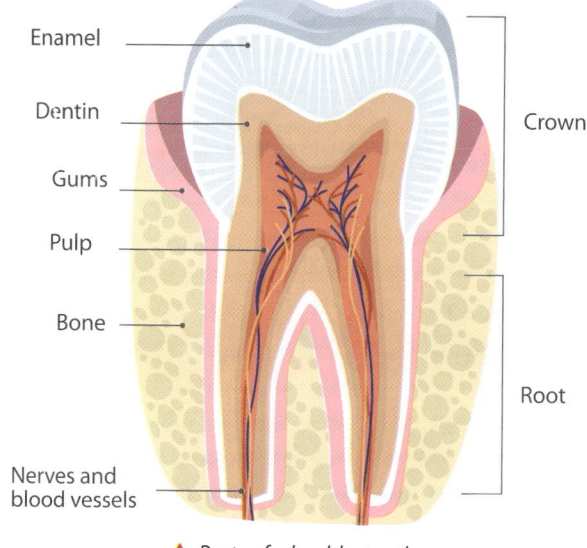

▲ Parts of a healthy tooth

In Real Life

The last molars on each side of the jaw are called wisdom teeth. They do not come out until you are 20 years or older. For many people, sadly, there is not enough space in the jaw for the wisdom teeth to come out properly, so they cause a lot of pain. Then these teeth have to be removed by the dentist.

▶ A wisdom tooth not growing properly

Tooth Decay

When you do not brush your teeth regularly or properly, a powdery, yellow plaque forms over your teeth. This is made of tiny bits of food and the bacteria that grow on them. If you allow the plaque to remain, the bacteria in them invade the gaps between your teeth and then their crowns, forming **caries**. This is the start of dental cavities. Once they form, brushing will not remove them. You have to go to the dentist, who will have to drill your teeth to remove the decay and put in a cement filling or remove the tooth altogether.

Isn't It Amazing!

A shark's teeth never stop growing. When they fall out, they are replaced by new teeth. Through its life, a shark may grow and lose up to 35,000 teeth.

▶ Sharks typically lose at least one tooth per week

Hunger & Thirst

Did you know that your brain plays a role in digestion? The brain is the part of our body which finds out that we are running out of energy and water. Then it tells our stomach to start rumbling or our throat to feel dry—and we start to feel hungry or thirsty!

Our body makes many chemicals called **hormones**. They travel through the blood to various organs. Together with nerves, these hormones tell our body to start or stop eating, to transfer digested food into the blood so that it reaches all tissues, and to turn them into energy.

▶ We can have a variety of foods and beverages to quench our hunger and thirst

 Feeling Hungry

There is a small region in the brain called the hunger centre. When the stomach is empty, it makes a hormone called ghrelin. Ghrelin passes through the blood to the brain's hunger centre, which then makes you feel hungry. This feeling makes your body start wanting food.

During an infection, the body sometimes switches off the hunger centre, so that you do not eat and allow the infectious pathogen to get food.

Feeling Full

It is very important to have some fat in your diet. Fat not only provides a lot of energy, but also helps the body know when to stop eating. Once you have started eating food, the intestine starts digesting food into the molecules that make it up—fats, proteins, sugars, vitamins, and minerals.

All these are taken into the blood, which gives it to the tissues. The tissue that takes up fat is called adipose tissue. When it has taken in enough fat, it makes a hormone called leptin. Leptin tells the brain's hunger centre to switch off and you start feeling full. The small intestine makes another hormone called incretin, which in turn tells the pancreas to make the hormone insulin. Insulin tells the brain that there is enough sugar in the blood.

If you eat too fast, your body does not have enough time to digest and know that you have eaten enough. You end up overeating! That is why we should eat slowly and chew our food. Some people have a medical condition in which they cannot switch off their hunger centres. They keep on eating and sometimes this leads to obesity.

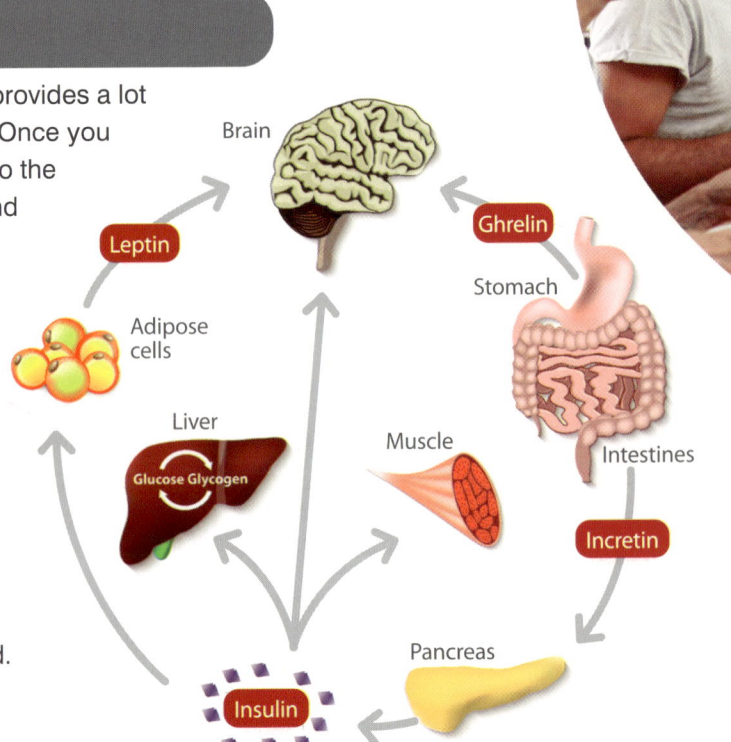

▲ The diagram represents the process from hunger to satiation

HUMAN BODY | STOMACH & DIGESTIVE SYSTEM

Feeling Thirsty

Alongside the hunger centre, the brain also has a thirst centre. When your body is running out of water, your blood becomes thicker and its pressure drops. This tells the brain that you need more water. It makes your throat feel dry, so you look for water.

Too much salt in your body also makes you feel thirsty. If you have not had enough water, the salt builds up in your body, and you start feeling dehydrated. This may make you dizzy. If you get dehydrated, take sips of water slowly, instead of drinking lots in one go.

▲ Your brain's thirst centre gets activated when your body is running low on water

▲ We should avoid overeating as it could lead to many health problems

Before eating — Ghrelin produced by cells in the gastrointestinal tract (Hunger)

After eating — Leptin hormone made by adipose cells (Satiety)

▲ Ghrelin is a hormone that makes us feel hungry, while leptin makes us feel full

In Real Life

Doctors and nutritionists recommend that we have a heavy breakfast and a light dinner. Yet most of us do the exact opposite. Scientists have found that this is because of our circadian rhythm that the brain processes that govern our daily behaviour. It makes us eat well since the brain expects that we will not get food for the next several hours. It also reduces hunger so that we do not wake up and spoil our rest because we feel hungry.

▲ Traditional English breakfast

Incredible Individuals

A hunger strike is a form of protest in which people go hungry for days till their protest is heard. Mahatma Gandhi of India was famous for going on hunger strikes to protest against the atrocities of British colonial rule.

The Stomach

Think of digestion and you immediately think of the stomach. It controls the pace of digestion and also makes a lot of hormones that communicate with the different parts of the body to either prepare to eat something, or to digest the eaten food and turn it into energy.

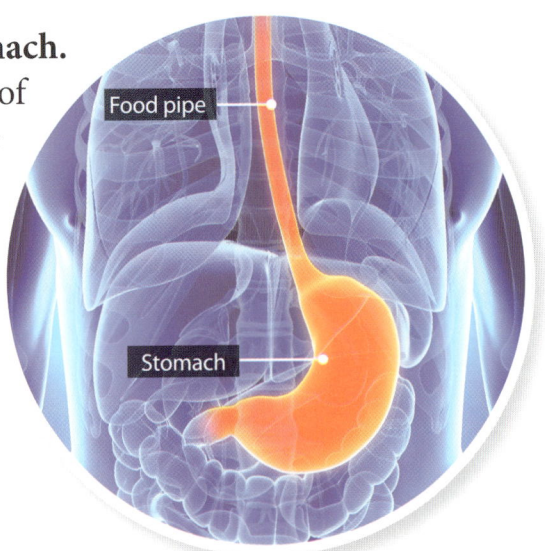

▲ The food pipe and the stomach

Getting Food to the Stomach

Between the throat and the stomach is a long pipe that passes between the lungs—the oesophagus, also called the food pipe. But food does not drop down it like water flows through a pipe. The oesophagus has to make the food go down, by squeezing and expanding, peristalsis. Between the oesophagus and stomach is a muscle called the sphincter, which prevents the stomach's acid from getting out.

Inside the Stomach

The stomach is a little like a balloon. It swells up as you eat food. But unlike a balloon, it has very tough walls so that it does not burst. The walls are made of three layers. The outer wall is made of muscles, the middle layer has arteries, veins, and nerves, while the inner wall is made of thousands of tiny folds called gastric pits.

The gastric pits make acid. This acid is released into the stomach when you start eating, and it breaks up the food into smaller chemical bits. The stomach also makes an enzyme called pepsin, which breaks up the proteins that you eat as part of your food into smaller bits called peptides.

The stomach muscles contract and expand when the stomach is full of food. This makes it churn, mixing up the food with the acid and pepsin and speeding up digestion.

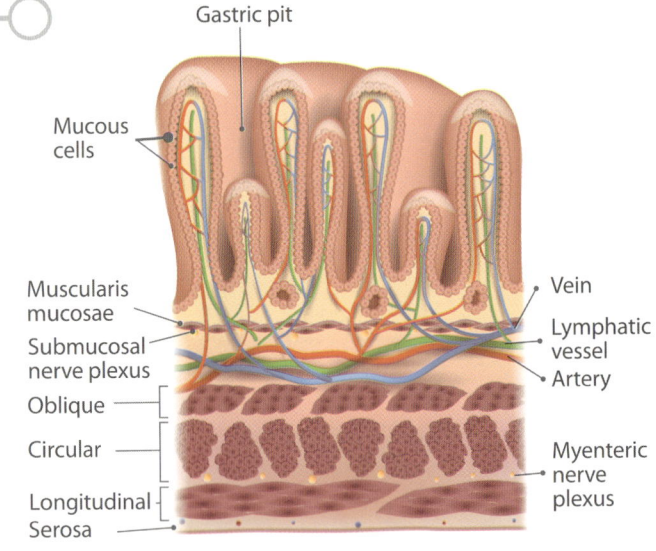

▲ The stomach does more than just digest food. It also makes many hormones that control digestion and hunger

Isn't It Amazing!

If you have ever visited a farm, you might have been told that a cow has four stomachs. Actually, the stomach of a cow is so large that it is divided into four parts. This is because the grass and leaves that cows eat are hard to break down and therefore, need a lot of time and space in order for them to be digested.

▶ Cows have a stomach divided into four parts that help them digest grass and leaves better

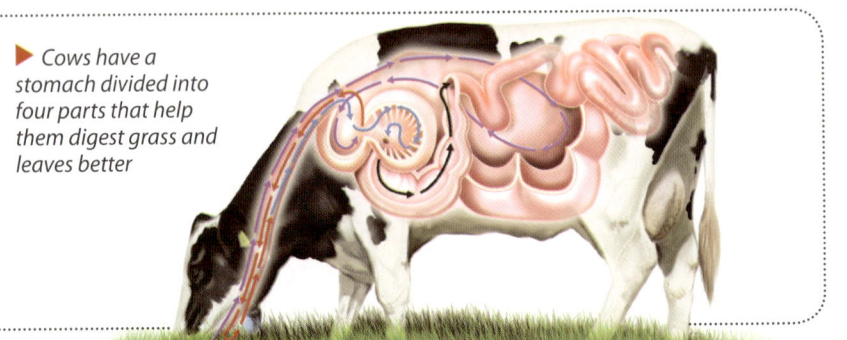

HUMAN BODY | STOMACH & DIGESTIVE SYSTEM

The Hormone Factory

The stomach makes a lot of hormones that help your body eat at the right time, eat enough, and avoid overeating.

- When food enters the stomach, the stomach makes gastrin which tells the gastric pits to release pepsin. It also makes histamine which tells the gastric pits to release acid and serotonin for the stomach muscle to start churning.

- Once you have eaten enough, somatostatin tells the stomach to stop and push the food into the small intestine. It also tells the pancreas and small intestine to stop performing their functions.

- When it is empty, the stomach makes ghrelin, the hormone that tells the brain to feel hungry. The stomach stops making it when food starts entering it.

- Gastrin also tells the small intestine to start its work when new food is coming and the large intestine to throw out the old, digested meal.

◀ *The stomach does more than just digest food. It guides the body to eat healthy*

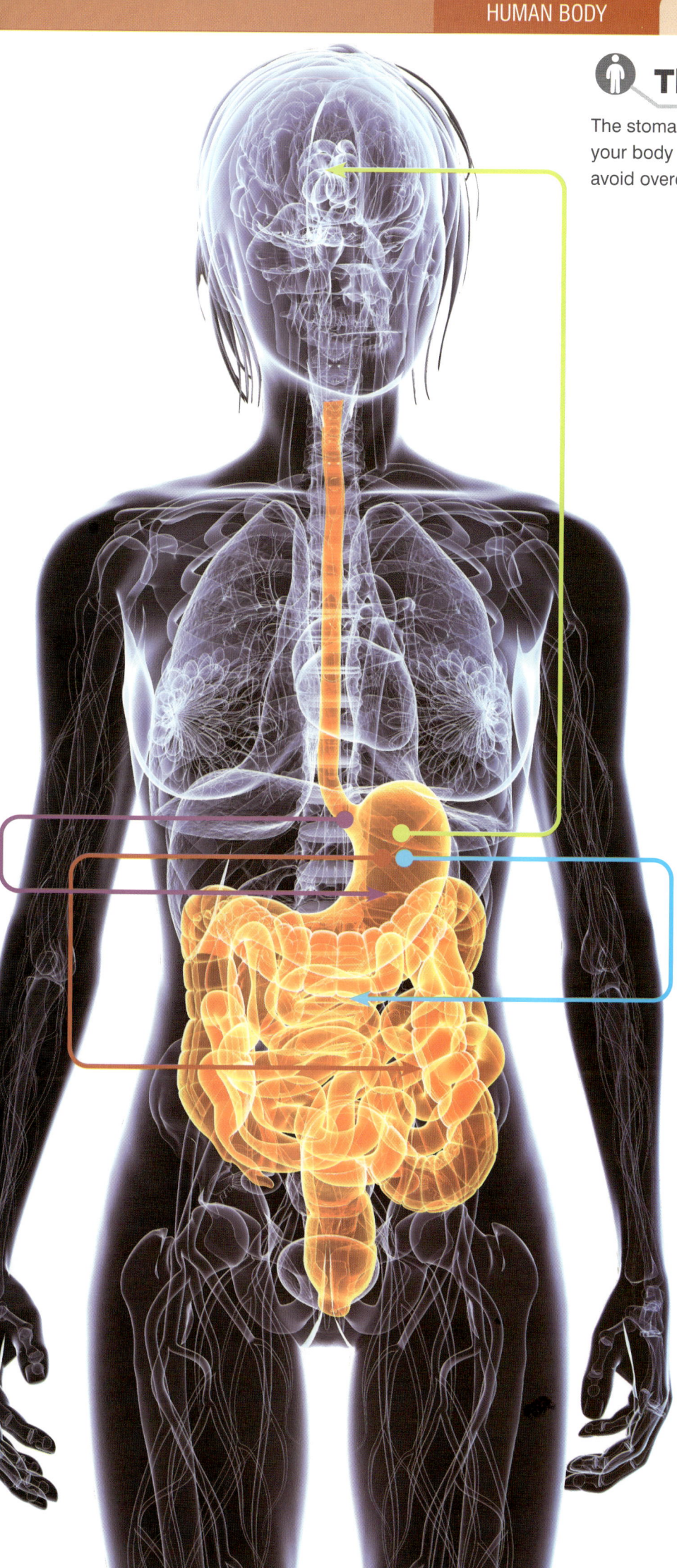

In Real Life

Somatostatin is also known as the growth hormone-inhibiting hormone. It regulates the endocrine system and is secreted by the D cells of the islets to obstruct the release of glucagon and insulin. It also prevents the release of the growth hormone when it is generated in the hypothalamus. So, there have been many studies conducted on the use of the somatostatin hormone on diseases like breast cancer and malignant lymphoma.

The Things that Make Us

30 per cent of our body is made up of proteins, carbohydrates, fats, vitamins, and minerals. The remaining 70 per cent is water. Those are exactly the things we need to eat and drink.

Proteins

This is what you need for building up muscle, repairing tissue, and growing up. You get a lot of it from meat, fish, eggs, cheese, beans, and lentils. Proteins are made of 20 kinds of amino acids. Our digestive system breaks up proteins into their amino acids and absorbs them. In the body, amino acids do a lot of things, including making the proteins that the body needs. You need to eat proteins every day as the body cannot store them.

Carbohydrates

You get them in milk, fruits, vegetables, and grains such as rice, corn, wheat, and oats. Our body needs them the most, because these are turned into our daily energy. Carbs are stored in the liver as glycogen, or turned into fats that can be stored in adipose tissue.

Fats

These are our body's energy source in an emergency, as they give more energy by weight compared to carbs. You also need them for absorbing some vitamins, for your nerves to work correctly and to make some important hormones. You get fat from butter, cooking oil, meat, and eggs. Fats are stored in adipose tissue, which is found in many tissues and organs in our body.

▼ Food items are grouped into different food groups

◀ Though people say we should drink eight glasses of water every day, this is not a fixed number. You should drink water whenever you feel thirsty

Vitamins

You need very little of these nutrients, but they are vital to our body's health. There are 13 of them and each does a different job:

Vitamin A, found in carrots, is needed for your eyes.

Vitamin B1, found in fish, eggs, and cabbage, helps your body burn sugar. The B-vitamins usually come together in most foods.

Vitamin B2 is needed to make healthy red blood cells (RBCs).

Vitamin B3 controls the amount of fat in the blood.

Vitamin B5 helps in general metabolism, along with **B7**.

Vitamin B6 is needed for making antibodies, haemoglobin, and for your nerves to work well.

Vitamin B9, found in peas and beans, is needed for repairing tissues and making new cells.

Vitamin B12 is needed for making healthy RBCs and keeping the nervous system healthy.

Vitamin C, found in citrus fruits, is needed for skin and bones.

Vitamin D, found in egg yolk and sea fish, helps your body absorb calcium and keep the immune system healthy.

Vitamin E, found in vegetable oils and nuts, helps your immune system.

Vitamin K, found in green vegetables and berries, keeps your bones strong and helps your blood clot properly.

Minerals

Minerals keep your body healthy in many ways. You need some in abundance, such as calcium—for the bones, phosphorus—for energy, magnesium—to maintain blood sugar levels, sodium and potassium—to help the nerves work properly, chloride—to digest food in the stomach, and sulphur—for healthy hair. Conversely, there are other minerals that are needed in small amounts, such as zinc—for immunity, iron—for haemoglobin, fluoride—for strong teeth, selenium—for healthy cells, manganese—for bone health, copper—for iron absorption, iodine—for making thyroid hormone, and cobalt—for Vitamin B12 that helps in producing RBCs.

Incredible Individuals

Elsie Widdowson (1906–2000) was a British nutrition scientist who worked out the chemical make-up of many foods. During the World War II, Britain faced a huge shortage of food. Soon, the British government found out that a simple diet of bread, potatoes, and cabbage could provide sufficient nutrition.

The Intestine

Though it is called the 'small' intestine, it is more than three metres long. It is taller than the tallest human that ever existed! To fit all this length, the small intestine coils and loops around itself, like a long rope would tangle around itself. The small intestine does the hard work of digesting all of your food.

Did you know that it also acts like a second tongue? It has the enterochromaffin cells which can sense chemicals that may be harmful to you. These cells send a message to the brain, which acts quickly to make you feel 'sick'. You stop eating and throw out what you have eaten.

▶ *The illustration shows the different internal organs of the digestive system. Notice the large and small intestines in detail*

In Real Life

Compared to chimps, our intestines are much shorter. Some scientists suggest that since we have been consuming cooked food for a very long time, our body has evolved to make shorter intestines. This is because cooking helps break down some food, making our body spend less energy on its digestion.

▲ *Chimpanzees eat things such as fruits, insects, leaves, and even smaller wildlife, so their digestion takes longer and consumes more energy*

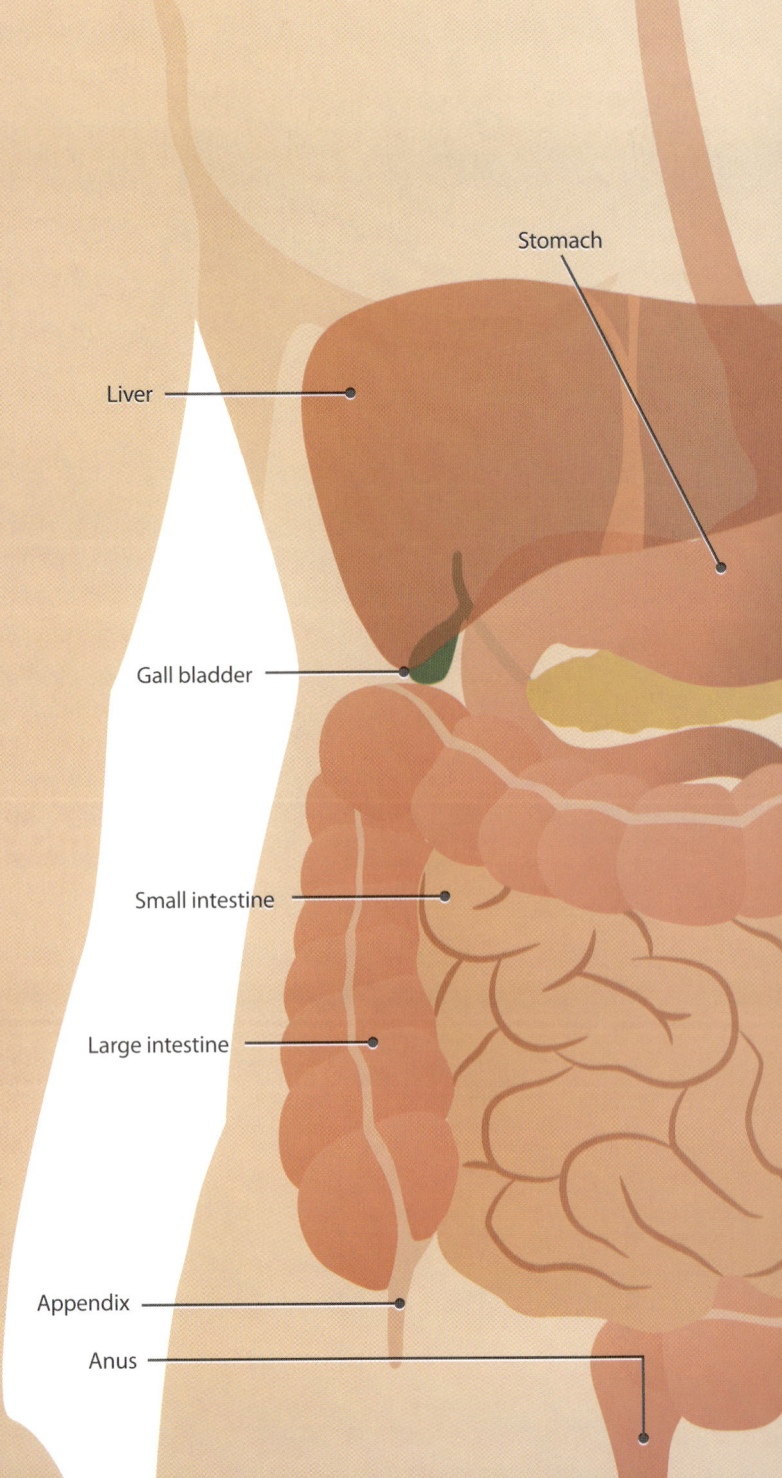

HUMAN BODY — STOMACH & DIGESTIVE SYSTEM

The Starter Course: Duodenum

The duodenum is the part that receives the partly digested food from the stomach. Two things happen here:

* Bile from the bile duct acts like soap and makes sure that the fat in your food mixes thoroughly with the rest.

* The pancreas pours a big mix of enzymes that digest the rest of the food. It keeps working till the food reaches the end of the small intestine.

The Main Course: Jejunum and Ileum

The jejunum starts digestion, and the ileum finishes it. The inner walls of both are made of thousands of tiny fingers called villi. Each of these villi are made of even tinier fingers called microvilli. They make the inside border of the intestine very large, so that enzymes can attack every part of the food. The walls of the intestine undergo peristalsis—contract and expand all the time— to keep pushing the food ahead. They also make some more enzymes.

◀ The intestinal wall is made up of thousands of villi and microvilli

▶ A microvillus with arteries and veins ready to absorb digested amino acids and sugars

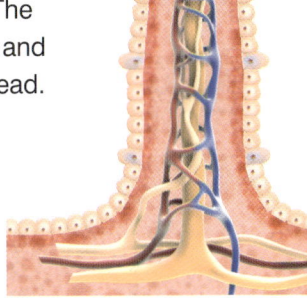

More Than Just a Digester

The intestine is busy with digestion all day long, but it does other things too. It has tied up with the immune system to create a protective tissue around it called MALT, which keeps bacteria from the intestine escaping into the blood. It also makes a lot of hormones that control digestion, just as the stomach does. One of these hormones is incretin, which tells the pancreas to make another hormone called insulin. Insulin tells the rest of the body to take up the sugar that the intestine just put into the blood.

Another hormone it makes is **serotonin**, which goes to the brain and spinal cord. It tells the nervous system to start the 'rest and digest' system. Serotonin makes the body feel sleepy and convinces the brain that the stomach is full.

Pancreas

★ Incredible Individuals

Travellers to Latin America often come down with a kind of diarrhoea called Montezuma's Revenge. This happens because of unhygienic hotel rooms or eating street food, where harmful bacteria irritate the intestine. It is named after Montezuma, the last king of the Aztecs who was killed by the Spanish conquerors.

▶ Montezuma fought wars all over Central America and doubled the size of the Aztec empire

The Liver

Did you know that the liver is the body's largest gland, weighing nearly 1.4 kg? It has to be, for it is the body's doctor. It is made of cells called hepatocytes. These cells make bile and also deal with a lot of chemicals that may enter the body—toxins in your food, the medicines you take for illness, and the wastes made by your tissues. The liver turns them into harmless chemicals that are gotten rid of through urine or bile.

A special vein—the hepatic portal vein—brings blood from the small intestine directly to the liver, where it can remove toxins that came from food and also take up glucose.

▲ Location of the liver in the body

▲ The liver does many things to keep our body healthy

 ## Storing Glucose

Once you have eaten, your blood suddenly has a lot of sugar in it. Most of this sugar goes to your cells where it is turned into energy. But there is still quite a lot left over. All this goes to the liver, where it turns them into long, stringy molecules called glycogen. If you miss a meal, your brain's hunger centre tells the liver to turn some of the glycogen back into sugar.

 ## Gall Bladder

The liver makes bile all day long, but it cannot store it. Instead, the bile is stored in the gall bladder, where it remains till the brain tells it that food has entered the stomach. Then it pours all the bile into the duodenum, so that the digestion can begin.

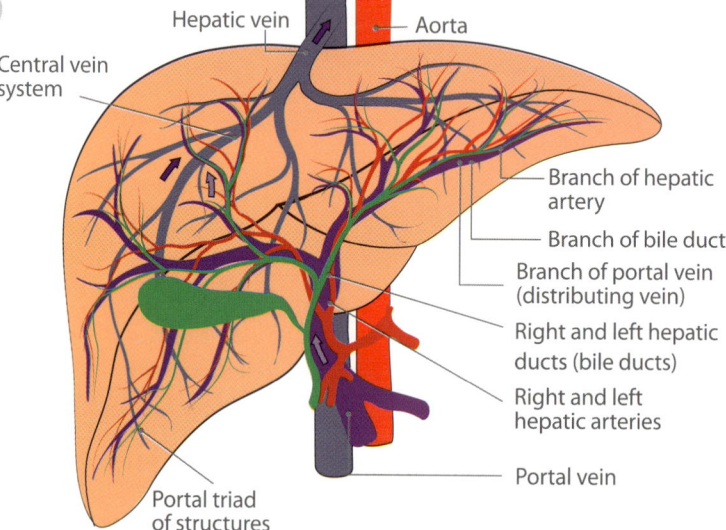

▲ The liver is full of veins that help it take up sugar from the blood very quickly

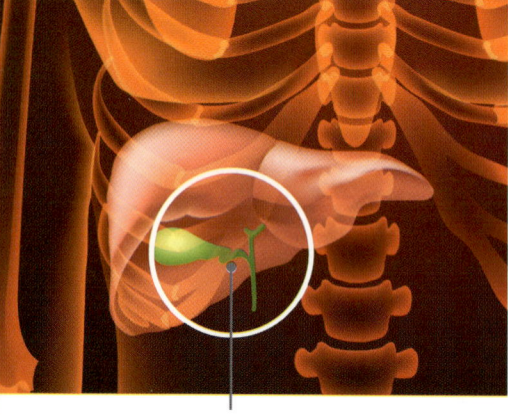

Gall bladder

▲ The gall bladder stores bile until needed. Bile helps in digesting fats

⊙ Incredible Individuals

The Ancient Greeks had a myth that the god Prometheus stole fire from the other gods and gave it to humans, who could use it to cook food and stay warm. For this, Prometheus was punished: he was chained to a rock and every day an eagle would peck out his liver. The liver would grow back in the night, and the eagle would come the next day.

▶ A painting showing an eagle pecking out Prometheus' liver. Human liver has the astonishing ability of repairing itself

The Pancreas

The pancreas is not one organ but two. Part of it works for the digestive system by making enzymes, while the rest of it makes hormones that help the body stay healthy and active.

The food coming from the stomach is full of acid, which stops the intestine's enzymes from working. Therefore, the pancreas makes a lot of sodium bicarbonate. This reacts with the stomach's acid to make salt, which is then absorbed into the blood. The carbonate is finally removed by your lungs as carbon dioxide.

$$HCl + NaHCO_3 \rightarrow NaCl + H_2O + CO_2$$

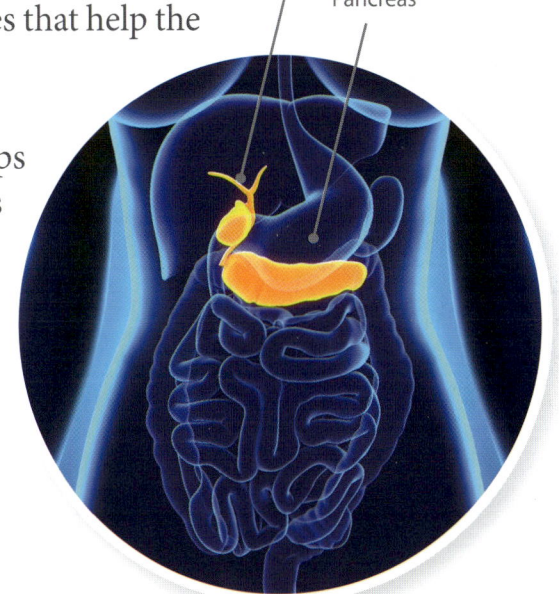

▲ *Pancreas and gall bladder*

Enzyme-making

The cells of the pancreas that make enzymes are called acinar cells. They make many different kinds of enzymes that can digest proteins, DNA, fats, and carbohydrates. The enzymes are collected in tiny pipes that finally flow into the main pancreatic duct.

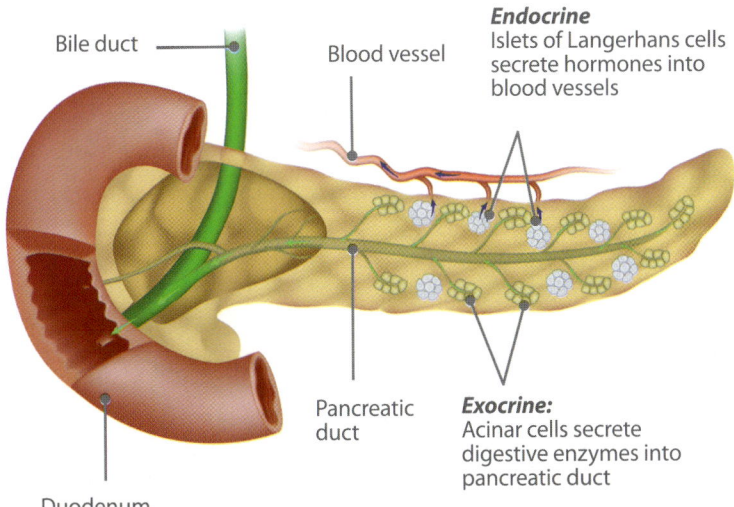

▲ *The pancreas makes both enzymes and hormones. Enzymes are poured into the duodenum, while hormones go into the blood*

Isn't It Amazing!

Diabetes is a disease we get if the body cannot make enough or proper insulin. Diabetic patients have to take regular injections to live normally. But did you know that most of the insulin they take, comes not from other human beings, but from specially modified bacteria?

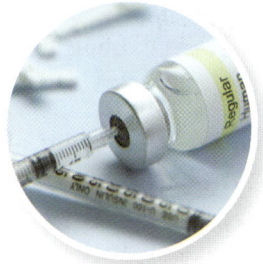

▶ *Artificial insulin made by genetically engineered bacteria*

Incredible Individuals

Insulin has got Nobel Prizes to more people than any other molecule.
* In 1923, F. G. Banting and J. J. R. Macleod got it for discovering insulin.
* In 1958, Frederick Sanger received the Nobel for finding out its molecular structure.
* In 1977, Rosalyn Yalow received one for finding a method to measure insulin in the body.

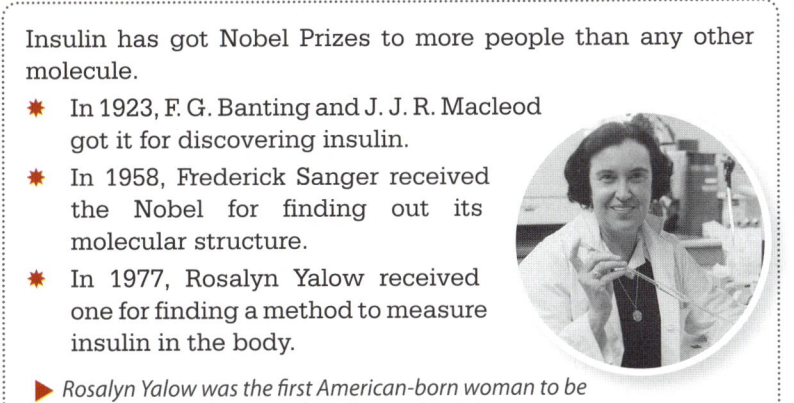

▶ *Rosalyn Yalow was the first American-born woman to be awarded the Nobel Prize in Physiology or Medicine*

Hormone-making

Hormones are made by a different kind of cell, the Islets of Langerhans. The pancreas makes two important hormones:

* Glucagon, which tells the liver to turn glycogen to glucose. This hormone is released when your blood has run out of sugar.
* Insulin, which tells your body's cells to take up sugar from the blood. It is released after you have eaten food.

Enzymes: How We Digest Food

We have talked a lot about enzymes, but what exactly are they? How do they digest food?

Our body makes several thousand different kinds of proteins that do many things. One kind are enzymes—proteins that help carry out biochemical reactions in the body. They do this by acting like tiny locks, each of which can only be opened by a matching key. In the intestine, these keys, called substrate are the food we eat, such as proteins, complex fats, and carbs. Each enzyme catches its own substrate and turns it into the product of digestion—amino acids, simple fats, and sugars.

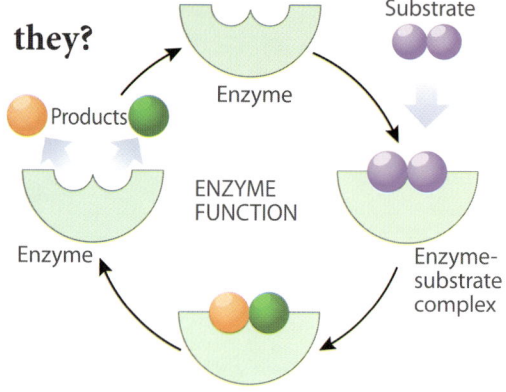

▲ *Each enzyme in the digestive system reacts with its own substrate, and nothing else*

Enzymes in the Mouth

Amylase is the first enzyme your food meets—in your drool or saliva. It turns starch into sugar. For instance, if you hold a thin slice of potato in your mouth for long enough, you can feel it turning sweet. Another enzyme is the lingual lipase, which starts digesting the fat in your food. Your saliva also has lysozyme, an enzyme that breaks up the walls of bacterial cells.

Stomach Enzymes

Pepsin is the main one in the stomach. It breaks proteins into smaller bits called peptides. The stomach also makes gastric lipase, which digests fats. Young children have another called rennin, which helps to digest the protein casein, found in milk and cheese.

◄ *After eating, the brain indicates that we rest, so digestion can happen properly*

Bile

To help you digest the fats you eat, the hepatocytes of the liver make bile. Bile is a yellowish-green soap-like liquid, made of bile salts and bile pigments. Bile salts have two parts—a fatty part that ties up fats in your food and a salt part that mixes them with the intestinal juice. Without bile, the fat would stick to the walls of the intestine and slow down digestion.

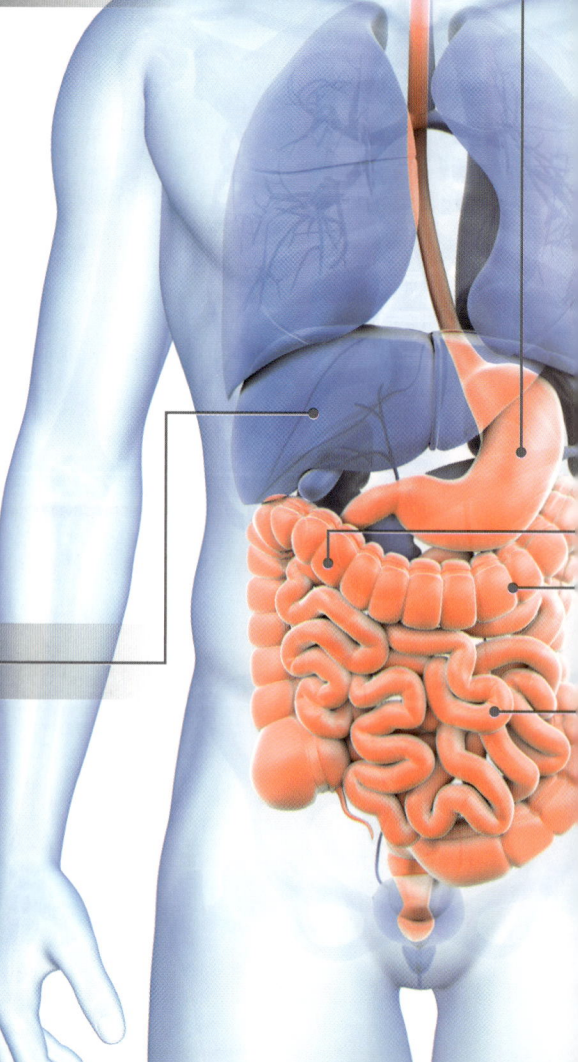

Pancreatic Juice

The pancreas makes a lot of enzymes, which it releases together into the duodenum as pancreatic juice. Here is what they do:

Enzyme	Acts on	End Product
Carboxy-peptidase	Peptides	Amino acids and peptides
Chymotrypsin	Proteins	Peptides
Elastase	Proteins	Peptides
Ribonuclease	RNA	Nucleotides
Deoxyribonuclease	DNA	Nucleotides
Pancreatic amylase	Starch	Sugars
Pancreatic lipase	Complex fats	Simple fats
Trypsin	Proteins	Peptides

 ## Enzymes from the Small Intestine

The small intestine's microvilli make the last few enzymes, which finish up digestion.

Enzyme	Acts on	End Product
α-Dextranase	Starch	Glucose
Lactase	Lactose	Glucose and galactose
Maltase	Maltose	Glucose
Sucrase	Sucrose (the most common sugar)	Glucose and fructose
Peptidases	Peptides	Amino acids
Enteropeptidase	Trypsinogen	Trypsin*
Pancreatic lipase	Complex fats	Simple fats
Trypsin	Proteins	Peptides

*The pancreas do not make trypsin directly, because it is a very powerful enzyme which can attack the intestine. So, it is made as trypsinogen, which another enzyme called enteropeptidase turns into trypsin.

 ## Lactose Intolerance

Lactose is the sugar present in milk, which is digested by lactase in the small intestine. Some people's bodies stop making lactase as they grow out of childhood. If they consume milk then, they cannot digest the lactose in it. Instead it goes into the large intestine, where bacteria turn it into gas. This is called being lactose intolerance.

 ## Incredible Individuals

Does the word pepsin sound familiar? In 1893, **Caleb Bradham** introduced a new soda to market as an aid to digestion. 'Pepsia' is Greek for digestion and cola comes from the name for kola-nut, which is used to make the drink.

Large Intestine and Appendix

Any picture of the digestive system will show it as a big tube curving around the abdomen. However, most of the food has been digested and absorbed in the small intestine. The stomach and the small intestine also make the hormones that tell the brain to get you to rest while you digest. So, what is left for the large intestine to do? Surprisingly, a lot!

Digesting Water

Did you know that every glass of water you drink, must travel the complete length of your digestive system, before it is absorbed? Why wait so long? That is because your digestive system needs all the water to keep the insides of the stomach and intestines moist till the enzymes have done their work.

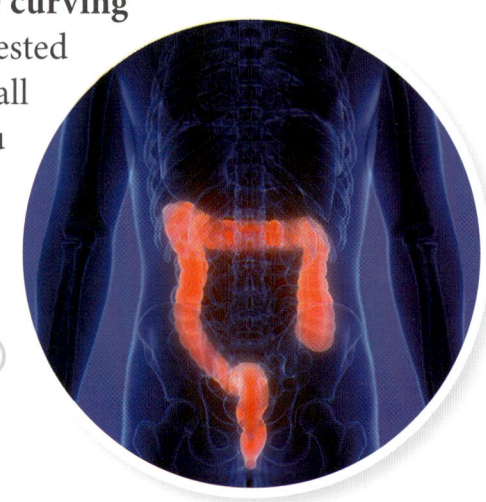

▲ The large intestine is also known as the colon

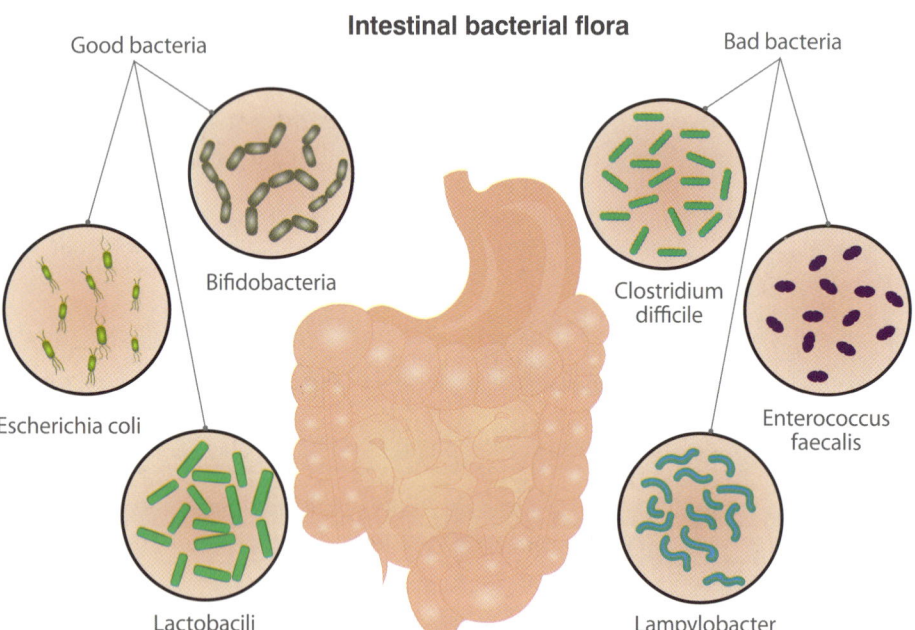

▲ Good bacteria make vitamins and keep bad bacteria away

Bacteria

Most of the digestive system is too dangerous for bacteria to live in. But did you know that trillions of them live in the large intestine, belonging to over 700 species? If you remove them, you would lose valuable friends who make some amino acids as well as Vitamins B7, B9, and K; who help you absorb minerals and break down some carbs. They also keep bad bacteria away.

The Mysterious Appendix

This is a tiny part of the large intestine, near where the small intestine meets it. In 2007, some scientists found out that it may act as a training centre for the immune system, and as a refuge for the bacteria that live in the intestine.

Isn't It Amazing!

A horse's appendix is nearly 4 feet long! In fact, this is true of all plant-eating animals, like cows, goats, and rabbits. The appendix is full of bacteria that help digest the complex carbs found in leaves and grasses.

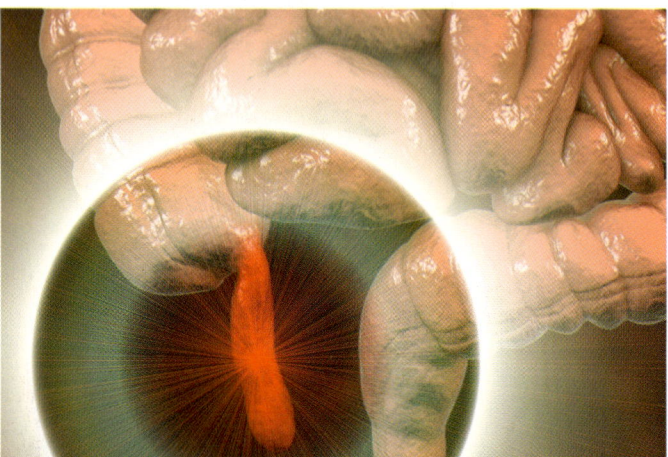

◀ The painful swelling of the appendix is called appendicitis

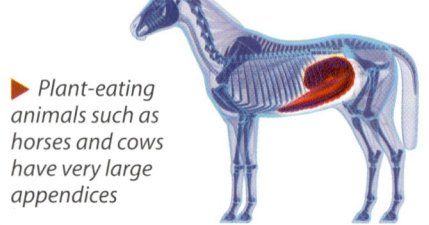

▶ Plant-eating animals such as horses and cows have very large appendices

Excretion: Eliminating Waste

Once the food has been consumed and all the protein, sugar, fat, vitamins, minerals, and water from it has been absorbed for the body's use, the remainder is completely unwanted, and must be eliminated. This is carried out through the process of excretion.

Regular exercise coupled with a timely eating and rest regimen helps in excreting wastes at regular intervals.

Keeping Toxins Out

This is actually the body's way of keeping itself safe from infectious bacteria, viruses, and fungi, or from some poisonous things in our food. The small intestine has cells that sniff out toxins. If they find any, the small and large intestinal walls contract very quickly, pushing the food to the anus. It is thrown out, undigested, and watery. Diarrhoea is a common symptom experienced by those suffering from cholera and dysentery.

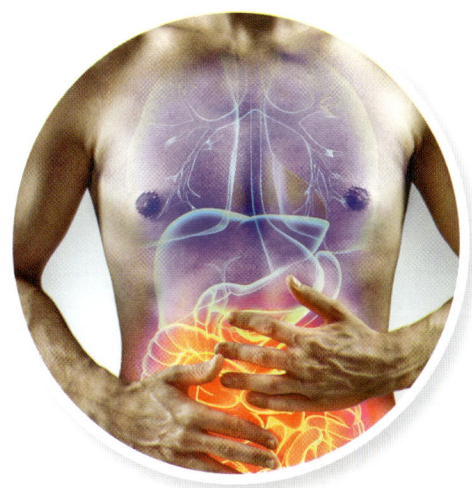

▲ *Diarrhoea causes pain, as the intestines try to shove out food that has gone bad or has bad bacteria in it*

Constipation

This happens if you have not had enough water. By the time food arrives in the large intestine, it becomes very dry, making it hard for the organ to move it along. It also happens if you do not have enough fibre in your diet. Fibre stimulates the intestines to keep shoving food forward. Add 5 to your age, that is the amount of fibre (in grams) that you need every day.

▲ *Grains and greens are rich in fibre, which prevents constipation*

Flatulence

If you have had lots of beans, you hope you do not get this in the classroom. Flatulence happens when the digestive system fills up with gas and has to throw it out, often making a whistling sound. Beans and fibre-rich foods have carbs that cannot be digested by the small intestine. But the bacteria in the large intestine can digest them, producing carbon dioxide. If your bacteria are not healthy, they might produce bad smelling gases.

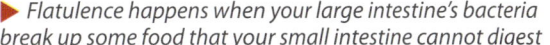 *Flatulence happens when your large intestine's bacteria break up some food that your small intestine cannot digest*

Taking Food to Those Who Need It

Once the digestive system has broken down all the food, it needs to get it to the rest of the body, which needs energy and raw materials for growth. But why do we need to digest food in the stomach and intestines before it goes into our bodies? That is because it has to be broken into chemicals like glucose and amino acids that are simple enough to be taken into the blood. This also makes sure that the body has control over what gets into it and what is excreted out.

Active Transport

Different nutrients are taken up by the body in different ways. Tiny little proteins in the walls of the intestine act like pumps to push sugars, minerals, and some vitamins into the blood. As the body needs energy to do this, it is called **active transport**.

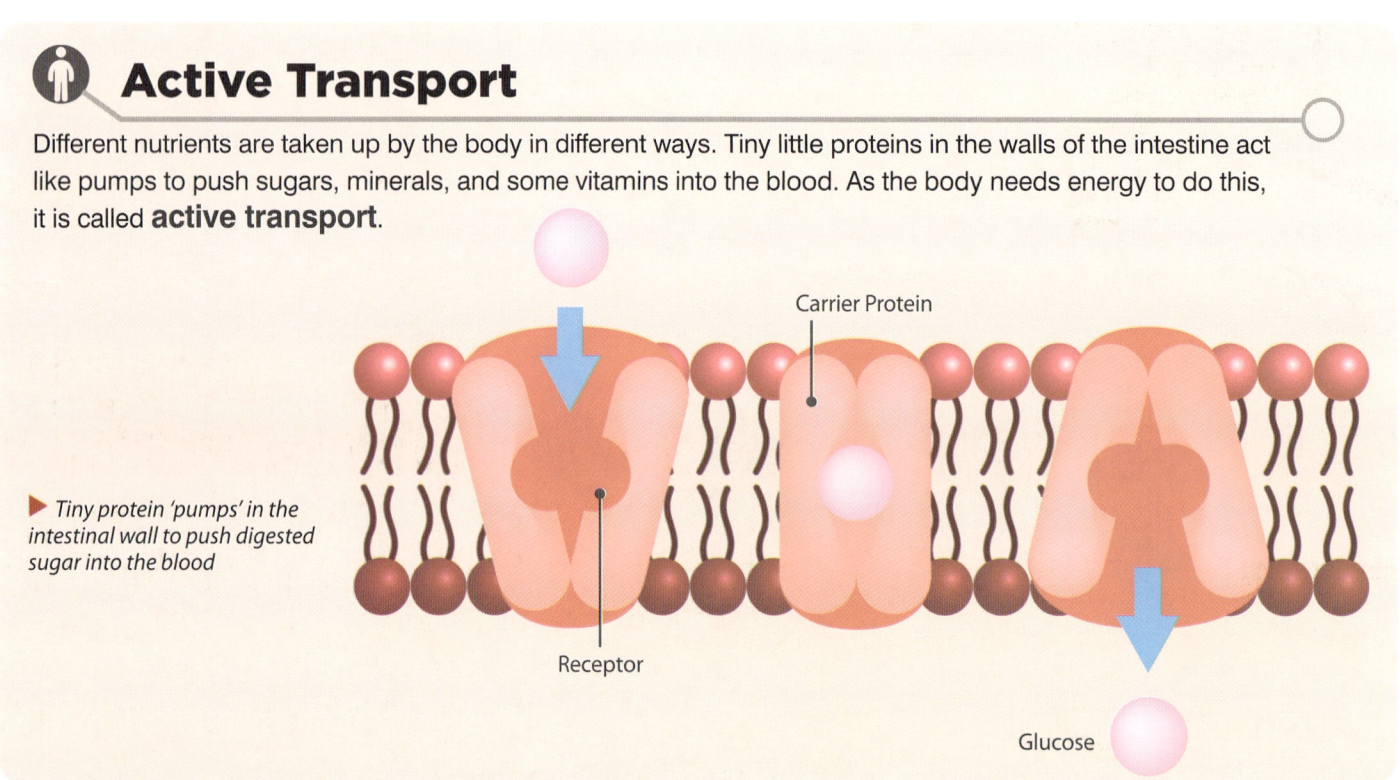

▶ Tiny protein 'pumps' in the intestinal wall to push digested sugar into the blood

Isn't It Amazing!

You know that pythons go hungry for many days and then suddenly catch one big meal. But did you know that digesting the meal can take up as much as 37 per cent of the energy the food gives? This makes the python go to a safe place and rest for a few days to digest all of its food.

▲ A python resting after eating a gazelle

Diffusion

This is how many fats and some vitamins are taken up. They pass through the intestine into the blood without any effort. That is why fatty foods seem to be digested faster than sugary foods.

Lipoprotein

Not all fats are allowed to go into the blood directly, as they can clog up your arteries causing plaque. Instead, they are tied with proteins to make **lipoproteins**, which act as escorts in the blood. If a plaque becomes too big, it can block the artery. The organ that gets blood from the artery begins to starve.

▶ Fats stick to the walls of arteries, causing plaque

Metabolism: Make It or Break It

What our cells do with the sugar, fats, and amino acids that they take in from blood is called metabolism. When they make things, it is called anabolism. For example, the amino acids are used to make proteins like haemoglobin, insulin, or collagen. The fats are used to make the membranes of cells. When they turn them into energy, it is called catabolism. One form of catabolism is cellular respiration.

Cellular Respiration

This is the way the body gets energy from glucose. Blood brings both glucose from the intestine, and oxygen from the lungs to each of the body's cells. They make the oxygen react with the glucose to form carbon dioxide and water.

$$C_6H_{12}O_6 + 6O_2 \rightarrow 6CO_2 + 6H_2O + \text{energy}$$

While doing this, a lot of energy is released. This energy, which is measured in calories can be used to:

* Make muscles work, so that you can move your arms and legs.
* Make energy to digest food.
* Grow new cells and tissues.
* Strengthen the immune system.

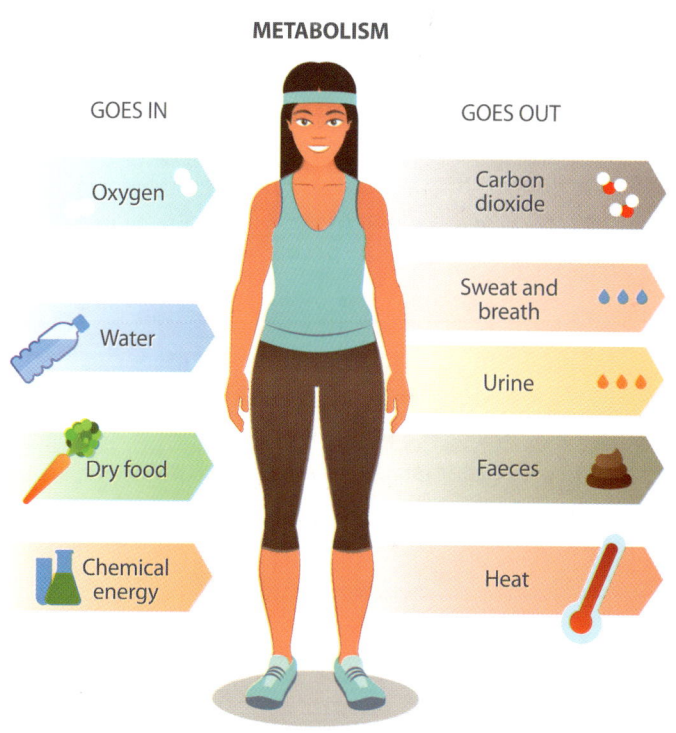

▲ A diagram showing what happens when you go running

In Real Life

During exercise or heavy work, our muscles use up a lot of energy. As blood cannot supply oxygen fast enough, the muscles use **anaerobic respiration** instead of cellular respiration. During this, the muscles release a lot of lactic acid into blood. Lactic acid builds up and causes pain. If you have overworked, the lactic acid may build up very fast to give you sudden pains called **cramps**.

▲ Pain due to lactic acid build up is temporary and is not a cause for a lot of concern

Burning Calories

Did you wonder why some people stay thin even though they eat a lot and some put on weight even though they eat little? That depends on the **basal metabolic rate**, the speed with which the body turns food to energy. If it uses up food very fast, the body is warmer and you feel more energetic. BMR differs from person to person and depends on your genes, diet, and overall health.

▲ Various exercises, including running, help us lose weight

The Urinary System

After your tissues have converted food into metabolites and energy, the remainder is waste. It is drained out by your blood and taken to the urinary system, which filters and removes it.

In a day, most people will pass 2,500 ml of urine and drink about that much of water. This is controlled by the brain and the pituitary gland. The brain makes sure that you do not need to pass urine when you are sleeping. The pituitary gland makes a hormone called the anti-diuretic hormone, which controls how much water you urinate, so that you do not become dehydrated.

 ## Kidneys

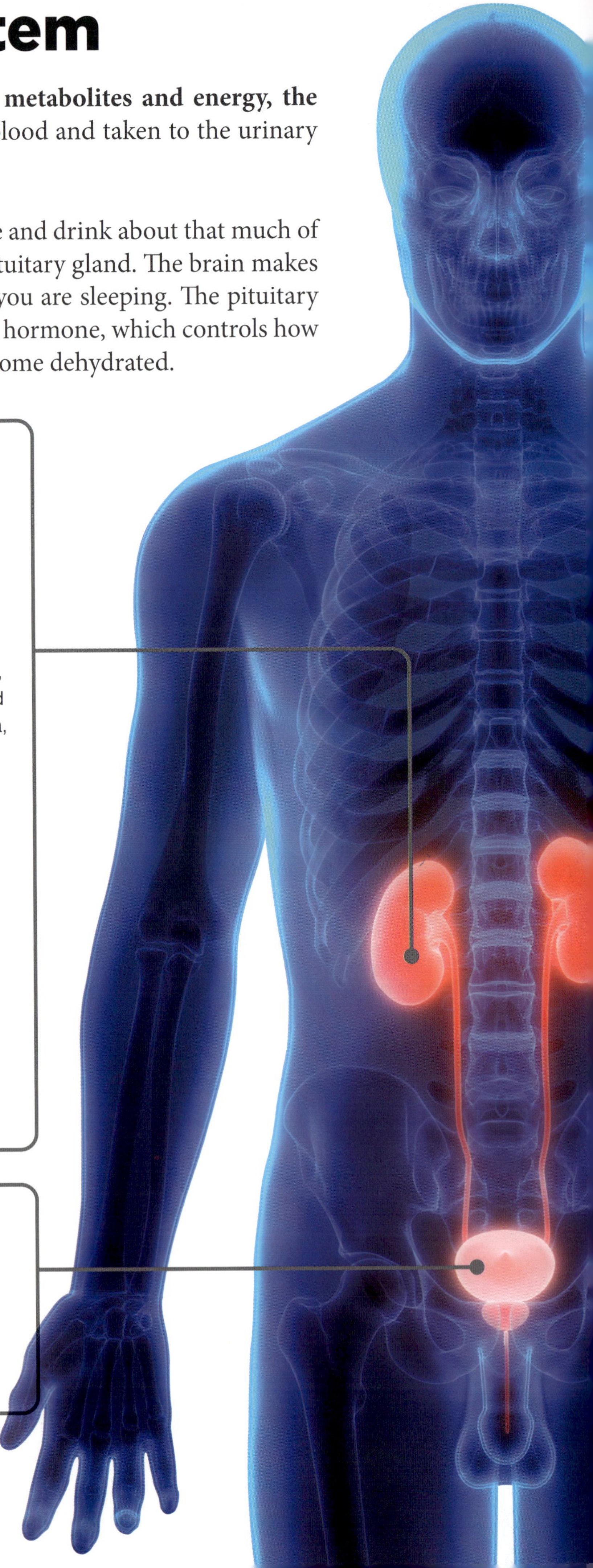

These are two bean-shaped organs near your intestines, which filter the blood. Each kidney is made up of thousands of tiny parts called **nephrons**. Each nephron is made of a tangle of blood capillaries, called glomerulus, which allows the plasma to seep out and enter the cup-shaped Bowman's capsule. The waste travels further through a long tube, shaped like a hairpin, the loop of Henle, where most of the water is reabsorbed by the blood. What remains finally is a thick soup of urea, other wastes, and a little water, which becomes urine. The urinary ducts merge at the centre of the kidney to become the **ureters**, which take urine to the bladder.

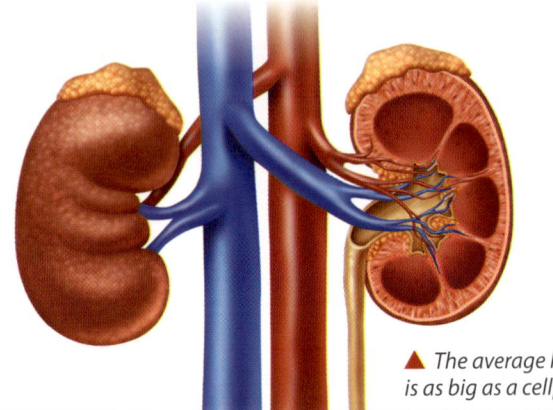

▲ *The average kidney is as big as a cellphone*

 ## Urinary Bladder

Did you know that the urinary bladder can hold up to half a litre of water? It collects the urine from the kidneys till enough has built up. Then the nerves of the bladder tell the brain that it is time to pass urine. The urine finally passes through the urethra out into the environment.

Isn't It Amazing!

Why do male dogs urinate so many times, unlike human beings who have to go all at once? That is because dogs use their urine to mark territory, by widdling on trees, car tyres, or streetlights. This is your dog's way of telling other dogs that he's the boss in this area. They can do this because they can control their bladders and let out only small amounts of urine each time.

▶ A male poodle marks a tree to say that this is his area; this is called urine-marking

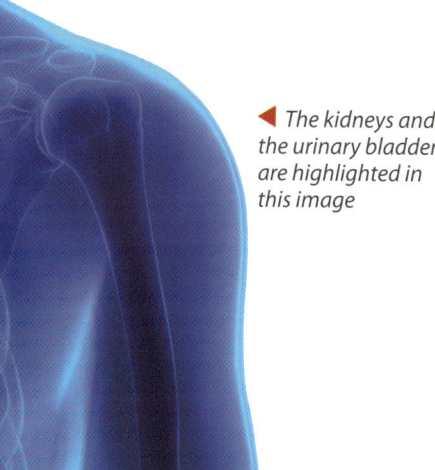

◀ The kidneys and the urinary bladder are highlighted in this image

▲ The colour of your urine says how healthy you are. Usually, a healthy person has clear urine.

Healthy and Unhealthy Urine

Did you know that the colour and smell of your urine can tell you how healthy you are? If you are healthy, have been eating regularly and drinking enough water, your urine should be clear, colourless, or pale yellow and there should be a lot of it. If you have thick urine which is dark yellow and smelly or cloudy, it means you are ill. If it is red, then your kidneys have been infected, because the red colour shows that blood cells have come into the urine.

Urine Testing

Doctors use a urine test to find out if you are ill. If there are germs in your blood, they may break through your kidneys and show up in urine. If you have jaundice, your urine will have bile. Too much stress may show up as creatine, a yellow-coloured substance. Incorrect metabolism may show many other biochemicals in urine.

 In Real Life

Urine is full of urea, which turns into ammonia over time. Stale urine is used to soak animal hides during the making of leather. The ammonia reacts with the hide and softens it and weakens the hair in it. In the past, urine was used as a source of ammonia for washing clothes and making gunpowder.

Tummy Troubles

Our digestive system sometimes goes wrong, giving us an upset stomach. It may be caused by disease, but often it is because we are not eating right. Good habits help keep tummy troubles away. Do not miss breakfast, eat slowly, and chew your food well. Ensure you are having fibre, which helps the intestines move the food along. Do not eat at odd hours, such as snacking while watching TV, or in the middle of the night, when your body is still digesting dinner. Do not eat when you are stressed, like when studying for an exam. Take a break instead.

▲ A little boy with indigestion. Healthy eating habits keep it away

Indigestion

Indigestion is also called **dyspepsia**. You know you have got it when you:

* have pain in the tummy (abdomen)
* keep belching or giving off wind
* do not feel hungry
* feel like vomiting
* have diarrhoea or constipation
* feel that your chest is burning (**heartburn**)

It usually happens when you are under stress due to an exam or are not eating properly on time, or are sensitive to some foods, such as nuts then this may induce indigestion. In Adults indigestion, maybe caused due to excessive smoking and very high consumption of coffee or alcohol.

Ulcers

Ulcers are serious ailments that happen when your stomach or duodenum's lining is eroded by the acid in it. These ulcers are not just painful but can also cause severe allergies. Ulcers happen because of stress, and also because of infection by a bacterium called **Helicobacter pylori**.

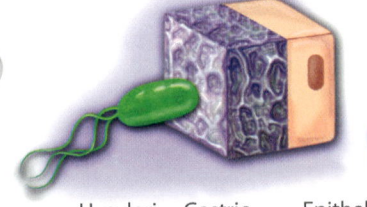

H. pylori   Gastric mucin gel   Epithelial cells

H. pylori raises pH, mucin de-gels

▲ The diagram shows how Helicobacter pylori reaches the stomach

Allergies

Some diseases of the digestive system may happen because of allergies. Coeliac disease, which damages the intestines, is one such disease. You get it if your body is allergic to gluten, a protein found in wheat. Others happen because of infections such as *Helicobacter pylori*, malnourishment, poor eating habits, or psychological conditions.

Incredible Inadividuals

Barry Marshall was an Australian scientist who was trying to show that Helicobacter pylori caused ulcers, but no one believed him because they thought nothing could live in the acidic environment of the stomach. Then one day, Barry Marshall drank a tube full of this bacterium. Within a week, he began showing the symptoms of ulcers, and could show that he was really infected. He got the Nobel Prize in Medicine in 2005.

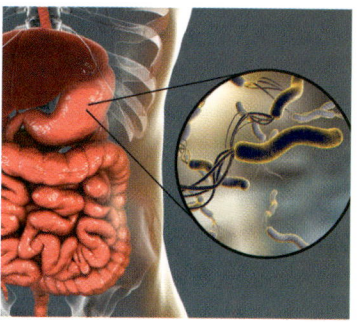

▲ Helicobacter pylori the main cause of peptic ulcers; and it can also cause gastritis and stomach cancer

Irritable Bowel Syndrome

Do you feel crampy and moody most of the time? You may have **irritable bowel syndrome**, which causes diarrhoea in some people and constipation in others, and sometimes both. This is because something affects your large intestine, but we don't know what yet. Eating regularly, consuming fibre, and keeping bad bacteria out keeps you healthy.

Beriberi and Scurvy

Not having enough vitamins can cause diseases. For example, not having enough Vitamin B1 causes **beriberi**, a disease in which you feel numb and weak, all the time. Lack of Vitamin C in the body causes bleeding gums, slow wound healing and tiredness. This disease is called **scurvy**.

Isn't It Amazing!

In the early 19th century, sailors in the British Navy were called limeys. This was because they were all forced to drink lemon juice, which contains lots of Vitamin C and prevents scurvy.

Eating Disorders

Some disorders cause people to eat too little or too much food. In anorexia nervosa, people eat too little because they feel they are fat all the time and become severely malnourished. Another disease is **binge-eating disorder**, where the brain's hunger centre does not work correctly, making people feel hungry all the time.

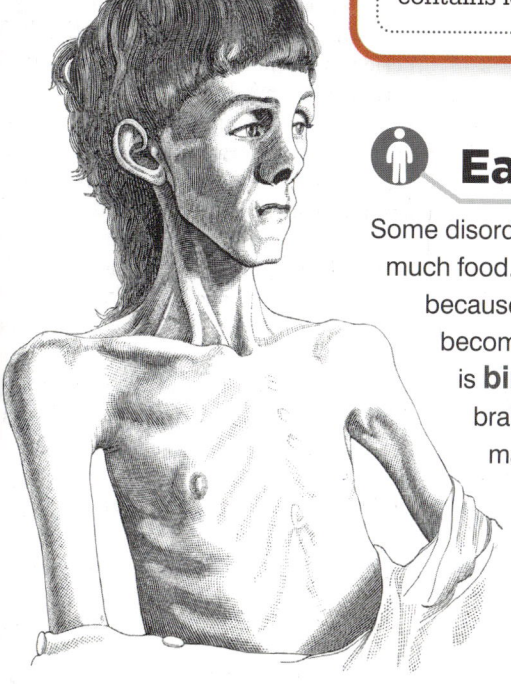

◂ The illustration shows a person suffering from anorexia nervosa

Kwashiorkor

Babies who do not get enough protein in their diet get **kwashiorkor**. In many countries, foods rich in protein, like meat, eggs, and milk are too expensive, and people instead eat food rich in carbs, such as grains and yams. It causes swollen bellies, thin limbs, and frequent diarrhoea.

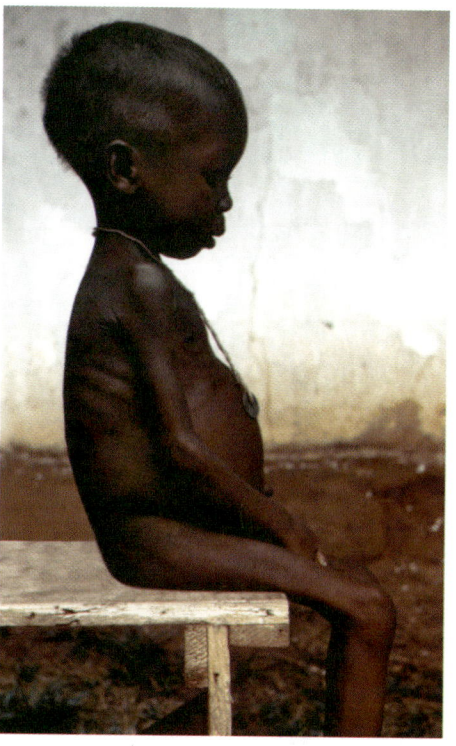

▲ Kwashiorkor is a disease brought on due to a severe dietary protein deficiency, and this child, whose diet fits such a deficiency profile, presented with symptoms including edema of legs and feet; light-coloured, thinning hair; anaemia; a pot-belly; and shiny skin

In Real Life

In many diseases or in cases of starvation, your body loses water and minerals very quickly. To replace them, the World Health Organisation (WHO) recommends an **oral rehydration solution**, which has sodium citrate, potassium chloride, and glucose. It was tested in refugee camps during the 1971 Bangladesh War by Indian doctor Dilip Mahalanabis. It is now used around the world.

A Healthy Mind & Healthy Body

A diet that has the right amounts of proteins, carbs, fats, vitamins, minerals, and water is a balanced diet. However, because of poverty, stress, bad eating habits, or even mental conditions, you may fall short of one or more of these nutrients. This is called **malnutrition**.

If a pregnant woman is malnourished, her baby can be born with many birth defects, including low intelligence, poor immunity, and other defects such as blindness or deafness. This is particularly true of micronutrients like vitamins and minerals. Malnutrition as a young child may also lead to low scores in school.

◀ One must always try to have a hearty and balanced breakfast

▶ Eating a light snack in the evening prevents us from feeling hungry before dinner

Breakfast is Essential... Even If You're in a Hurry

Late for school and not interested in breakfast cereal? But you shouldn't skip it, because breakfast is the most important meal. When you wake up, your body has digested its dinner and both your brain and muscles need fresh energy. A good breakfast does that—it gives you the energy to stay awake and alert in class, and the energy to play in the break.

Snacks are Good for You

Children eat less in meals and like having a snack now and then. Tell mum and dad that that's right for you. It keeps your energy levels up. Fries, crisps, and chocolate are high in sugar—have a little if you've not had a meal in a long time. But the best are fruits, muffins, bread-and-butter, or nuts.

In Real Life

Athletes go bananas for bananas! That's because bananas are packed with sugar and vitamins and give an instant dose of healthy energy. They are also rich in fibre and stop constipation. Go ahead, have a banana, and then another.

Eat Less, but Eat More

Puzzled? We mean to say, eat less in each meal, but have more meals. Research says that fewer meals make you overeat, while more meals help you eat smaller, healthier portions. A 3-ounce bag of chips actually has 3 servings, so don't eat it all at once. Share it with friends or eat it over 3 days.

▶ Smaller portions make for healthier meals

Sugar and Sleepiness

After eating foods rich in proteins and carbohydrates, like meat and rice, we feel sleepy. This is because the sugar and amino acids entering the blood make the pancreas produce more insulin. Insulin pushes sugars and amino acids into the tissues, except for one amino acid called tryptophan. This instead goes into the brain, where it is turned into serotonin. Serotonin tells the brain to slow down and make the body go to sleep. Sleep is good for digestion as it needs energy. If you don't get good sleep, you can have problems digesting food.

◀ A quick afternoon nap helps digest your lunch

Coffee is Good... in Small Amounts

Some adults love drinking coffee, sometimes up to 10 cups a day. The caffeine in it keeps them awake and refreshed. It helps them pass urine and maintain water balance. But it is good only as long as they drink under 400 mg of caffeine—which is about four cups of coffee—in a day. Too much caffeine causes more acid to be released in the stomach and stops your body from absorbing calcium. So if mum or dad have a lot of coffee, tell them to go easy!

▶ The rule for coffee—do not drink more than four cups in a day

Incredible Individuals

Gerty Cori (1896–1957) got the Nobel Prize in 1947, with her husband, for finding out what happens when muscles work very hard. They make lactic acid, which goes to the liver to be made into glucose again. During this time you feel tired and need to rest. Gerty Cori did most of her work when men were still opposed to women doing scientific research, but she did not get discouraged and kept working.

Gall & Kidney Stones

Diseases like liver cancer, infections, or anaemia cause the gall bladder to form gallstones. Similar things happen in the kidney too, where **kidney stones** are formed. Doctors know you have got them if:

* you feel a lot of pain in your belly or your back;
* there is blood when you urinate or excrete;
* you get fever and shivering;
* your urine looks cloudy or your excreta looks pale;
* your urethra burns when you pass urine.

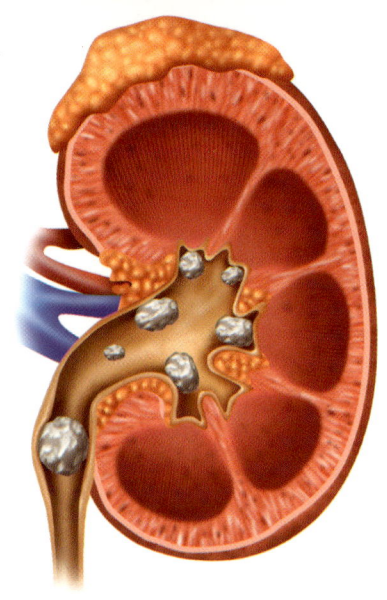

▲ *Kidney stones are more common in summer and in hotter climates*

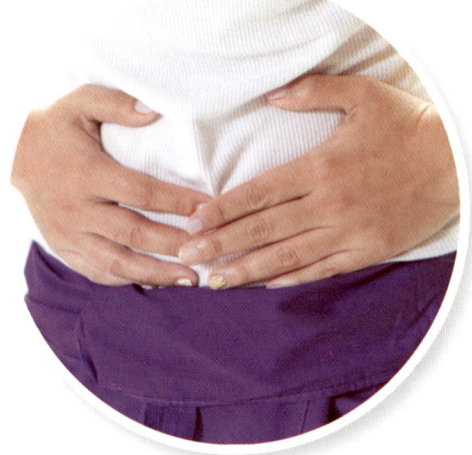

Gallstones

A gallstone is formed when bile becomes very thick due to lack of water, and the bile salts begin to crystallise. These tiny crystals then block the bile duct, causing a lot of pain and indigestion, because bile does not reach the intestine. Gallstones may often have to be removed by surgery.

◀ *Gallstones cause lasting, severe pain and indigestion*

▶ *Kidney stones taken out after surgery. They can grow upto 5 mm wide*

Kidney Stones

Drinking a lot of water every day usually stops kidney stones from forming. If there's too little water in the body, the salts and urea filtered from the blood into the kidney begin to crystallise. These stones block the ureters and can cause a lot of pain.

Ultrasound

Certain sounds with an ultrasonic frequency are inaudible to the human ear. Such sounds are known as ultrasound. The vibrations caused by this sound can help break up kidney stones and gall stones in the body into tiny pieces that come out easily and unblock the kidneys or gall bladder. This helps you avoid surgery.

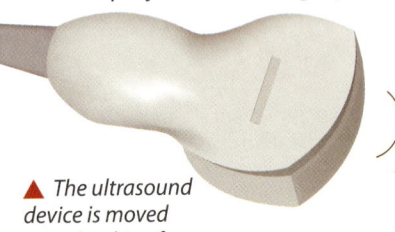

▲ *The ultrasound device is moved over the skin of the stomach*

Incredible Individuals

Alexander, the Greek conqueror, died in great pain when he was just 33. Some medical historians believe that the symptoms of his disease described by the writers of his time, point to an inflammation of the gall bladder, **cholecystitis**, which is usually caused by gallstones.

▶ *Alexander of Macedon is commonly known as Alexander the Great*

When Organs Give Up

Our kidneys and digestive system are at work throughout our lives, without any rest. Sometimes, because of stress, poor nutrition, disease, or old age, they may fail completely. If the artery that supplies blood to the organ is clogged with a plaque, the organ may be starved of nutrients and oxygen, and its cells begin to die.

Kidney Failure

When one or both kidneys fail, they are no longer able to filter out toxins from your blood. These toxins then accumulate, causing all kinds of damage to your tissue. Most people with kidney failure have to go to the hospital for **dialysis**, every week. This is a medical procedure by which your blood is filtered by a machine and the clean blood is put back into you.

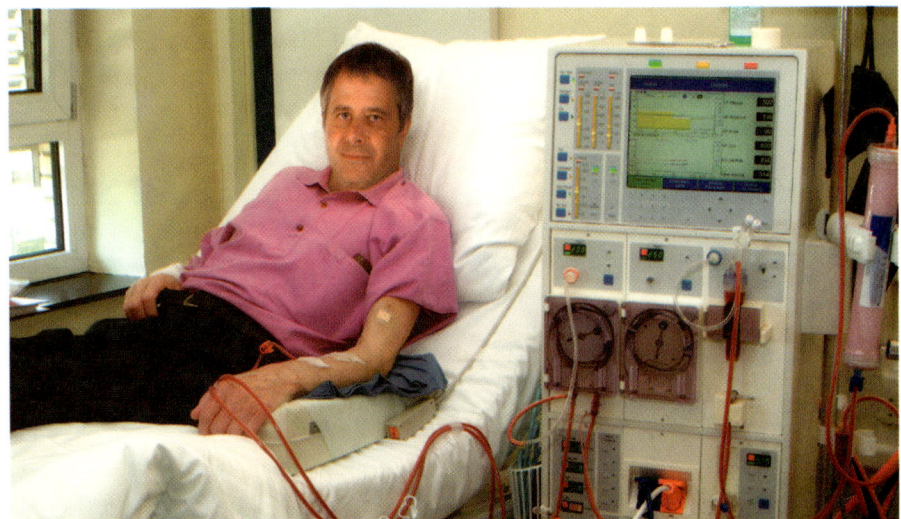

▶ Early treatment of kidney stones can prevent chronic kidney disease or renal failure that would require dialysis treatment or a kidney transplant

Organ Transplants

Did you know that you can have your large intestine, appendix, or spleen removed without much harm? But some organs, like the liver, heart, or kidneys cannot be removed without putting in a new one. This is called an organ transplant.

You need a healthy human **donor** who can give you one of their organs if your kidney or liver fails. But for a donation to work, the donor must match you as closely as possible genetically, such as a parent, sister, or brother. Otherwise, your immune system will treat the new organ like it treats germs and destroy it.

In Real Life

Scientists are trying to find out how you can grow your organ again, rather than get it from another person. They do this by studying the **stem cells**, which are cells that grow in all organs, and can turn themselves into any kind of cell. If the right biochemical instructions are given, stem cells can become different tissues and finally make a new organ.

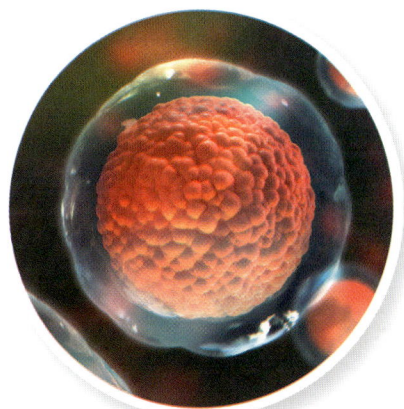

▲ The photograph shows a close-up of a cell in the human body. Stem cells are studied in the field of cellular therapy and regeneration

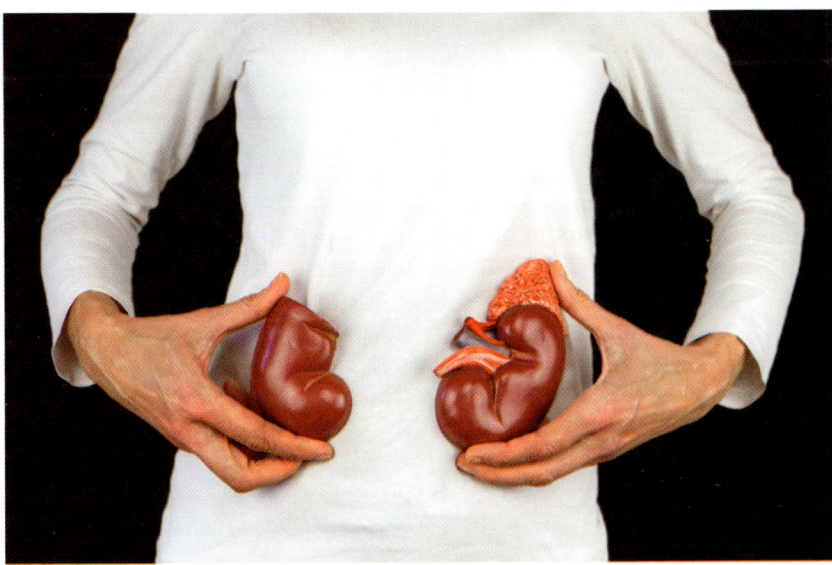

▲ Living kidney donation help in getting a kidney faster

Word Check

Brain & Nervous System

Amygdala: It is the part of the brain that controls emotions and memory.

Atrophy: It is the shrinking of a tissue due to illness.

Axon: Part of a neurone that takes the action potential to the next neurone

Basal ganglia: It is the nuclei in the inner brain that connect with the higher brain and help with decisions.

Cone Cells: They are the cells in your eye that make out different colours.

Cortex: It is the part of the cerebrum that does higher functions such as calculations after making sense of stimuli.

Dendrites: The parts of a neurone that take the action potential from the previous neurone or sensory organ

Dopamine: It is a neurotransmitter that helps regulate the movements a person makes as well as their emotional responses to a stimuli.

Ganglia: They are the clumps of neurone cell bodies in the brain and spinal cord.

Grey matter: It is the part of the brain and spinal cord that is made of the cell bodies of neurones.

Gustatory receptor cells: The cells in the tongue which pick up taste and relay it to the sensory nerves

Gyri: They are the folds of the cerebral cortex full of grey matter.

Hippocampus: Part of the temporal lobe responsible for memory

Homunculus: It represents the way the sense of touch from each part of the body is placed on the parietal lobe.

Hormones: They are the chemical messengers that travel through blood.

Interneurones: They are neurones in the spinal cord that control reflexes.

Involuntary movements: They are the movements within your body, such as heartbeats, which you cannot control.

Lower motor neurones: They are the neurones that carry messages from the spinal cord to the muscles.

Medulla oblongata: It is the part of the brain that connects it to the spinal cord.

Melanin: It is a natural skin pigment that gives our eyes, skin, and hair their colour.

Meninges: It forms the protective covering of the brain and the spinal cord.

Mirror neurons: They are the neurones that make the body do what it sees.

Myelin sheath: It is the protective covering around axons of neurones.

Neurology: It is a field of medicine that studies how the nerves work and what can go wrong with them

Neuromuscular junctions: They are the synapses between neurones and muscles.

Neurones: They are the cells that take messages from one part of the body to another.

Neurotransmitters: They are the chemicals in synapses that turn up or turn down neurones.

Nodes of ranvier: They are the gaps in the myelin sheath covering an axon.

Non-REM sleep: It is the kind of sleep when you are in deep sleep or sleepwalking.

Noradrenaline: It is a hormone and a neurotransmitter.

Nuclei: They are the clumps of grey matter in the brain and spinal cord.

Olfactory bulb: It is the part of the brain that deals with smell.

Olfactory receptor cells: They are the cells in your nose that catch smelly chemicals.

Papillae: They are the taste organs in your tongue.

Photoreceptor cells: They are the cells in the retina that are sensitive to light, made up of rod cells and cone cells.

Pituitary gland: It is the gland that makes many hormones, it is controlled by the hypothalamus.

Rapid eye movement sleep: It is the part of sleep when you are dreaming.

Retina: It is a thin membrane which lines the back wall of the eye.

Rod cells: They are the cells in the retina that sense whether it is light or dark.

Schwann cells: They are the cells that make myelin and cover the axons of neurones.

Sulci: They are the grooves between folds (gyri) of the cerebrum.

Synapses: They are the connections between neurones that act as switches.

Synaptic vesicles: They are the bags at the end of axons filled with neurotransmitters.

Ventricle: It refers to a chamber of the organ, in this case, the brain.

White matter: It is the part of the brain and spinal cord made of the axons and dendrites of neurones.

Heart & Circulatory System

Anaemia: It is a condition that results in weakness caused by lack of red blood cells and/or haemoglobin.

Aneurysm: It is caused by a bulge or weakness in the wall of the arteries.

Arrhythmia: It is when the heart beats too fast, too slow, or irregularly.

Arteries: They are the blood vessels which carry blood away from the heart.

Atria: They are the upper chambers of the heart.

Blood Pressure: It is the pressure maintained in your blood vessels by the regular pumping of the heart.

Blood transfusion: It is the process of receiving blood from another person.

Capillaries: They are the smallest blood vessels.

Cholesterol: It is a complex fat molecule present in many cooking oils and is required, in small amounts, by the body to make hormones.

Circulatory system: It comprises of the

heart and the blood vessels—arteries, veins, and capillaries.

Continuous capillaries: They let fluids in and out at the joints between cells in their walls.

Deoxygenated blood: It is the blood that is low on oxygen.

Diastole: It is the phase of the cardiac cycle in which heart muscles relax to let the heart fill up with blood.

Fenestrated capillaries: They are porous and allow more fluids to enter and leave the bloodstream.

Haemoglobin: It is a special protein in the RBCs that contains iron.

Heart attack: It occurs when the flow of oxygen-rich blood to the heart is not enough.

Heart failure: It means that the heart is not pumping enough to meet the body's requirements.

Heart Transplant: It is a medical procedure in which surgeons replace a sick heart inside the body, with a healthy one.

Homeostasis: It refers to the ability of birds and mammals to maintain constant body temperature.

Hypertension: It refers to a condition where blood pressure is above the normal range.

Hypotension: It refers to a condition where blood pressure is below the normal range.

Leukaemia: It is a type of cancer that affects the white blood cells.

Lymph: It is fluid from the blood, without red blood cells, which flows through the lymphatic system.

Lymphatic system: It helps carry fluids around the body. Therefore, it is helpful to the circulatory system. It also helps to protect our body from diseases and infections.

Macrophages: These are immune cells that eat and destroy pathogens in the body.

MALT: Mucosa-Associated Lymphatic Tissue is an organ of the lymphatic system which is associated with mucous tissue like the mouth, throat, intestines, etc.

Oxygenated blood: It is the blood that is rich in oxygen.

Pacemaker: It is a medical device that regulates the heart's contractions by sending it signals.

Pericardium: It is the sac that protects the heart.

Plasma: It is the only liquid component in the blood. It helps facilitate blood flow in our body.

Platelets: They help to clot out blood whenever we have a cut or a wound.

Red blood cells: It is the red component of the blood.

Sinusoid capillaries: They are enlarged; their cell-sized openings allow blood cells to enter and leave the bloodstream.

Sphygmomanometer: It is an instrument to measure your blood pressure.

Stroke: It is when the arteries or blood vessels to the brain are narrowed or blocked. This way, enough blood does not reach the brain.

Systole: It is the phase of the cardiac cycle in which heart muscles contract to pump blood.

Veins: They are the blood vessels which transport blood from the lungs and tissues back to the heart.

Ventricles: They are the lower chambers of the heart.

Vertebrates: Animals including mammals, fish, amphibians, birds, and reptiles that possess a backbone or spinal column.

White blood cells: They are the cells in our blood that help defend our body from any sort of infection.

Immune System & Common Diseases

Allergens: They are the things that cause allergies.

Antibiotics: They are the chemicals made by fungi to protect themselves from bacteria.

Antibodies: These are proteins made by the B-cells that stick to specific antigens and get the rest of the immune system to destroy them.

Antigens: They are parts of foreign bodies used by the immune system to recognise an infection/allergy and destroy it.

Autoimmune disease: It is a kind of disease in which our immune system attacks cells of our own body.

Bacteriophage: It is a virus that infects bacteria.

Complement pathway: It is the number of proteins that react with each other to make the blood clot in case of an injury.

Dendritic Cells: They are the phagocytic cells that enter tissues from the blood. They swallow pathogens and present antigens to T-cells and B-cells.

Dermis: It is the inner layer of the skin, made of live cells, hair follicles, and sebaceous glands.

Diarrhoea: It is the defence against infection which makes the intestines rapidly throw out bad food.

Epidermis: It is the outer layer of skin that acts like a waterproof wall against pathogens and allergens.

Fibrin: It is a protein in the blood that makes tiny fibres that seal a wound during clotting.

Gut Microflora: They are the bacteria that live in our large intestines, which make vitamins and keep bad bacteria from infecting us.

Hair Follicles: They are the cells in the dermis which give rise to hair.

Helminths: They are a class of microscopic worms, some of which infect us.

Histamines: They are the biochemicals which trigger an inflammation or allergy in response to an infection or allergen.

Hypodermis: It is the third layer of the skin. It is a deeper subcutaneous tissue made up of fat and connective tissue.

Immunoglobulin: It is the medical term for antibodies.

Inflammation: It is the body's reaction to insect bites, or some infections marked by redness and pain in the affected part of the body.

Keratin: It is the protein that makes hair and nails.

Langerhans Cells: They are the phagocytic cells in the skin.

Lymph Nodes: They are the glands that filter lymph and act as resting and training centres for WBCs.

Lymphatic System: It is the body's second circulatory system after blood, which drains tissues and hosts WBCs.

Lysosomes: They are the bags of enzymes inside each human cell that contain germ-destroying enzymes.

Macrophages: They are the WBCs that spot and kill harmful pathogens.

Melanin: It is the dark pigment that gives our skin its colour and keeps us safe from UV rays.

Mucus: It is the slime that covers the epithelium and keeps germs from getting into the body.

Natural Killer Cells: They are the WBCs full of granzymes and perforins that kill infected cells.

Pathogens: They are the germs that attack us, like bacteria, fungi, viruses, protozoa, and helminths.

Phagosome: It is a sac inside a cell which contains the swallowed pathogen. It will merge with a lysosome.

Rhinorrhoea: It is a defence against infection which makes the lungs rapidly throw out mucus from the nose, along with germs trapped in it.

Sebaceous glands: They are the glands in the skin that make sebum.

Sebum: It is the oil made by the skin that keeps our body waterproof.

Sweat Glands: They are the glands in the skin that make sweat. Sweat evaporates from the skin and keeps us cool.

Lungs & Respiratory System

Acclimatise: It is the practice of making the body get used to different weather conditions, especially high altitude.

Alveolar macrophage: It is an immune cell in the alveoli that destroys germs.

Alveolar pressure: It is the pressure of the mix of gases in the alveoli.

Alveoli: They are the functional units of the lungs where gas exchange happens.

Anaerobic respiration: They are the chemical reactions in cells that do not require oxygen to produce energy from glucose.

Bohr effect: It is the effect of acidity on the binding of oxygen to haemoglobin.

Boyle's Law: It is the law of physics that governs the relationship of a gas' volume and pressure.

Bronchi: They are the branches of the trachea leading into the lungs.

Bronchial tree: It is the network of bronchi and bronchioles.

Bronchial-associated lymphatic tissue (BALT): It is the lymphoid tissue in the bronchial tree that helps fight germs.

Bronchioles: They are the branches of the bronchi leading into the pulmonary lobes.

Bronchitis: It is the inflammation of the bronchi and bronchioles leading to difficulty in breathing.

Bronchospasms: It is a sudden contraction of the bronchi or bronchioles in asthma or COPD.

Carbamino-haemoglobin: It is the haemoglobin chemically attached to carbon dioxide.

Coelom: It is the body cavity of invertebrates, filled with haemolymph.

Conchae: They are also called turbinates. They are the folds of the skull that swirl the air in your inner nose.

Conducting zone: It is the part of your respiratory system that takes air from the atmosphere to the lungs.

Congenital disease: It is a disease that happens to newborn babies, sometimes caught while still a foetus.

Diaphragm: It is a giant muscle attached to the ribs, sternum, and spine that helps in breathing.

Emphysema: It is damage to lung tissue.

Epiglottis: A movable piece of cartilage that closes the wind pipe while swallowing food.

Eupnea: It means to breathe without making a conscious effort.

Goblet cells: They are the cells in the respiratory epithelium that make mucus.

Haemocyanin: It is the oxygen-carrier molecule in invertebrates.

Haemoglobin: It is the oxygen-carrier molecule in vertebrates.

Hilum: It is the point where bronchi, the pulmonary nerve, and blood vessels enter the lungs.

Hyperpnea: It means to breathe with a conscious effort.

Hyperventilation: It means breathing very fast when worried or excited, till the person collapses.

Influenza: It is a disease of the respiratory system that causes cold and sneezing.

Intercostal muscles: They are the skeletal muscles between your ribs, used in breathing.

Lamina propria: It is the tissue covering the bronchi, which has BALT.

Laryngopharynx: It is the part of the pharynx above the larynx that acts as a resonating chamber.

Laryngospasm: It is a sudden closure of the larynx to stop fluid from entering the lungs.

Lingual tonsils: It is the lymphoid tissue under the tongue.

Meditation: It is an Asian method of calming the mind and body by breathing slowly and in a controlled manner.

Metastatic: This is a phase of cancer when the cancerous cells spread to other tissue.

Nasal-associated lymphoid tissue (NALT): It is the lymphoid tissue in the nose that helps fight germs.

Nasopharynx: It is the part of the pharynx between the nose and the mouth.

Operculum: It is the skin covering a fish's gills.

Oropharynx: It is the part of the pharynx where the food pipe and the wind pipe cross.

Oxyhaemoglobin: It is the haemoglobin that is chemically attached to oxygen.

Palatine tonsils: It is the lymphoid tissue in the oropharynx.

Pandemic: It is used to refer to a disease that spreads across the world.

Paranasal sinuses: They are the hollow, air-filled spaces in the skull that help clean air.

Partial pressure: It is the pressure exerted by a gas in a mixture of gases.

Pharyngeal tonsils: Also called adenoids, it is the lymphoid tissue in the nasopharynx.

Phlegm: It is the thick, germ-filled mucus that is coughed out.

Pleura: It is the bag that cushions the lungs.

Pollutant: It is anything in air or water that should not be there naturally.

Pulmonary arteries: They are the arteries that take deoxygenated blood from the heart to the lungs.

Pulmonary Function Test: It is a test that diagnoses lung condition by measuring how hard it is to breathe.

Pulmonary lobes: They are the internal divisions of the left and right lungs, further divided into segments, lobules, and alveolar sacs.

Pulmonary oedema: It is a condition caused by plasma infiltrating the lungs.

Pulmonary surfactant: It is a soap-like liquid in the lungs that helps diffuse oxygen.

Pulmonary veins: They are the veins that take oxygenated blood from the lungs to the heart.

Pulmonary ventilation (PV): It is also called the respiratory cycle. It is a sequence of one inspiration and one expiration.

Respiratory bronchiole: It is the part of the bronchiole that ends in an alveolus.

Respiratory epithelium: It is the lining of the conducting zone of your respiratory system.

Respiratory zone: It is the part of your respiratory system that exchanges gases with blood.

Shock of life: It is the first breath taken by a newborn baby.

Trachea: It is the part of the respiratory system between the larynx and the lungs.

Trachealis muscle: It is the muscle that contracts the trachea.

Uvula: It is a flap of tissue that stops food going up into the inner nose.

Skeletal & Muscular System

Aerobic respiration: It refers to the conversion of glucose to carbon dioxide.

Anaerobic respiration: It refers to the conversion of glucose to lactic acid in muscles.

Articular cartilage: It is the cartilage that cushions the bones in a joint.

Biceps brachii and triceps: They are the muscles that flex the forearm.

Bursitis: It is the infection of the synovial sac.

Cardiac muscles: It is a muscle that keeps the heart pumping.

Carpals: They are the bones of the wrist.

Collagen: It is a protein that forms threads easily. It is seen in tendons and ligaments, as well as hair and nails.

Cranial Sutures: They are the immovable joints between skull bones, except the mandible.

Endoskeleton: An overarching supportive structure or shell inside an organism.

Exoskeleton: A tough exterior structure or shell that protects an organism.

Femur: It is the thigh bone.

Fracture: It is a break in the bones caused by stress exerted upon them.

Gluteus: They are the muscles that flex the legs.

Humerus: It is the bone of the upper arm.

Intervertebral disc: It is a piece of cartilage that cushions vertebra from each other. Joints: They are the junctions between bones. Joints may be movable or immovable.

Ligaments: They are the collagen sheets that help with the movement of the joints.

Mandible: It is the bone of the lower jaw.

Marrow: It is the fleshy tissue within bones.

Maxilla: It is the bone of the upper jaw.

Metacarpals: They are bones found in the palm of the hand.

Metatarsals: They are the bones of the sole.

Myocardium: It is the muscular part of the heart wall.

Myofibril: It is the functional unit of a muscle.

Neuromuscular junction (NMJ): It is a connection between muscles and motor neurones.

Orbit: It is the socket in the skull where the eye sits.

Osteoblasts: They are cells that make collagen.

Osteoclasts: These are the cells that resorb the bone's matrix to remodel it.

Osteoporosis: It is a disorder where a person suffers from a loss of minerals from bones, making them weak.

Pectoral girdle: It is the part of the skeleton to which the forelimbs are attached.

Pelvic girdle: The part of the skeleton to which the hindlimbs are attached.

Phalanges: They are the bones of the fingers and toes.

Radius and Ulna: They are the bones of the lower arm.

Ribs: They are the C-shaped bones in the chest.

Skull: It is the collective name for facial bones and the bones that protect the brain.

Smooth Muscle: It is a muscle that carries out involuntary movement.

Sternum: It is a tie-shaped bone in front of the chest.

Striated muscles: It is a muscle that carries out voluntary movement.

Synovial Fluid: It is the fluid that lubricates joints.

Tarsals: They are the bones of the ankle.

Tendons: It is a tissue that attaches muscle to the bone.

Thoracic basket: Also called the rib cage, it is made by the ribs, sternum and the spine, which encase the heart and lungs.

Tibia and Fibula: They are the bones of the lower leg.

Vertebra: It is a unit of the vertebral column.

Vertebral column: It is the chain of 33 bones in the back that supports the skeleton. It is also called the spine.

Stomach & Digestive System

Active transport: It is the way by which some food is pumped from your intestines to blood.

Anaerobic respiration: It is the way in which muscles work without enough oxygen, making a lot of lactic acid.

Basal metabolic rate: It is the speed with which your body turns food into energy.

Beriberi: It is a disease caused by the lack of Vitamin B1 in one's diet.

Binge-eating disorder: It is a disorder that makes you overeat because your brain's hunger centre isn't working properly.

Caries: It is the medical name for tooth cavities.

Cavities: It is a disease of the teeth caused by bacteria growing on the sugar on your teeth.

Cholecystitis: It is the inflammation of the gall bladder.

Cramps: It is used to refer to sudden pain in the muscles due to a build-up of lactic acid.

Dialysis: It is the way blood is filtered outside the body.

Digestive system: The digestive system breaks nutrients into small parts for the body to absorb and use for energy, growth, and cell repair.

Donor: It is a person who can give a healthy organ to a person whose organ has failed.

Dyspepsia: It is the medical word for indigestion.

Gag reflex: It is the throat's defence against choking by making you spit out what you've swallowed.

Heartburn: A burning feeling in the chest caused by indigestion.

Helicobacter pylori: It is a bacterium that infects the stomach and causes ulcers.

Hormones: They are the different kinds of biochemicals that act as messengers between organs.

Irritable bowel syndrome: It is a set of conditions that make you uncomfortable, with either diarrhoea, constipation, or both.

Kidney stones: They are crystals of urea or salts that form in your kidneys.

Kwashiorkor: It is a disease caused by not having enough protein in one's diet.

Lipoproteins: They are the proteins that help escort fat to tissues through the blood and stop them from sticking to the walls of your arteries.

Malnutrition: It is a disorder caused by not having enough food or certain nutrients in food.

Mandible: It is the lower jaw, which can move.

Maxilla: It is the upper jaw, which cannot move.

Metabolism: It is the way in which your body turns food into energy, or uses it to make tissue parts.

Nephrons: They are the tiny parts of the kidney that filter blood.

Oral rehydration solution: It is a mixture of sodium citrate, potassium chloride, and glucose dissolved in water and given to patients suffering from dehydration.

Pharynx: It is the inside of your throat, where the windpipe and food pipe cross each other.

Root canal operation: It is a medical procedure in which dentists remove rotten tissue from inside your teeth.

Scurvy: It is a disease caused by lack of Vitamin C in food.

Serotonin: It is the chemical in your brain that induces sleep.

Stem cells: They are the cells in the body that can turn themselves into any kind of cell and help repair organs.

Thyroid gland: It is an organ in the throat that makes the thyroid hormone.

Thyroid hormone: It is a hormone that helps the body make energy from food, helping you grow.

Ulcers: They are wounds inside the lining of the digestive system, often caused by Helicobacter pylori.

Ureters: Tubes that take urine from the kidneys to the urinary bladder.

HUMAN BODY — IMAGE CREDITS

a: above, b: below/ bottom, c: centre, f: far, l: left, r: right, t: top, bg: background

Cover:
Shutterstock Front: Chanon saguansak; SciePro; MattLphotography; Nerthuz; DM7; Lightspring; Designua; jacksparrow007; Nerthuz; BlueRingMedia; sciencepics; Designua; Lightspring; eranicle;
Back: joshya; Tijana Moraca; Lightspring; Schira
Wikimedia Commons Front: File:Blausen 0909 WhiteBloodCells.png

Brain & Nervous System
Insides:
Dreamstime: 12cr/Andreykuzmin; 14&15c/Nataliia Prokofyeva; 26&27c/Jose Manuel Gelpi Diaz; 31cr/Inna Volodina; 31b/ Evgeniia Kuzmich;

Shutterstock: 3b/patrice6000; 4l/Sebastian Kaulitzki; 4cr/bogadeva1983; 4br/hidesy; 4br/itor; 5tr/458677222/moncrub; 5tr/458677222/Christos Georghiou; 5br/Sirintra Pumsopa; 5bl/Simona Chira; 5cl/Sakurra; 6cl/decade3d - anatomy online; 6bl/stihii; 6r/ilusmedical; 7cr/AlexLMX; 6&7 bg/pro500; 7cl/De udaix; 8cl/Michael Braham; 8&9c/Atthapon Raksthaput; 9tr/Syda Productions; 10tr/lzf; 10cl/AJP; 10br/BUY THIS; 10cr/Luis Louro; 11cr/fizkes; 11br/ Alila Medical Media; 11bl/Alila Medical Media; 12tr/pathdoc; 12cl/Artorn Thongtukit; 12cr/AJP; 12br/Tracy Whiteside; 13tl/Sari ONeal; 13br/chrisdorney; 13cr/Designua; 13tr/Estrada Anton; 14cl/pablofdezr; 14br/Antonio Guillem; 15tr/solar22; 15tl/Anastasia Petrova; 15cr/Stas B; 15br/eveleen; 16&17c/ Alex Mit; 16br/Christina Li; 16bl/Designua; 17cr/extender_01; 18bl/annalisa e marina durante; 18r/Sebastian Kaulitzki; 19tr/ducu59us; 19cr/decade3d - anatomy online; 19bl/Holger Kirk; 20bg/KateStudio; 20bl/Monkey Business Images; 22tr/Alila Medical Media; 22c/Blamb; 22br/Africa Studio; 22bl/ vitstudio; 23tr/Iwona Wawro; 23bl/CLIPAREA l Custom media; 23br/CLIPAREA l Custom media; 24cr/Olga Bolbot; 24bl/Blamb; 25cr/CLIPAREA l Custom media; 25bl/yomogi1; 26&27c/StockImageFactory.com; 26cl/Pressmaster; 27br/japansainlook; 27tr/Axel Bueckert; 28tr/Kateryna Kon; 28c/Leremy; 29cr/ Intellistudies; 29bl/maxim ibragimov; 29br/Designua; 30bl/j.chizhe; 30bc/Ekaterina Markelova; 30tl/Antonina Vlasova; 30br/Keep Smiling Photography; 31tr/Africa Studio; 31cr/David Tadevosian; 31bl/Alena Ozerova;

Wikimedia Commons: 7br/File:Moniz.jpg PD-Sweden-photo/wikimedia commons; 7br/File:FreemanII.jpg/wikimedia commons; 9br/Luigi Novi / Wikimedia Commons / CC BY (https://creativecommons.org/licenses/by/3.0); 17br/Luigi Novi / Wikimedia Commons / CC BY (https://creativecommons.org/licenses/by/3.0); 21cr/Camillo_Golgi/wikimedia commons;

Heart & Circulatory System

Shutterstock: 3b/MDGRPHCS; 4bl/MDGRPHCS; 4&5c/Liya Graphics; 5t/Eviart; 6&7c/GraphicsRF; 6bl/Lightspring; 7cr/Marochkina Anastasiia; 8c/ NelaR; 8&9bg/Elena Paletskaya; 9t/Designua; 9cr/Neveshkin Nikolay; 9bl/fusebulb; 10cr/Kateryna Kon; 11tr/VGstockstudio; 11br/metamorworks; 12cl/457789036/Irina Strelnikova; 12cl/457789867/Irina Strelnikova; 12cl/561563713/Lyudmyla Ishchenko; 12cl/1145448170/Chompoo Suriyo; 13tr/ ilusmedical; 13cl/xpixel; 13cr/vlastas; 13br/Ling Stock; 14b/Dmitry Kalinovsky; 15b/Design_Cells; 15c/Vectorism; 15tr/ShadeDesign; 16t&17br/ Interior Design; 16cl/Standard Studio; 17tl/cenksns; 18br/Roman Zaets; 18bl/Room98; 18t/Designua; 19cl/Visual Generation; 19cr/Schira; 19cr/ awsome design studio; 19br/Elnur; 20t/Marina9; 20b/RTimages; 21br/Alexander Raths; 21br/Love the wind; 22br/diy13; 22tr/Double Brain; 22cl/ Jfanchin; 23cl/Sebastian Kaulitzki; 23cr/ilusmedical; 24cl/Anatomy Insider; 24r/S K Chavan; 25br/Alila Medical Media; 25cl/Nerthuz; 23,24,25bg/ Anatomy Insider; 25tl/Magic mine; 26tl/Double Brain; 26bl/Victor Josan; 26cr/solar22; 27br/TisforThan; 27tr/joshya; 27cl/decade3d - anatomy online; 28b/VectorMine; 29tr/fizkes; 29br/GoMixer; 29cr/Maridav; 29bl/Tukaram.Karve; 30&31b/Victor Metelskiy; 30cl/TijanaM; 31cr/pcruciatti; 31tr/Zerbor;

Wikimedia Commons: 5cl/Blausen_0724_PericardialSac/Blausen Medical Communications, Inc. / CC BY (https://creativecommons.org/licenses/by/3.0); 5b/2004_Heart_Wall_gl/OpenStax College / CC BY-SA (https://creativecommons.org/licenses/by-sa/3.0); 9r/ile:SS-harvey.jpg/wikimedia commons; 10b/2104_Three_Major_Capillary_Types/OpenStax College / CC BY-SA (https://creativecommons.org/licenses/by-sa/4.0); 11bl/St Jude Medical pacemaker in hand.jpg/Steven Fruitsmaak / CC BY (https://creativecommons.org/licenses/by/3.0); 11lc/Wilson Greatbatch.jpg/wikimedia commons; 14cl/ Heart_transplant/wikimedia commons; 14bl/Christiaan_Barnard_1969/Mario De Biasi (Mondadori Publishers) / Public domain; 16br/1GZX_Haemoglobin/ Zephyris at the English language Wikipedia / CC BY-SA (http://creativecommons.org/licenses/by-sa/3.0/); 17c/Blausen_0909_WhiteBloodCells/wikimedia commons; 21tr/James Harrison.jpg/wikimedia commons;

Immune System & Common Diseases

Shutterstock: 3b/Lightspring; 4bl/Dean Drobot; 4cr/Jose Luis Calvo; 5tr/Werayuth Tes; 5cr/Richard E Einert; 5b/Creations; 6&7c/S K Chavan; 6tc/ sciencepics; 6b/sciencepics; 7tl/BlueRingMedia; 7cl/Timonina; 7bl/ O2creationz; 7br/Surapol Usanakul; 8&9c/Lightspring; 8tl/717464182/BLACKDAY; 8tl/1132982033/CHEN I CHUN; 8cr/Kateryna Kon; 8cl/Veronika Zakharova; 8bl/Christoph Burgstedt; 9tr/Veronika Zakharova; 9tc/Veronika Zakharova; 9c/Meletios Verras; 9cr/Design_Cells; 10c/Designua; 10br/Kateryna Kon; 11tr/sciencepics; 11cl/Juan Gaertner; 11b/Christoph Burgstedt; 12cl/Liya Graphics; 12cr/Tatiana Shepeleva; 12bl/Kateryna Kon; 13tl/24Novembers; 13tr/Design_Cells; 13cr/Christoph Burgstedt; 13cl/Kateryna Kon; 13br/ nobeastsofierce; 14tr/Butsaya; 14bl/Lightspring; 15tr/TeraVector; 15cl/gritsalak karalak; 15cr/royaltystockphoto.com; 15b/Naeblys; 16tr/SvetaZi; 16cl/paulaphoto; 16br/Oksana Kuzmina; 17tr/Kateryna Kon; 17cl/Andrii Bezvershenko; 17bl/Jun MT; 17br/Indian Food Images; 18&19c/Designua; 18cl/royaltystockphoto.com; 18bl/Designua; 19tr/Lebendkulturen.de; 19cr/Designua; 20&21bc/434566348/sciencepics; 20&21bc/614705354/Sky vectors; 20&21c/marina_ua; 20br/ibreakstock; 20tr/Vladimir Zotov; 20&21tc/Kateryna Kon; 21tl/Kateryna Kon; 21bl/Mirror-Images; 21br/Karel Gallas; 22tr/Sebastian Kaulitzki; 22cl/Designua; 22cr/Alexandr III; 22bl/REDPIXEL.PL; 23tr/Evan Lorne; 23bl/goldeneden; 23br/New Africa; 24tr/imaagio. stock; 24cl/Illustration Forest; 24br/fototip; 24bl/Nelli Syrotynska; 25cl/517957894/BlurryMe; 25cl/571889917/Proxima Studio; 25c/232029478/ Ocskay Mark; 25c/1170304156/korn ratchaneekorn; 25cr/142857037/Jo Ann Snover; 25cr/757210732/one photo; 25br/MNStudio; 26bl/Gelpi; 26cr/Emily frost; 26br/Sergey Novikov; 27tl/Giovanni Cancemi; 27cr/BioMedical; 27br/sirtravelalot; 28&29c/Africa Studio; 28cl/Fomin Serhii; 29tr/

Wikimedia Commons: 14cr/Photographie de Leopold Duc d'Albany -Leopoldalbany.jpg/wikimedia commons; Africa Studio; 29clRost9; 29br/Everett - Art; 30tl/Tunatura; 30br/Kallayanee Naloka; 31cr/Daxiao Productions; 31l/Rawpixel.com; 31br/VaLiza;

Lungs & Respiratory System

Cover Image Credits: Shutterstock- MDGRPHCS;

Shutterstock: 3b/MDGRPHCS; 4tr/Tursk Aleksandra; 4b/Alila Medical Media; 5cl/YAKOBCHUK VIACHESLAV; 5bl/Designua; 5br/Reimar; 6tl/komokvm; 6cr/BravissimoS; 6bl/stockshoppe; 7bl/Yuliya Evstratenko; 7cl/Blamb; 7cr/Jen Watson; 7br/Africa Studio; 8tr/Alex Mit; 8cr/Aldona Griskeviciene; 8br/Lightspring; 8bl/Sakurra; 9cr/Vecton; 9bl/u3d; 10br/Susan Schmitz; 10bl/Steve Cymro; 10cr/Sakurra; 11cl/Dirk Ercken; 11cr/ArchMan; 11br/Normana Karia; 12tr/Pete Pahham; 12cl/Blamb; 12cr/VectorMine; 12br/fizkes; 13cl/gritsalak karalak; 13br/Halfpoint; 14&15bg/Charcompix; 14bl/MJTH; 15b/Fahkamram; 15cr/Designua; 16cr/VectorMine; 16tr/Master1305; 16br/Umomos; 17bl/zhengzaishuru; 17br/SNeG17; 17cr/normaals; 18tl/Tappasan Phurisamrit; 18b/KrutSolt; 19cr/Alila Medical Media; 19bl/Lizmyosotis; 19br/risteski goce; 20&21 bg/Sergey Nivens; 20&21tc/Koldunova Anna; 20cl/vchal; 21bc/Anton Watman; 21bl/VGstockstudio; 21br/aslysun; 22b/antpkr; 22&23bc/Maridav; 22tr/begun1983; 23tl/Rost9; 23cl/didesign021; 23cr/luanateutzi; 24tr/Kateryna Kon; 24bl/Alexander Raths; 24cr/Miriam Doerr Martin Frommherz; 25bl/Marko Aliaksandr; 25tr/Puwadol Jaturawutthichai; 26bl/Robert Kneschke; 26cl/fizkes; 26tr/Virales; 27tr/Designua; 27c/VectorMine; 27br/freeskyline; 28br/VectorMine; 28tr/VectorMine; 28bl/Luis Molinero; 30bl/narin phapnam; 30c/joshya; 31br/Bangkokhappiness; 31bl/pixelaway; 31tr/Michael Pervak;

Wikimedia Commons: 5tr/Kleopatra-VII.-Altes-Museum-Berlin1/Louis le Grand / wikimedia commons; 11t/Kleine_Bonaire-Underwater_life-js/Michal Strzelecki, Wojtek Strzelecki i Jerzy Strzelecki / CC BY (https://creativecommons.org/licenses/by/3.0); 18cr/Tibetan Woman (5278082263)/Paul Hamilton / CC BY-SA (https://creativecommons.org/licenses/by-sa/2.0); 23cr/A_nurse_vaccinates_Barack_Obama_against_H1N1/White House (Pete Souza) / Maison Blanche (Pete Souza); 25ltl /John_Keats_by_William_Hilton/William Hilton; 25ltc/Percy_Bysshe_Shelley_by_Alfred_Clint/After Amelia Curran / Public domain; 25cr/William_Henry_Harrison_daguerreotype_edit/Albert Sands Southworth (American, 1811–1894) and Josiah Johnson Hawes (American, 1808–1901). Edited by: Fallschirmjäger and subsequently by Emiya1980 / Public domain; 31br/Pablo picasso 1/Argentina. Revista Vea y Lea/ Public domain;

Skeletal & Muscular System

Shutterstock: 3bc/Horoscope; 4tl/Arcady; 4b/Sakurra; 4&5c/design36; 5cr/Sathit; 6cl/Corona Borealis Studio; 6cr/corbac40; 6&7b/sportpoint; 6&7c/Sebastian Kaulitzki; 7tr/Prostock-studio; 7cr/Ellen Bronstayn; 7br/lovepetch; 8bl/Ruslan Huzau; 8cr/Blamb; 8&9c/1267474846/TreesTons; 8&9c/145154638/decade3d - anatomy online; 9tr/Javid Kheyrabadi; 9cr/Blamb; 9r/ancroft; 10l/Alex Mit; 10cr/Designua; 10bc/wildestanimal; 10br/eranicle; 10&11c/sciencepics; 11tr/sciencepics; 12l/wavebreakmedia; 12tr/AM-STUDiO; 12br/BlueRingMedia; 13tl/Salienko Evgenii; 13tc/Sky Antonio; 13tr/wavebreakmedia; 13br/stihii; 13cl/Maridav; 14tr/Roop_Dey; 14l/Nerthuz; 14bl/Alila Medical Media; 14br/Sebastian Kaulitzki; 15cl/wavebreakmedia; 15cr/Natee Jitthammachai; 16l/Monika Wisniewska; 16cr/Double Brain; 17r/Sebastian Kaulitzki; 17bl/joshya; 18&19c/Valentyna Chukhlyebova; 18cr/Fer Gregory; 18br/Anan Kaewkhammul; 19br/Designua; 20tr/Sebastian Kaulitzki; 20cr/Heiko Kiera; 20c/Artemida-psy; 20bl/Designua; 21tr/Nerthuz; 21c/Tefi; 21br/madesapix; 22l/Ramona Kaulitzki; 22cr/Sakurra; 22bl/Elana Erasmus; 22br/Teguh Mujiono; 23tr/Sakurra; 23cr/Designua; 23cl/Designua; 23br/Designua; 24l/Choksawatdikorn; 24tr/Oleksandr Zamuruiev; 24cr/joshya; 25tr/Blamb; 25cl/Blamb; 25br/fizkes; 26tr/ancroft; 26cr/sciencepics; 26cl/1027255258/Richman Photo; 26cl/1055560397/AlexLMX; 26b/Allies Interactive; 27tl/wavebreakmedia; 27cr/kwanchai.c; 27r/Skumer; 28&29c/Robert Adrian Hillman; 28tl/evrymmnt; 28cr/Designua; 28bl/Alila Medical Media; 29tr/GraphicsRF; 29cr/Tyler Olson; 30tr/Beate Panosch; 30cl/Designua; 30bl/Artemida-psy; 30br/Evan Lorne; 31tr/VectorMine; 31cl/Alohaflaminggo; 31cr/ChiccoDodiFC; 31bl/Jo Ann Snover;

Wikimedia Commons: 15br/Peter Paul Rubens181.jpg/Peter Paul Rubens / Public domain/Wikimedia; 26br/Portrait of William Einthoven, head and shoulders. General Collections Keywords: Willem Einthoven/See page for author / CC BY (https://creativecommons.org/licenses/by/4.0)/Wikimedia;

Stomach & Digestive System

Cover Image Credits: Shutterstock- Magic mine;

Dreamstime: 8&9c/Monkey Business Images; 23bl/Ljupco, 23br/Arne900; 28&29c/Boarding1now; 29br/Roman Stetsyk;

Shutterstock: 3b/Sebastian Kaulitzki; 4&5c/StockImageFactory.com; 4cl/Thomas M Perkins; 4bl/Automobus; 5tr/Peter Hermes Furian; 5cr/Dmytro Vietrov; 6&7c/Gelpi; 6cl/Roman3dArt; 7tr/Torrenta Y; 7cr/Xcages; 7br/Martin Prochazkacz; 8tr/Nattakorn_Maneerat; 8br/Designua; 9tl/greenaperture; 9cl/Zurijeta; 9cr/Designua; 9br/etorres; 10tr/Sebastian Kaulitzki; 10cr/Tefi; 10br/Nicolas Primola; 11l/Sebastian Kaulitzki; 12&13c/StockImageFactory.com; 12bl/Anatoliy Sadovskiy; 12tr/margouillat photo; 12cr/Oleksandra Naumenko; 12br/margouillat photo; 13cl/marilyn barbone; 13tl/210587053/ronstik; 13tl/158485127/gresei; 13bl/1233254824/ZuKIN Art Studio; 13bl/81072301/V.S.Anandhakrishna; 14bl/Eric Isselee; 14&15c/metamorworks; 15tl/nobeastsofierce; 15cr/Akadet Guy; 16tl/Magic mine; 16tr/Designua; 16cr/piyapon chantra; 16bl/Elen Bushe; 16br/Hoika Mikhail; 17tr/Nerthuz; 17cl/Alila Medical Media; 17cr/Sherry Yates Young; 18&19c/Sebastian Kaulitzki; 18tr/Designua; 18cl/Anastasia Tveretinova; 20tr/Magic mine; 20cl/Olena758; 20br/decade3d - anatomy online; 20bl/Kateryna Kon; 21tr/Romariolen; 21cl/marilyn barbone; 21br/Aaron Amat; 22cr/gritsalak karalak; 22bl/Alberto Loyo; 22br/Victor Josan; 23cr/VectorMine; 23bl/Fresnel; 23br/Jacob Lund; 24&25c/Magic mine; 24cl/Lightspring; 25tr/ThamKC; 25cr/plenoy m; 26t/kwanchai.c; 27tl/Kateryna Kon; 28cl/CKP1001; 28r/Arvind Balaraman; 28bl/Pixfiction; 29tr/BW Folsom; 29cl/StockImageFactory.com; 29br/StockImageFactory.com; 30tr/Lightspring; 30cl/BlurryMe; 30cr/Evan Lorne; 30bl/joshya; 30br/giannimarchetti; 31tr/gopixa; 31bl/Ben Schonewille; 31br/Yurchanka Siarhei;

Wikimedia Commons: 15br/Motzume/wikimedia commons; 17bl/Rosalyn_Yalow/wikimedia commons; 26br/Ulcer-causing_Bacterium_(H.Pylori)_Crossing_Mucus_Layer_of_Stomach/wikimedia commons; 27cr/Starved_girl/wikimedia commons; 27bl/Miss_K._R-_aged_14,_before_treatment_for_anorexia_Wellcome_L0073694_(cropped)/wikimedia commons;